ACOUSTO-ULTRASONICS
Theory and Application

ACOUSTO-ULTRASONICS
Theory and Application

Edited by

John C. Duke, Jr.

Virginia Polytechnic Institute and State University
Blacksburg, Virginia

Springer Science+Business Media, LLC

Library of Congress Cataloging in Publication Data

Acousto-ultrasonics: theory and application / edited by John C. Duke, Jr.
p. cm.
"Proceedings of a workshop on acousto-ultrasonics, theory and application, held July 12–15, 1987, at the Virginia Polytechnic Institute and State University in Blacksburg, Virginia"—T.p. verso.
Includes bibliographical references and index.

ISBN 978-1-4757-1967-3 ISBN 978-1-4757-1965-9 (eBook)
DOI 10.1007/978-1-4757-1965-9

1. Acoustic emission testing—Congresses. 2. Ultrasonic testing—Congresses. I. Duke, John C., 1951–
TA418.84.A265 1988 88-23507
620.1'1274—dc19 CIP

Proceedings of a workshop on Acousto-Ultrasonics: Theory and Application,
held July 12–15, 1987, at the Virginia Polytechnic Institute
and State University in Blacksburg, Virginia

Originally published by Plenum Press, New York in 1988.

PREFACE

Finding and sizing cracks and other crack-like discontinuities has been the center of attention for scientists and engineers developing and using nondestructive evaluation (NDE) technology. However, with advanced materials being "engineered" and used in critical structural components, a new challenge for NDE has emerged. Whereas many traditional engineering materials fail due to the initiation and self-similar propagation of a crack, reinforced composite materials degrade and fail in a manner more analogously to the collapse of a structure. Consequently the NDE of such materials involves assessing the combined effect of the material's damaged condition rather than identifying and sizing single critical imperfection.

In 1979 Alex Vary, seeking to address the challenge confronting the NDE of advanced fiber reinforced composite materials began work on a new method of materials characterization. Focusing on the problem of evaluating graphite fiber reinforced/epoxy laminated plates; Vary used a piezoelectric transducer to excite a mechanical disturbance in a plate and, with a sensitive piezoelectric transducer monitored the disturbance on the same surface of the plate. (Placing the transducers on the same surface was primarily for practical purpose but their displacement in the direction of anticipated service load was of fundamental significance!) To quantify this observation, he counted the number of excursions, of the resulting electrical signal, above a arbitrary voltage threshold; a procedure frequently used for acoustic emission signal analysis.

Because the disturbance was introduced in a manner similar to that used in ultrasonics and the detected signal was analyzed in a manner similar used in acoustic emission monitoring the name "acousto-ultrasonic measurements was coined. Vary and coworkers were encouraged, immediately, by the close correlation observed between acousto-ultrasonic measurements, quantified through a procedure which yields a stress wave factor (SWF), and mechanical properties of advanced composite materials.

Since that time proponents and opponents of the use of the method have considered its application to an enormous variety of situations. Concerns have been expressed about the difficulty in reproducing measurements and the issue of whether or not the resulting parameters are physically meaningful. Excitement has resulted as good correlation between acousto-ultrasonic measurements and strength, and mechanical properties for wood, advanced composites, bonded components and so on, have been obtained. Other workers have extract aspects of the original approach for application to some special situations, while still others have explored alternative ways of quantifying and explaining the phenomena.

Against this backdrop an international symposium was held at Virginia Polytechnic Institute and State University in Blacksburg, Virginia form July 12-15, 1987; this volume is the proceedings of that conference - Acousto-Ultrasonics: Theory and Application. The organizing committee included: Edmund G. Henneke, II, VPISU, USA; Alex Vary, NASA/LeRC, USA; James H. Williams, Jr., MIT, USA; Alain Lemascon, CETIM, France; Karl Schulte, DFVLR, Germany; C. R. L. Murthy, IIS, India; Yapa Rajapakse, ONR, USA; Robert E. Green, Jr. JHU, USA; Ramesh Talreja, TUD, Denmark; John C. Duke, Jr., VPISU, USA (chairman).

Many of the papers in this volume were critically reviewed and revised. Appreciation is expressed to: C. Anderson, J. Liu, W. Sachse, R. Kriz, J. W. Wagner, C. Teller, E. Papadakis, Y. Bar-Cohen, and all of the authors for reviewing manuscripts; S. W. Bartlett and M. A. Horne for proofreading several of the galleys; and Ms. Shelia Lucas for typing a large portion of the volume and proofreading many of the final galleys. Finally a word of thanks to my family, BJD, ECD, and JCDIII, for sharing me with this project.

John C. Duke, Jr.

Blacksburg, Virginia 1988

CONTENTS

KEYNOTE

THEORY - WAVE PROPAGATION FOR ACOUSTO-ULTRASOUND

DATA INTERPRETATION

SOURCES/DETECTORS

CALIBRATION AND METHOD IMPLEMENTATION

APPLICATIONS - GENERAL

APPLICATIONS - COMPOSITE MATERIALS

APPLICATIONS - BONDING

THE ACOUSTO-ULTRASONIC APPROACH

Alex Vary

National Aeronautics and Space Administration
Lewis Research Center
Cleveland, OH 44135

SUMMARY

This paper reviews the nature and underlying rationale of the acousto-ultrasonic approach, suggests needed advanced signal analysis and evaluation methods, and discusses application potentials. The term acousto-ultrasonics denotes an NDE technique that combines some aspects of acoustic emission methodology with ultrasonic simulation of stress waves. The acousto-ultrasonic approach uses analysis of simulated stress waves for detecting and mapping variations of mechanical properties. Unlike most NDE, acousto-ultrasonics is less concerned with flaw detection than with the assessment of the collective effects of various flaws and material anomalies. Acousto-ultrasonics has been applied chiefly to laminated and filament-wound fiber reinforced composites. It has been used to assess the significant strength and toughness reducing effects that can be wrought by combinations of essentially minor flaws and diffuse flaw populations. Acousto-ultrasonics assesses integrated defect states and the resultant variations in properties such as tensile, shear, and flexural strengths and fracture resistance. Matrix cure state, porosity, fiber orientation, fiber volume fraction, fiber-matrix bonding, and interlaminar bond quality are factors that underline acousto-ultrasonic evaluations.

INTRODUCTION

Conventional NDE

The main purpose of nondestructive evaluation (NDE) is to provide a basis for determining whether a structure will perform reliably when placed in service. This is usually done with conventional flaw detection techniques and it depends on defining which flaws are harmful. Obviously, it is unrealistic to discard a part simply because it contains minor flaws or anomalies that normally arise during fabrication. Certainly, some flaws are likely to be more harmful than others because of their size, location, and proximity to each other. But, in some materials (e.g., fiber reinforced composites) the harmful effects of certain types of flaws cannot always be determined unambiguously.

Although composites can have broken fibers, delaminations, local porosity, resin rich areas, etc., the overall effect may be benign. The combined effect of these defects is often difficult to assess analytically. More theoretically valid analysis models are needed to incorporate the complex fracture process that occur in composites.[1] On the other hand, a composites structure may be free of readily detectable flaws but still exhibit low strength and be unsuitable for use. Low composite strength may be due to poor bonding between plies or between the fibers and matrix.

Mechanical Property Assessment

The preceding observations suggest a need for NDE methods that can assess variations in mechanical properties rather than merely detect minor flaws whose relevance is questionable. These methods should at least quantify relative strength and fracture resistance by measuring the combined effect of all deficiencies due to subcritical flaw populations and microstructural anomalies in a volume of material. Then, fracture analysis and analytical life prediction methods would be supplemented by NDE approaches that verify properties that contribute to structural integrity and reliability.[2]

Acousto-Ultrasonics

The acousto-ultrasonic approach addresses the above-mentioned requirements.[3] It complements other NDE approaches to materials characterization[4] and offers advantages that make it the preferred one to use in some cases. Accordingly, the purpose of this paper is to describe the acousto-ultrasonic approach, its genesis, rationale, signal analysis methods, and applications.

BACKGROUND

Terminology

The term acousto-ultrasonics was coined to express the close relation with acoustic emission. "Acousto-ultrasonics" may be taken as a contraction of "acoustic emission simulation with ultrasonic sources." Acousto-ultrasonic waves like acoustic emissions are generally ultrasonic in nature. The acoustic emission method depends on loading to excite spontaneous stress waves such as those accompanying plastic deformation and crack growth.[5] Acousto-ultrasonics differs mainly in that the ultrasonic waves are benign and are generated externally by a pulsed source (usually a piezotransducer).[6]

Objective

The objective in acousto-ultrasonics is to simulate stress waves that resemble acoustic emission waves but without disrupting the material.[7] One launched inside the material sample, the waves are modified by stochastic processes like those that affect spontaneous acoustic emissions from internal sources during stressing, deformation, etc. Moreover, acousto-ultrasonic waves are launched periodically at predetermined times and with predetermined repetition rates.

In contrast to acoustic emission practice, the idea in acousto-ultrasonics is to keep the nature and location of the source of ultrasonic radiation known and fixed. Then, the inverse problem is not concerned with source location and characterization but with characterization of the material medium between the source and receiver.[8]

Alternative Approaches

Materials characterization can, of course, be accomplished by other ultrasonic methods,[9] for example, by pulse-echo ultrasonic measurements of velocity and attenuation. These measurements can usually be correlated with variations in microstructure (e.g., grain size, porosity), elastic constants, strength, toughness, etc. Pulse-echo ultrasonics is based on knowing the wave propagation path, avoiding multiple reflections and overlapping echoes, and by accounting for dispersion and similar effects.

The acousto-ultrasonic approach is among the alternatives that should be considered for material property characterization. It is specifically designed for cases where constraints imposed by pulse-echo and similar conventional ultrasonic approaches are impractical.

METHODOLOGY

Probe Configuration

In acousto-ultrasonics, the sender and receiver proves are usually coupled to the same side of the test object. (In many applications only one side will be accessible.) The probes are coupled at normal incidence to the surface of a test piece. The receiver is displaced by a fixed distance from the sender. In the limit of zero separation, this configuration reduces to conventional pulse-echo ultrasonics. The send-receive transducer pair is moved about as a unit and the test object is scanned to map material property variations, Fig. 1.

Direction Criterion

The rationale of the acousto-ultrasonic probe configuration may be understood by considering the materials characterization problem posed by fiber-reinforced composite laminates. For laminates it is desirable to measure properties with ultrasonic energy that has propagated laterally (parallel to the lateral surfaces). Transducers coupled to edges could send and receive the

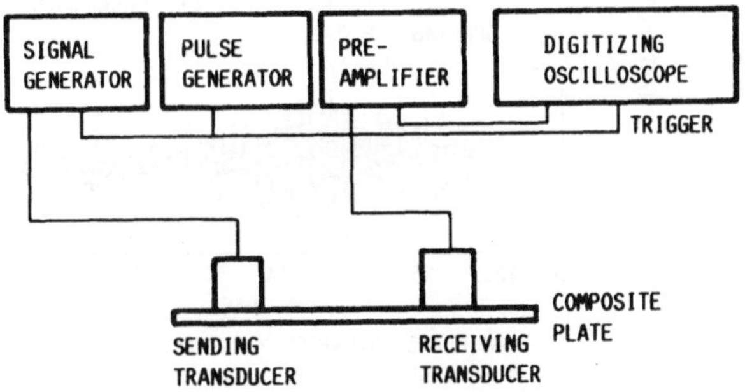

Fig. 1. Basic probe configuration and instrumentation for acousto-ultrasonic measurements. Digitizing oscilloscope is used to feed time domain signals (waveforms) a computer for analysis and evaluation of acousto-ultrasonic parameters.[45]

signals, but although thick laminates may allow effective coupling to edges, thin laminates pose a problem. (Laminate thicknesses range from a fraction of a millimeter to several centimeters.) In large laminated sheets the signal would probably be lost before reaching an opposite edge because composites are usually highly attenuating.

In laminated composite structures, some or all edges may be inaccessible (e.g., cylinders, vessels, fuselages). Other examples include wood fiber boards, paper products, ropes and cables, and plate/sheet/strip stock. The need to use energy that has propagated parallel to major surfaces arises whether the laminates are polymer, metal, or ceramic matrix fiber reinforced composites.

Echo System

The acousto-ultrasonic configuration assures that numerous wave interactions occur in the volume of material interrogated. Instead of consisting of a series of isolated echoes, as in pulse-echo ultrasonics, the acousto-ultrasonic signal will be much more complex. In test objects like laminated panels that have a thickness less than the spacing between probes the received signal will consist of overlapping echoes, Fig. 2. For objects having greater thicknesses, echoes will tend to separate due to delayed arrivals of individual echoes. Despite its complexity the acousto-ultrasonic signal can be readily analyzed to provide information about lateral property variations. (Signal analysis methodologies are discussed later.)

Transducers

In current practice piezoelectric probes are coupled directly to the surface of a test object. An alternative, based on contactless laser probes will be discussed later. It is important that the sending transducer have a spectral bandwidth sufficient to excite all frequencies needed to interrogate the

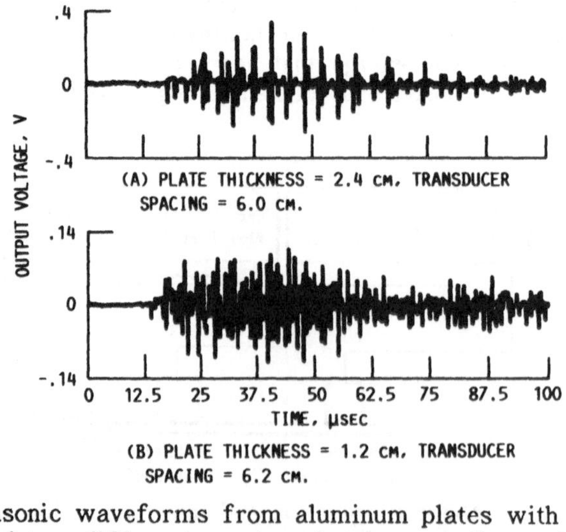

Fig. 2. Acousto-ultrasonic waveforms from aluminum plates with two different thickness and two different spacings between sending and receiving transducers. The sender and receiver had bandwidths of approximately 1 MHz at center frequencies in the range from 1.3 to 1.7 MHz.[26]

material. The character of the acousto-ultrasonic signal and the data that can be extracted will depend rather heavily on the spectral response and sensitivity of the receiving probe.

For many graphite fiber polymer matrix composite laminates both the sender and receiver may be broadband 2 MHz (megahertz) transducers. A bandwidth of about 1 MHz may be adequate to characterize these materials. Experimental determination of appropriate transducer properties should be made for each new structure. In highly attenuating materials the receiver may need to be a sensitive transducer of the type used for detecting acoustic emissions.

Coupling

The acousto-ultrasonic technique depends strongly on establishing reproducible probe coupling. Otherwise, signal modulations due to material variations become confused with those due to coupling variations. Probes are usually coupled to a surface with a thin film of fluid (glycerin, gel, silicone grease, and shear wave couplant have been found useful.) Enough pressure has to be applied to eliminate unwanted reverberations within the couplant. In addition, deleterious effects of surface roughness must be overcome. Similar precautions are needed with dry coupling which usually involves a thin elastomer buffer (e.g., silicone rubber) bonded to the transducer wearplate.

Extended Scanning

Scanning is done by intermittently lifting and recoupling the probes. There are obvious disadvantages to the intermittent contact required with direct coupled (dry or liquid) probes. The potential need to scan large surface areas demands probes that can be readily moved about. Using probes with the piezoelectric crystal mounted in the hub of a rubber-rimmed wheel allows continuous rolling-contact scanning. Even then, great care has to be taken to ensure that coupling variations are insignificant.

Laser Methods

The prospect of contactless laser probes, particularly for scanning large complex surfaces, is very attractive.[10,11] Use of lasers to excite and acquire acousto-ultrasonic signals does not eliminate potential coupling problems because the signals are still influenced by surface roughness and also emissivity, reflectivity, and other thermal and optical factors.

Probe Accommodation

Certain accommodations may be required for either contact or laser probes to overcome coupling and surface access problems. This can entail surface preparation or even changes in part geometry and design. Insertion of sufficient laser thermal energy may ultimately depend on adding sacrificial layers (e.g., fluids, plating). Extraction of signals by means of interferometric or other methods may depend on affixing reflecting layers or echelle (or blazed) diffraction gratings.[12,13]

Alternative Methods

There are alternatives to the acousto-ultrasonic configuration, for example, the use of closely-spaced angle beam transducers in contact with the surface.[14]

Similar measurements can be made with immersion scans based on leaky Lamb waves[15] that arise from radiation that propagates laterally within thin laminates. Experience and results with the acousto-ultrasonic method suggest that angle beam transducers are not necessary and immersion needed by leaky waves can be avoided.

FUNDAMENTAL HYPOTHESIS

Conditional Statement

The ultimate purpose of the acousto-ultrasonic approach is to rate relative efficiency of stress wave energy propagation in a material. If the material is subject to brittle or quasi-brittle failure but exhibits efficient stress wave energy transfer, then it will exhibit higher extrinsic strength and fracture resistance. This does not necessarily apply to materials than can sustain plastic deformation or slow crack growth.[1]

For many materials, such as fiber reinforced composites, better stress wave energy transfer means better transmission of dynamic strain, better load distribution, greater strength, and greater fracture resistance. In these cases the hypothesis is that increased energy flow (either stress or strain energy) corresponds to increased strength and fracture resistance, especially when precursor conditions for fast, brittle fracture exist. This hypothesis is based on the "stress wave interaction" concept which holds that spontaneous stress waves at the onset of fracture promote rapid crack growth unless their energy is dissipated in other ways.[16,17] Prompt and efficient flow of stress wave energy away from crack nucleation sites is desired when the energy cannot be absorbed locally without cracking.

Stress Wave Evaluation

Acousto-ultrasonic measurements are made by means of a stress wave factor (SWF). The SWF is used to quantify acousto-ultrasonic signals for comparison with variations in mechanical properties like strength and fracture resistance. The SWF will indicate regions where strain energy is likely to concentrate and result in crack nucleation and fracture.

SWF and Attenuation

Lower values of the SWF correspond to regions of higher attenuation. Indeed, the dominant effect measured in acousto-ultrasonics is relative attenuation. When properly measured, any magnitude variations of the acousto-ultrasonic signal will depend primarily on material factors that govern attenuation: microstructure, morphology, porosity, bond quality, cure state, microcracks, and so on. The acousto-ultrasonic approach assumes that these factors similarly affect the natural stress waves that arise during dynamic loading, deformation, and crack nucleation.

Stress Wave Simulation

The nature and configuration of acousto-ultrasonic probes are selected to simulate the frequency content of spontaneous stress (strain) waves that arise at the onset of microfracture, crack nucleation, deformation, etc. If this is accomplished, then the acousto-ultrasonic approach should measure the effects of factors that govern relative efficiency of strain energy transfer. Accordingly, regions that exhibit high values of the SWF would also exhibit enhanced stress wave energy flow. Conversely, low values of SWF would indicate places where the dynamic strain energy is not effectively dissipated or redistributed (with resultant deformation or fracture).

STRESS WAVE FACTOR (SWF)

General Note

The SWF may be defined in a variety of ways. The ones mentioned here are based directly on acoustic emission practice (e.g., ringdown count, peak voltage, energy).[5] This is appropriate because acousto-ultrasonic signals are usually quite similar to acoustic emission signals. The formulation of an expression for the SWF depends on which features in a waveform are most relevant to a given probe configuration, material, or structure geometry.

Ringdown SWF

Acousto-ultrasonic signals that resemble acoustic emission bursts are readily characterized by a ringdown count or count rate. In the former case the SWF is formulated as,

$$SWF = E_c = PRC \qquad (1)$$

where, P is the repetition rate of an ultrasonic pulser, R is the reset time, and C is the digital counter output. A threshold voltage setting is the basis for counting the number of ringdown oscillations per waveform. Defined this way, the SWF measures relative signal strength. The threshold voltage is usually set at just above noise level. The pulse repetition rate, P, is set so that each signal rings down below the threshold before a new one starts. The reset time, R, allows averaging a predetermined number of signals into the count, C.

Ringdown oscillations toward the end of a signal might be more characteristic of the transducer than the material (i.e., ringdown in undampened piezoelectric transducers). Trailing oscillations might also be reflections from regions just outside the volume between the sending the receiving probes. Therefore, it may be necessary to increase threshold level slightly or to truncate the ringdown counting time zone. In either case, threshold and reset criteria should be based on experimental feedback.

Peak Voltage SWF

By using peak detection the SWF may be defined as,

$$SWF = -E_v = V_{max} \qquad (2)$$

where, the SWF is base on the maximum (max) voltage swing. This assumes that dominant oscillations always represent any material variations. But, smaller oscillations that precede or follow may be more representative. An appropriate alternative based on peak voltage might be the measurement of signal rise time or signal decay time (as in acoustic emission practice).

The pulse repetition rate, P, and reset time, R, settings mentioned above for ringdown SWF still apply for measuring peak voltage SWF and also for the energy SWF, described next.

Energy SWF

The relative energy of the acousto-ultrasonic signal can be defined in the time domain as,

$$SWF = E_t = (V_{rms})^2 = (1/T) \int_{t_1}^{t_2} v^2 \, dt \qquad (3)$$

where, the SWF is based on root mean square (rms) voltage, T is a time interval (t_1 to t_2), t is time, and v is time-varying voltage. An equivalent frequency domain definition of the SWF in terms of the root mean square of the power spectrum is,

$$SWF = E_f = (S_{rms})^2 = (1/F) \int_{f_1}^{f_2} s^2 \, df \qquad (4)$$

where, F is a frequency interval (f_1 to f_2), f is frequency, and s is a function of frequency. Although it is unnecessary to set a threshold voltage, it is still necessary to specify the size and location of the interval (i.e., T or F) in the time or frequency domain that will most closely associate with material variations. This suggests the need to experimentally determine specific time or frequency intervals for each new material, structure, and probe configuration.

Using the previous definitions of the SWF, eqs. (1) to (4), it is usually better to normalize values of quantities like E_c, E_t, or E_f for comparison with material property variations. It has been found practical to normalize against the maximum asymptotic value found for these quantities for a given material, structure, and probe configuration.[7,18]

SIGNAL ANALYSIS

General Observations

The envelope of the acousto-ultrasonic time domain waveform usually exhibits complicated amplitude variations, Fig. 3, while the corresponding frequency spectrum usually exhibits numerous prominent frequency components. These time and frequency domain features are related to one or more material and structural factors: velocity, dispersion, attenuation, dynamic vibration modes, plate waves, etc. These factors are influenced in turn by the nature of the transducers, probe configurations, coupling, instrumentation, etc.

There is a need for better understanding of wave propagation factors that underlie acousto-ultrasonic waveforms. Accordingly, the problem of predicting and analyzing waveforms for various material conditions (shape, texture, isotropy) has been broached in order to satisfy this need.[19,20] Some insights that have been gained and improved signal analysis methods that have been proposed are discussed below.

Natural Modes

Experimental evidence suggests that natural vibration modes and associated nodal lines on the surface of a solid will certainly affect acousto-ultrasonic measurement.[21] Evaluation of the SWF can be misleading if specimen resonances are ignored or misinterpreted. Resonant frequencies and their corresponding nodal patterns can have an enormous effect on the acousto-ultrasonic waveform and on the evaluation of the SWF. Resonant frequencies due to natural vibration modes of a structure can dominate the spectral content. This is

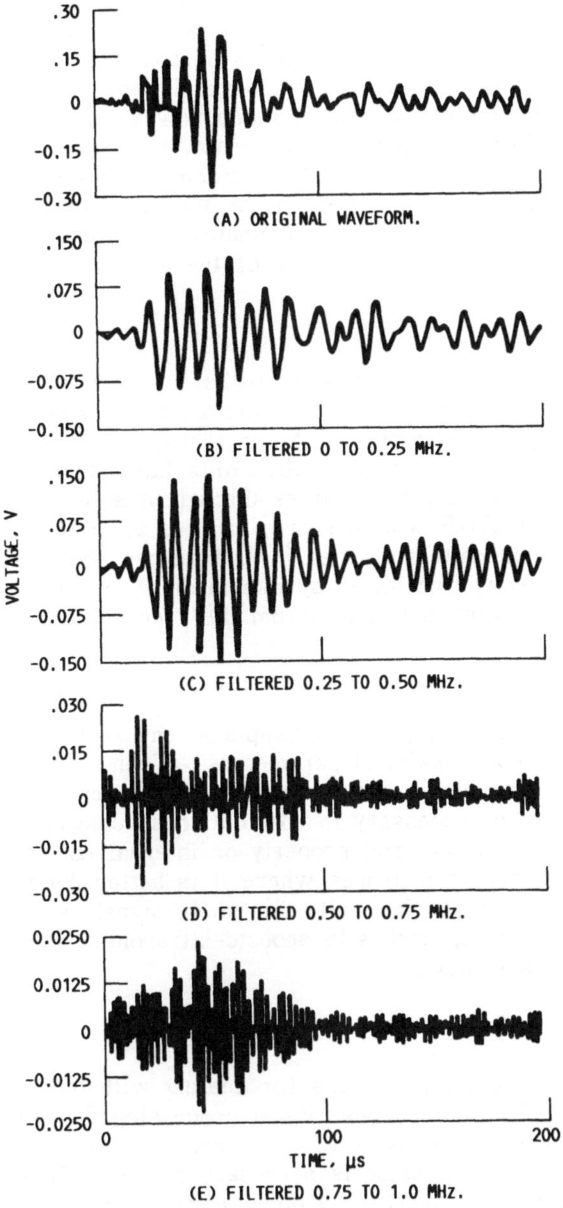

Fig. 3. Acousto-ultrasonic waveforms from a filament wound composite bend bar with thickness of 0.65 cm and 5 alternating cross-plies. The original waveform of the received signal, A), was passed through four filters, B) through E), to illustrate diverse frequency and modal components present.[25]

especially true when the sending transducer inputs energy at the resonant frequencies. Location of the receiver in relation to nodal lines further influences the spectrum of the acousto-ultrasonic waveform.

Dispersive Modes

Although the sending transducer inputs longitudinal waves perpendicular to the specimen surface, the energy radiated into the material will produce oblique reflections and shear waves. Interactions of shear and longitudinal waves with plate boundaries and with plies in laminates produced various dispersive wave modes.[14] When laminate or ply thicknesses of composite structures are comparable to the ultrasonic wavelengths, several Lamb modes with different speeds propagate simultaneously. Since Lamb waves are dispersive, their phases and group velocities depend on frequency, laminate thickness, bond quality (adhesion) between plies, etc.[22,23] Some Lamb modes will be excited and others will be extinguished in accordance with boundary conditions and the interlaminar bond quality in composite laminates. Factors that govern these wave modes also determine the character of the acousto-ultrasonic waveform.

Wave Paths

It is useful to attempt to trace wave paths, to track multiple reflections, and to gauge depths of penetration. This proves to be particularly difficult in most laminated composite structures. Some investigators have inferred that the SWF is typical only of the first few plies of a laminate.[24] Others have held that the SWF was influenced by all plies throughout a laminate. In the latter cases it appeared that SWF was sensitive to plies at the opposite surface even in relatively thick composite laminates.[25] Ambiguities about wave paths are bothersome because waves reflected by intermediate layers and bondlines are likely to be more relevant than those from the free surface of a structure.

Problem Summary

The acousto-ultrasonic approach attempts to ensure that interrogating waves interact freely with many material parameters. Then, a major signal analysis problem in acousto-ultrasonics is the need for separating interactive waveform components. It becomes necessary to sort out those components that are most relevant to the particular material property or internal condition to be assessed. Alternatively, there may be instances where it is better not to separate waveform components but to deal with the whole signal as a multivariate data set. Several powerful approaches to acousto-ultrasonics signal analysis along both lines are suggested next.

Homomorphic Processing

One of the most promising means for dealing with the signal deconvolution problem is provided by homomorphic signal processing.[26] It has been successfully applied to the analysis of audio acoustic, speech, and seismic signals. Homomorphic processing is necessary for waveforms with components whose Fourier transforms overlap. Homomorphic filters provide the means for deconvolving such waveform components, Fig. 4. A detailed discussion of homomorphic processing methodology is given elsewhere.[27] It is sufficient here to note that the method should lead to improved evaluation of the SWF by isolating key components in waveforms that consist for superimposed multiple echo systems (as in thin composite laminates). Homomorphic processing should be particularly useful in identifying significant wave modes and their paths.

Partition-Regression Method

Another methodology that has proven to be useful is based on partitioning waveforms and their spectra followed by regression analysis.[28] One procedure is to divide the time and frequency domain records into a large number of equal segments. The SWF for each segment is regressed against a material property of interest to find corresponding time and frequency segments that correlate with that property. The procedure is iterated until the correlation is optimized. The basic assumption of the partition-regression method is that relevant signal components and their frequency bands can be separated with simple linear filters.[29] This is in contrast to conditions that indicate homomorphic processing.[26]

When used with either eqs. (3) or (4) the partition-regression method found corresponding time and frequency domain intervals that gave the best correlation with a given material property (i.e., filament-wound composite interlaminar shear strength), Fig. 5.[30] An interesting outcome was that only the least prominent part of the spectrum gave the greatest correlation coefficient with shear strength. That is, while lower frequency components dominated the spectrum, only the higher frequencies varied significantly with interlaminar shear strength variations. In the experiment cited higher frequencies corresponded to the initial oscillations of the acousto-ultrasonic signal while lower frequencies were associated with trailing oscillations.

Finite Element Vibration Analysis

Among factors that are likely to influence the SWF are boundary conditions that govern resonant frequencies of natural vibration modes.[21] The prominent low frequency components in the previously-cited case provide an example of this situation. This indicates a need for independent means for identifying resonant frequency components. But, in heterogeneous, anisotropic materials with uniform properties this need poses a problem.

Because flaw systems in composite materials can be quite complex, finite element analysis may be needed to help define inspection parameters. Accordingly, the use of finite element models has been suggested to study the effect of material anomalies and possible damage (type, severity, area) on boundary conditions and on SWF measurements.[31] Finite element predictions of resonant frequencies and mode shapes can help optimize the location, configuration, and sensitivity requirements for acousto-ultrasonic probes. Developing simple analytical models that incorporate appropriate parameters should reduce trail and error while providing a rational basis for performing SWF measurements.

Method of Moments

Simple structural geometries and boundary conditions combined with judicious selection of transducers and probe configurations may eliminate or substantially reduce ambiguous and irrelevant signal components. In these cases the homomorphic and partitioning methodologies may be unnecessary. Then, with simple noise elimination, the entire acousto-ultrasonic waveform can be retained and analyzed as a whole. For example, excellent correlations have been obtained with SWFs based on essentially raw total waveforms.[18,32]

The method of moments invokes the use of parameters such as those used to describe distribution functions, namely location, scale, and shape parameters.[33] In the frequency domain, in addition to the mean square definition given

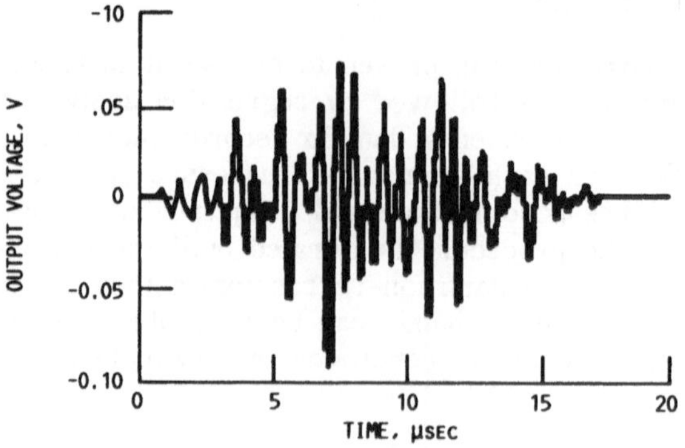

(A) WINDOWED WAVEFORM.

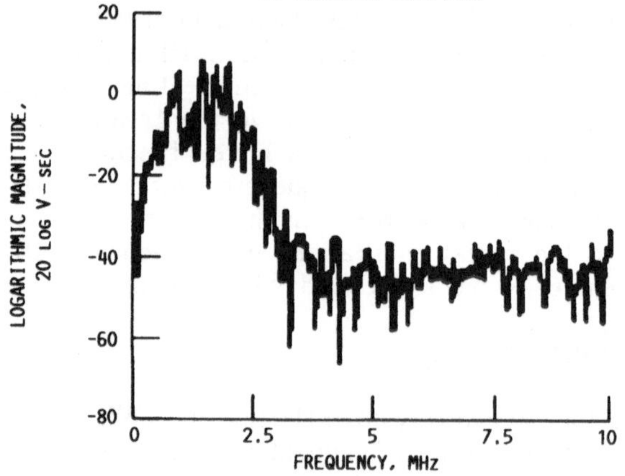

(B) LOG SPECTRUM.

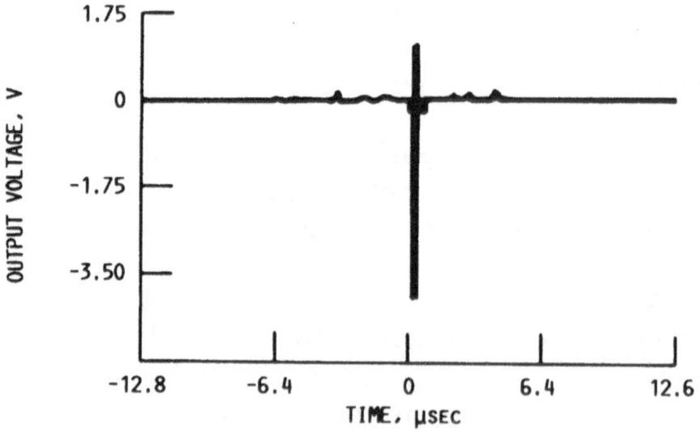

(C) CEPSTRUM WINDOW.

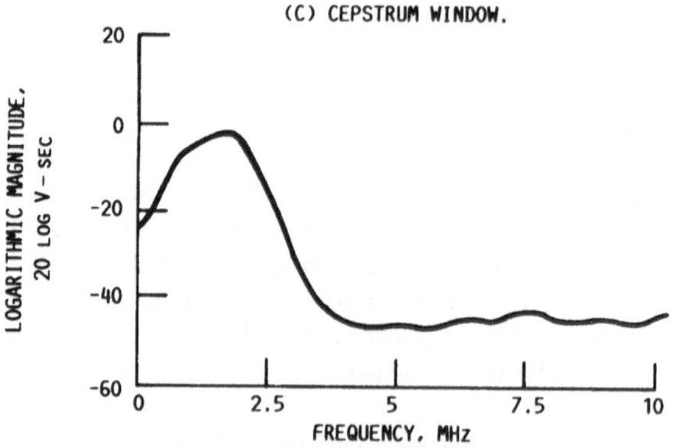

(D) SMOOTHED SPECTRUM.

Fig. 4. Homomorphic analysis of an acousto-ultrasonic waveform: A Hamming window is used to select a portion, A), of the original waveform for analysis. Rapid variations in the log spectrum, B) are removed by cepstrum windowing, C), to get the smoothed log spectrum, D). The procedure is repeated to characterize log spectra of various reflections comprising the waveform, the analysis proceeds by comparison with characteristic log spectra of homomorphic signal components with know modes and ray paths in companion sample.[26]

in eq. (4), the SWF can be defined from full waveform power spectra in terms of various moments and ratios of moments: the centroid, kurtosis, skewness, standard deviation, variance.[28] These quantities can be readily evaluated using digital fast Fourier transforms.[29] This stratagem acknowledges the stochastic nature of attenuating interactions that shape the waveforms and also the validity of a statistical treatment of their spectra.

Pattern Recognition Method

There are a rather large number of possible mechanical properties and morphological conditions that can effect a correspondingly large number of

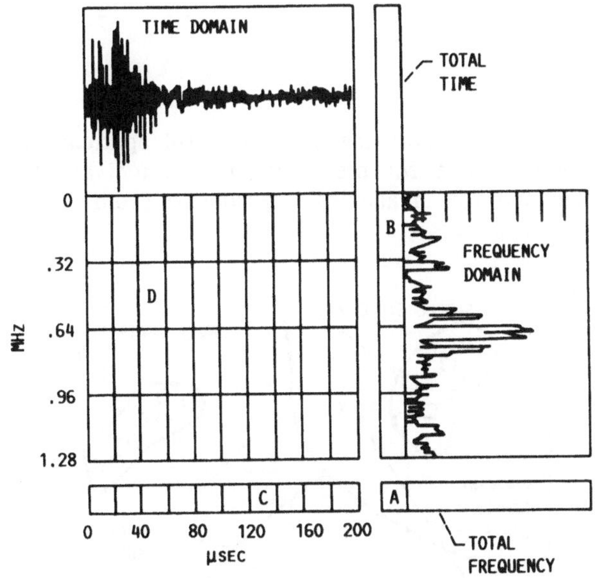

Fig. 5. Waveform partitioning scheme for stress wave factor (SWF) calculations and format for presenting regression coefficents. Block "A" holds the regression coefficent basedon the SWF for the total waveform and spectrum. Block "B" is for the frequency interval 0 to 0.32 MHz and total waveform. Block "C" is for the time interval 120 to 140 μsec and total spectrum. Block "D" represents partitions of both the waveforms and spectra. Blocks are filled with coefficients obtained by regressing SWF values within the intervals against a material property to find the best correlation with that property.[30]

acousto-ultrasonic signal parameters. This suggests determining the state of a material with pattern recognition methodology.[34,35] This methodology involves statistically-based data generation, feature extraction, and classification. By using samples containing known states (i.e., training samples) significant discriminatory pattern vectors can be defined and used for feature extraction, Fig. 6. Then, as described in detail elsewhere,[34] the most likely states in an unknown material can be identified and associated with particular acousto-ultrasonic signal parameters. This "adaptive learning" methodology has been successfully used for analyzing acoustic emission signals and may be a practical alternative to the a priori modeling demanded by the finite element approach mentioned above.

Auxiliary Measurements

Additional ultrasonic measurements may occasionally be needed to corroborate or clarify acousto-ultrasonic measurements. Velocity or attenuation measurements using conventional time-of-flight or pulse-echo methods, respectively, are likely to be quite useful for independently evaluating certain material parameters (e.g., elastic constants).[36,37] Combined with acousto-ultrasonic measurements they can help remove ambiguities and aid in signal analysis. As an example, SWF values combined with surface-parallel velocity measurements gave better correlations with composite laminate shear strength than either the SWF or velocity alone.[18]

EXEMPLARY FINDINGS

General Findings

There are now abundant examples of successful applications that show the viability of acousto-ultrasonics for measuring mechanical strength variations in structural composites and for measuring degradation from cyclic fatigue and

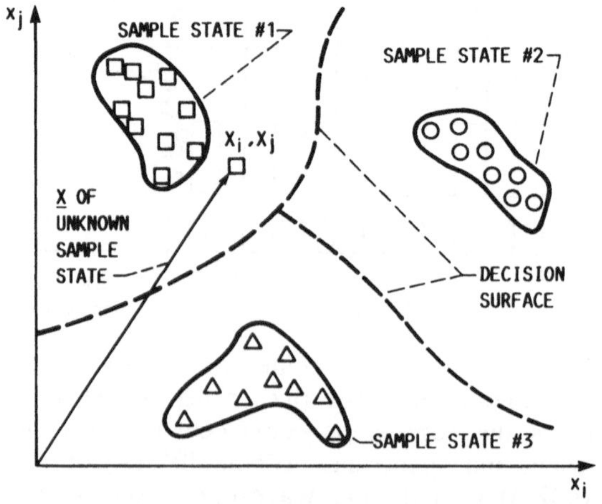

Fig. 6. Schematic illustration of decision surfaces separating feature space into regions corresponding toi distinct sample states arrived at by adaptive learning methodology. Pattern recognition analysis is used to classify an unknown sample as belonging to a particular sample state.[34]

impact damage. Acousto-ultrasonics has also been used to measure adhesive bond strength,[38,39] polymer composite cure state,[40] filler content in wood and paper products,[41] wire rope strength,[42] and porous metal diffusion bond quality.[43] Details for some relevant experimental findings are reviewed next.

SWF Hypothesis

Perhaps the best demonstration of the previously-stated hypothesis regarding SWF measurement of relative strain energy transfer and enhanced relative strength is obtained with unidirectional composite laminates. It is easy to show that the SWF is sensitive to fiber direction and that the SWF is greatest in the fiber direction and least perpendicular to the fiber direction. This agrees with the fact that ultimate tensile strength is also greatest in the fiber direction. The SWF measured perpendicular to the fiber direction will have a low value corresponding to low ultimate strength in that direction (because the load is then sustained only by the matrix.)

When fiber orientations are mixed, the magnitude of the SWF will vary accordingly, Fig. 7.[44] In a laminate consisting of alternating cross-ply layers the 0° plies will transmit stress wave energy more efficiently than 45° plies in turn are more efficient than 90° plies. The SWF will rank laminates according to the proportion of 0° plies aligned with the load direction.

Failure Site Location

Mapping SWF variations along the load axis of tensile specimens is an effective way to identify weak regions.[44,45] These regions will usually have the lowest SWF values. This has proved to be true in a variety of composite laminate tensile specimens having mixed fiber orientations. It is noteworthy that the composites subjected to acousto-ultrasonic evaluation the potential failure loci gave no prior indication of overt flaws such as delaminations. Any flaws, if present, were so minute or diffuse that they were sensed only as low values of the SWF (or greater ultrasonic attenuation).

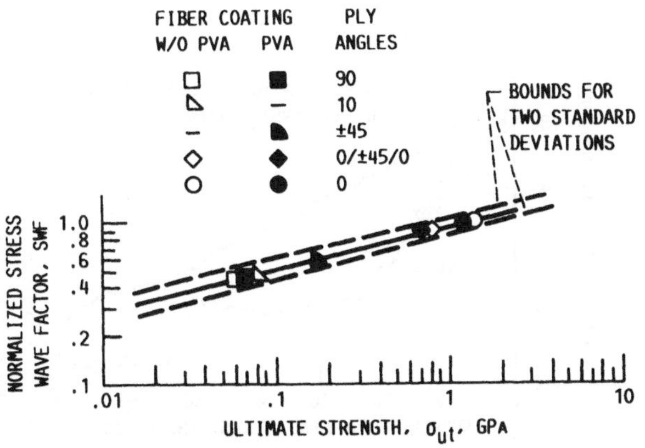

Fig. 7. Normalized SWF as a function of ultimate strength for a series of graphite fiber, epoxy matrix composite laminate specimens. One set of specimens had (polyvinyl alcohol) PVA-coated fibers. Each specimen had eight plies with fiber orientations as indicated. The correlation coefficient for the fitted curve is 0.996.[44]

Apparently, there is a close correspondence between the SWF and stiffness in composite laminates. Moiré interferometry has shown that variations in strain during axial loading agree rather closely with SWF variations along the load axis. Local regions of low SWF correspond to high local displacement in moire fringe patterns and also to failure loci when specimens are loaded to failure.[24]

Degradation Assessment

Among the most useful aspects of acousto-ultrasonics is the ability to assess degradation states in a material. In fiber reinforced composites degradation can result from moisture absorption, chemical attack, cyclic fatigue, or impact damage. Fatigue and impact damage although subtle can significantly change material properties by reducing stiffness, strength, and ultimately, curtail the service life of composite structures.

The SWF is sensitive to progressive degradation in composite laminates subjected to cyclic fatigue. A close relation between decreasing SWF and decreasing stiffness due to cyclic fatigue of composite laminated has been demonstrated.[32] Indeed, SWF appears to be sensitive indicator of local fatigue damage when compared to overall stiffness reduction as measured by the secant modulus, Fig. 8. Apparently, changes in the SWF are related to the accumulation of damage primarily in the form of matrix crazing and fiber breakage distributed throughout the volume being fatigued.[36,46] It has been noted that systematic shifts in certain spectral components of the acousto-ultrasonic waveform tend to accompany fatigue.

The SWF is also a sensitive indicator of accumulated degradation in composite laminates subjected to impact damage, Fig. 9.[47] Changes in the SWF correspond closely to linear reduction in elastic modulus, ultimate tensile strength, and toughness exhibited by unidirectional laminates that sustained increasing number of impacts. In laminates subjected to combined fatigue and impact, it was found that resultant reductions in fatigue life were accompanied by abrupt reductions in the magnitude of the SWF.[48] Correspondences were found between the SWF, the extent and nature of impact or fatigue damage

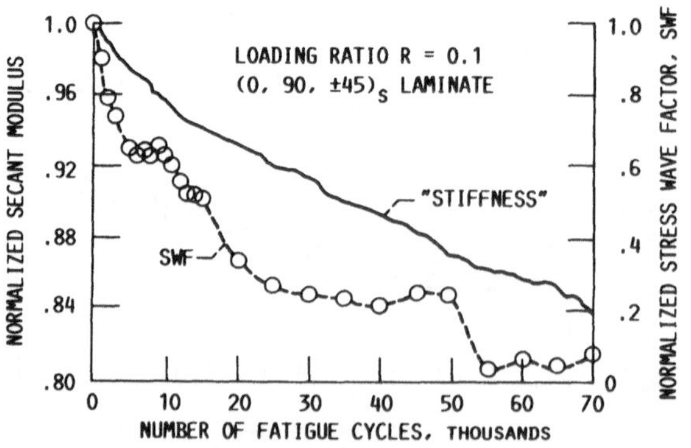

Fig. 8. Convariation of normalized SWF and longitudinal secant modulus with fatigue damage in a graphite/epoxy fiber composite laminate. The difference in slopes is because secant modulus is for full gauge length while SWF is for a local portion of the test specimen.[32]

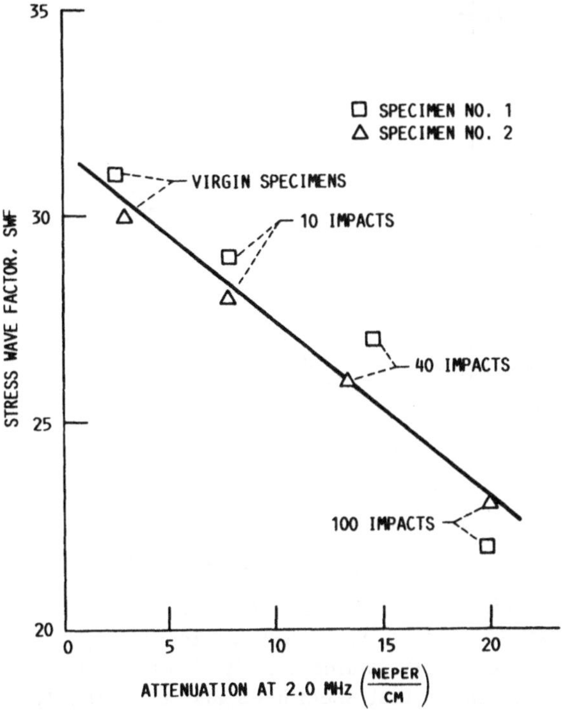

Fig. 9. Stress Wave Factor (SWF) versus through-transmission ultrasonic attenuation at the impact site of a 10-ply unidirectional graphite/ epoxy composite laminate specimen.[47]

(ranging from microcrack populations to fractured fibers), and changes in load response and life.

Interlaminar Strength

High interlaminar strength in composite laminates and high adhesive bond strength in joints are pivotal to structural integrity. Nondestructive measurement of these interface properties is very difficult with conventional NDE approaches because poor bond quality does not always manifest itself as sets of discrete flaws or discontinuities. Usually, conventional NDE methods give only qualitative results.

The acousto-ultrasonics approach has proven particularly useful for quantifying relative bond strength. Excellent correlations have been obtained between interlaminar shear strength in composite laminates as measured by mechanical bend tests and the SWF, Fig. 10.[18,30] Similarly, good correlations have been obtained between the peel-test strength of adhesive bonds and the SWF.[38] Apparently, the acousto-ultrasonic wave interactions are sensitive to the properties of interfaces.

Flaw Detection

Although the emphasis here is on materials characterization, nothing precludes using acousto-ultrasonics for detecting delaminations, hidden impact damage areas, or other overt flaws.[49,50] Indeed, acousto-ultrasonics or other

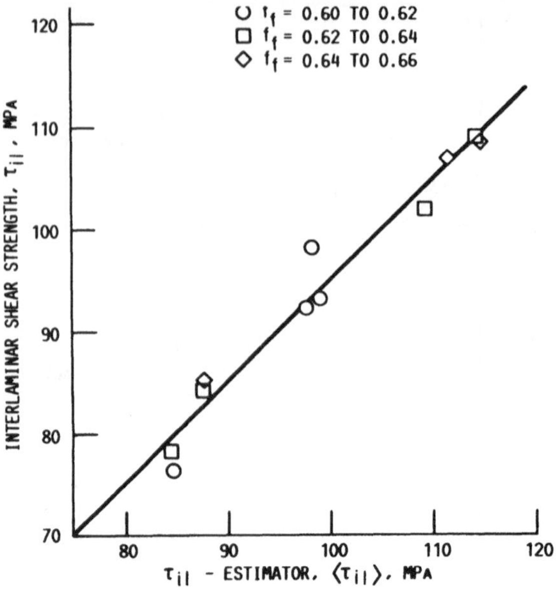

Fig. 10. Correlation between measure interlaminar shear strength and its acousto-ultrasonic estimator for unidirectional graphite/polyimide composite 12-ply laminates. Each data point is the average for ten bend-shear test specimens and f_f is the fiber fraction. The estimator is defined in terms of an equation that combines stress wave factor, E_c, and surface-parallel velocity, v_s, measurements.[18]

NDE methods should be used to identify and help eliminate items that contain obviously harmful flaws.

CONCLUSIONS

The sensitivity of the acousto-ultrasonic approach for detecting and quantifying subtle but significant variations in the strength and fracture resistance of fiber reinforced composites has been experimentally demonstrated. This is somewhat remarkable because, for the most part, it has been accomplished with relatively unsophisticated signal processing and analysis procedures. Nevertheless, the viability of the acousto-ultrasonic approach certainly warrants the development of more advanced instrumentation and signal analysis methodology, along the lines suggested in this paper. Results obtained thus far with polymer matrix composites indicate that the acousto-ultrasonic approach should prove useful for other composite materials, e.g., fiber reinforced metal and ceramic matrix composites. This view follows from similarities in the nature of these materials. Based on the concepts and findings discussed in this paper, the acousto-ultrasonic approach should be considered whenever there is a need to assess integrated defect states and related strength and fracture resistance in composites. The approach merits special consideration to quantify and damage or property degradation after composites are exposed to hostile environments.

REFERENCES

1. M. F. Kanninen and C. H. Popelar, Advanced Fracture Mechanics, Oxford University Press, New York (1985).
2. A. Vary, A Review of Issues and Strategies in Nondestructive Evaluation of Fiber Reinforced Structural Composites, in: "New Horizons -Materials and Processes for the Eighties," SAMPE, Azusa (1979).
3. A. Vary and K. J. Bowles, An Ultrasonic-Acoustic Technique for Nondestructive Evaluation of Fiber Composite Quality, Poly Engrng & Sc., 19:373 (1979).
4. "Nondestructive Methods for Material Property Determination," C. O. Ruud and R. E. Green, Jr. eds., Plenum Press, New York (1984).
5. "Acoustic Emission," J. R. Matthews, ed., Gordon and Breach Science Publishers, New York (1983).
6. A. Vary, Acousto-Ultrasonic Characterization of Fiber Reinforced Composites, Matls Eval. 40:650 (1982).
7. A. Vary, Concepts and Techniques for Ultrasonic Evaluation of Material Mechanical Properties, in: "Mechanics of Nondestructive Testing," W. W. Stinchcomb, ed., Plenum Press, New York (1980).
8. R. E. Green, Jr., Basic Wave Analysis of Acoustic Emission, "Mechanics of Nondestructive Testing," W. W. Stinchcomb, ed., Plenum Press, New York (1980).
9. A. Vary, Ultrasonic Measurement of Material Properties, in: "Research Techniques in Nondestructive Testing, Vol. 4," R. S. Sharpe, ed., Academic Press, London (1980).
10. J-P. Monchalin and R. Heon, Laser Ultrasonic Generation and Optical Detection with a Confocal Fabry-Perot Interferometer, Matls Eval. 44:1231 (1986).
11. D. A. Hutchins, R. J. Dewhurst, S. B. Palmer, and C. B. Scruby, Laser Generation as a Standard Acoustic Source in Metals, Appl Phys. Ltrs. 38:677 (1981).
12. A. Sarrafzadeh-Khoee and J. C. Duke, Jr., Noncontacting Detection in Ultrasonic Nondestructive Evaluation of Materials: Simple Optical Sensor and Fiber-Optic Interferometric Application, Rev of Sc Instr. 57:2321 (1986).
13. A. Sarrafzadeh-Khoee and J. C. Duke, Jr., Small In-Plane/Out-of-Plane Displacement Measurement Using Laser-Speckle interferometry, Exp Tech. 10:18 (1986).
14. J. Krautkramer and H. Krautkramer, "Ultrasonic Testing of Materials, 2nd Edition," Springer-Verlag, New York (1969).
15. Y. Bar-Cohen and D. E. Chimenti, Detection of Porosity in Composites Using Leaky Lamb Waves, in: "Proceedings of the Eleventh World Conference on Nondestructive Testing Vol.III," Taylor Publishing, Dallas (1985).
16. H. Kolsky, "Stress Waves in Solids", Dover, New York (1963).
17. A. Vary, "Ultrasonic Nondestructive Evaluation, Microstructure, and Mechanical Property Interrelations, NASA-TM-86876," NASA, Cleveland (1984).
18. A. Vary and K. J. Bowles, Ultrasonic Evaluation of the Strength of Unidirectional Graphite/Polyimide Composites, in: "Proceeding of the Eleventh Symposium on Nondestructive Evaluation", Southwest Research Institute, San Antonio (1977).
19. J. H. Williams, Jr., H. Karagulle, and S.S. Lee, Ultrasonic Input-Output for Transmitting and Receiving Longitudinal Transducers Coupled to Same Face of Isotropic Elastic Plates, Matls Eval., 40:655 (1982).

20. J. H. Williams, Jr., S. S. Lee, and H. Karagulle, Input-Output Characterization of an Ultrasonic Testing System by Digital Signal Analysis, in: "Analytical Ultrasonics in Materials Research and Testing, NASA CP-2383," A. Vary, ed., NASA, Cleveland (1986).
21. J. H. Williams, Jr., E. B. Kahn, and S. S. Lee, Effects of Specimen Resonances on Acoustic-Ultrasonic Testing, Matls Eval., 41:1502 (1983).
22. J. H. Hemann and G. Y. Baaklini, The Effect of Stress on Ultrasonic Pulses in Fiber Reinforced Composites, SAMPE Journal 22:9 (1986).
23. S. I. Rokhlin, Adhesive Joint Evaluation by Ultrasonic Interface and Lamb Waves, in: "Analytical Ultrasonics in Materials Research and Testing, NASA CP-2383," A. Vary, ed., NASA, Cleveland (1986).
24. A. K. Govada, J. C. Duke, Jr., E. G. Henneke, II, and W. W. Stinchcomb, "A Study of the Stress Wave Factor Technique for the Characterization of Composite Materials, NASA CR-174870," NASA, Cleveland (1985).
25. H. E. Kautz, "Ultrasonic Evaluation of Mechanical Properties of Thick, Multilayered, Filament Wound Composites, NASA TM-87088," NASA, Cleveland (1985).
26. H. Karagulle, J. H. Williams, Jr., and S. S. Lee, Application of Homomorphic Signal Processing to Stress Wave Factor Analysis, Matls Eval. 43:1446 (1985).
27. A. V. Oppenheim and A. S. Willsky with I. T. Young, "Signals and Systems," Prentice-Hall, Englewood Cliffs (1983).
28. R. A. Johnson and D. W. Wichern, "Applied Multivariate Statistical Analysis," Prentice-Hall, Englewood Cliffs (1982).
29. R. N. Bracewell, "The Fourier Transform and Its Applications," McGraw-Hill, New York (1978).
30. H. E. Kautz, "Acousto-Ultrasonic Verification of the Strength of Filament Wound Composite Material, NASA TM-88827," NASA, Cleveland (1986).
31. C. J. Rebello and J. C. Duke, Jr., Factors Influencing the Ultrasonic Stress Wave Evaluation of Composite Material Structures, J. of Comp Tech & Res. 8:18 (1986).
32. J. C. Duke, Jr., E. G. Henneke, II, W. W. Stinchcomb, and K. L. Reifsnider, Characterization of Composite Materials by Means of the Ultrasonic Stress Wave Factor, in: "Composite Structures, 2," I. H. Marshall, ed., Applied Science Publishers, London (1984).
33. A. Govada, E. G. Henneke, II, and R. Talreja, Acousto-Ultrasonic Measurements to Monitor Damage During Fatigue of Composites, in: "1984 Advances in Aerospace Sciences and Engineering," U. Yuceoglu and R. Hesser, eds., American Society of Mechanical Engineers, New York (1984).
34. J. H. Williams, Jr. and S. S. Lee, Pattern Recognition Characterization of Micromechanical and Morphological Materials States via Analytical Quantitative Ultrasonics, in: "Analytical Ultrasonics in Materials Research and Testing, NASA CP-2383," A. Vary, ed., (1986).
35. H. C. Andrews, "Introduction to Mathematical Techniques in Pattern Recognition," Wiley-Interscience, John Wiley and Sons, New York (1972).
36. J. H. Williams, Jr. and B. Doll, Ultrasonic Attenuation as an Indicator of Fatigue Life of Graphite Fiber Epoxy Composite, Matls Eval. 38:33 (1980).
37. D. W. Fitting and L. Adler, "Ultrasonic Signal Analysis for Nondestructive Evaluation," Plenum Press, New York (1981).

38. H. L. M. dos Reis, L. A. Bergman, and J. H. Bucksbee, Adhesive Bond Strength Quality Assurance Using the Acousto-Ultrasonic Technique, Brit J NDT 28:357 (1986).
39. H. L. M. dos Reis and H. E. Kautz, Nondestructive Evaluation of Adhesive Bond Strength Using the Stress Wave Factor Technique, J Acous Em. 5:144 (1986).
40. R. J. Hinrichs and J. M. Thuen, "Control System for Processing Composite Materials," U. S. Patent No. 4,455,268 (1984).
41. H. L. M. dos Reis and D. M. McFarland, On the Acousto-Ultrasonic Characterization of Wood Fiber Hardboard, J Acous Em. 5:67 (1986).
42. H. L. M. dos Reis and D. M. McFarland, On the Acousto-Ultrasonic Non-Destructive Evaluation of Wire Rope Using the Stress Wave Factor Technique, Brit J NDT 28:155 (1986).
43. A. Vary, P. E. Moorhead, and D. R. Hull, Metal Honeycomb to Porous Wireform Substrate Diffusion Bond Evaluation, Matls Eval. 41:942 (1983).
44. A. Vary and R. F. Lark, Correlation of Fiber Composite Tensile Strength with the Ultrasonic Stress Wave Factor, J Test & Eval. 7:185 (1979).
45. J. C. Duke, Jr., E.G. Henneke, II, and W. W. Stinchcomb, "Ultrasonic Stress Wave Characterization of Composite Materials, NASA CR-3976," NASA, Cleveland (1986).
46. J. H. Williams, Jr., H. Yuce, and S. S. Lee, Ultrasonic and Mechanical Characterization of Fatigue States of Graphite Epoxy Composite Laminates, Matls Eval. 40:560 (1982).
47. J. H. Williams, Jr. and N. R. Lampert, Ultrasonic Evaluation of Impact-Damaged Graphite Fiber Composite, Matls Eval. 38:68 (1980).
48. N. Nayeb-Hasemi, M. D. Cohen, J. Zotos, and R. Poormand, Nondestructive Evaluation of Graphite/Epoxy Composite Materials Subjected to Combined Fatigue and Impact, in: "Proceedings of the International Conference and Exposition on Fatigue, Corrosion Cracking, Fracture Mechanics, and Failure Analysis," American Society for Metals, Cleveland (1986).
49. J. E. Green and J. Rodgers, Acousto-Ultrasonic Evaluation of Impact-Damaged Graphite Epoxy Composites, in: "Materials Overview for 1982," SAMPE, Azusa (1982).
50. J. E. Green, J. D. Carlyle, and P. Kukuchek, Impact Damage Epoxy Composites: Impact Testing and NDT Evaluation, in: "RP/C'83: Composite Solutions to Material Challenges," Society of the Plastics Industry, New York (1983).

INTERACTION OF ULTRASONIC WAVES WITH LAYERED MEDIA

Adnan H. Nayfeh and Timothy W. Taylor

Dept. of Aerospace Engrg. & Engrg. Mechanics
University of Cincinnati
Cincinnati, OH 45221

ABSTRACT

A unified theoretical treatment is presented for the interaction of ultrasonic waves with multilayered media. The wave is supposed to be incident from water, at an arbitrary angle, upon a plate consisting of an arbitrary number of different material layers. The composite plate is supported from the bottom by a solid half-space. It is assumed that all solid interfaces are rigidly bonded. Reflection and transmission coefficients are derived for the total system. By examining the behavior of the reflection coefficient, all of the propagating modes are identified. Numerical results are given in order to delineate the influence of the plate material orderings on the propagation process.

INTRODUCTION

In three recent papers[1-3] Nayfeh and Chimenti considered, both analytically and experimentally, the reflected waves from liquid-solid half-spaces interfaces separated by a solid layer of different elastic material in rigid contact to the solid half-space. In the latter[2,3] we derived exact expressions for the reflection coefficients where, as in our earlier paper,[1] only an approximate expression was reported. As was discussed in these papers all physical effects of the reflected beam can be explained by examining the behavior of the appropriate reflection coefficient. Generally speaking, the inclusion of the layer was found to give rise to dispersive effects in both the surface wave speed and the lateral shifting (displacement) of the beam. It was also found[3] that the specific influence of the layer depends highly upon its material properties as compared to that of the substrate. Specifically, the layer can either load or stiffen the substance. The loading situation occurs when the layer's properties (especially its shear wave speed) are smaller than those of the substrate, whereas stiffening occurs for layers stiffer than the substrate.

In this paper we present a generalization of our single layer results to the case of an arbitrary number of layers separating the fluid and solid half-spaces. A more extended version of the analysis is found elsewhere.[4] Among its many applications it can be used to examine the combined. influence of loading and softening layers. This is of particular importance in advanced device applications where there is a need for characterizing material properties of material deposition (in the forms of thin layers) on host substrates.

In order to generalize our single layer result to the multilayer case we use the matrix transfer technique introduced originally by Thomson[5] and somewhat later on by Haskell[6] and others[7-10] for applications in the geophysics, acoustic and electromagnetic fields. According to this technique we construct the propagation matrix for a stack of arbitrary number of layers by extending the solution from one layer to the next while satisfying the appropriate interfacial continuity conditions.

In the "Theoretical Developments" section we summarize analysis for the multilayered system with all rigidly bonded interfaces. This will result in algebraic expressions for the reflection and the transmission coefficients. In the "Numerical Results" section, we present a wide variety of numerical results to delineate the utility of the models.

THEORETICAL DEVELOPMENTS

Formulation of the Problem

Consider a laminated plate consisting of an arbitrary number, n, of elastic isotropic layers rigidly bonded at their interfaces. This plate is assumed to be rigidly attached to an elastic isotropic solid half-space separating it from a fluid half-space. The problem then is to study the reflected beam from the fluid-plate interface for an incident beam originating in the fluid at an arbitrary angle from the normal to the interface.

Guided by our single layer plate analysis,[2] in order to facilitate the present analysis, we shall use two sets of two-dimensional coordinate systems (x,z), as illustrated in Fig. 1. One system is global which has its origin at the substrate-plate interface such that x denotes the propagation direction and z is normal to the interfaces. Here the layered plate will then occupy the space $0 < z < d$ where d denotes the total thickness of the plate. The second system is local for each layer of the plate. Since the plate is made of n layers, the kth layer will then have its local coordinates x and $z^{(k)}$ with local origin at the interface between layers k-1 and k. Hence layer k occupies the space $0 < z^{(k)} < d^{(k)}$, where $d^{(k)}$ is its thickness. In Fig. 2 we display a representative layer k with its appropriate coordinates and boundary field variables.

With this choice of coordinate systems all motions will be independent of the y-direction and the relevant elastodynamic equations for each solid (including each layer and the substrate) consist of the momentum equations

$$\partial\sigma_x/\partial x + \partial\sigma_{xz}/\partial z = \rho\, \partial^2 u/\partial t^2 \quad (1)$$

$$\partial\sigma_z/\partial z + \partial\sigma_{xz}/\partial x + \rho\, \partial^2 w/\partial t^2 \quad (2)$$

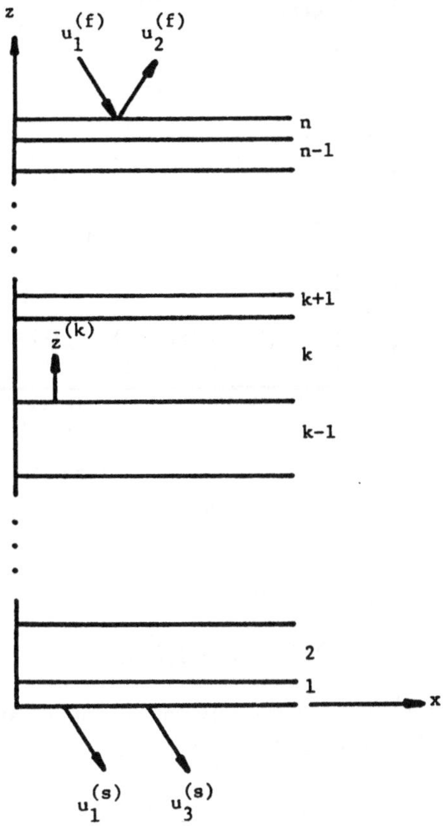

Fig. 1. The laminated plate model.

and the constitutive relations

$$\sigma_x = (\lambda + 2\mu)\partial u/\partial x + \lambda \partial w/\partial z \tag{3}$$

$$\sigma_z = (\lambda + 2\mu)\partial w/\partial z + \lambda \partial u/\partial x \tag{4}$$

$$\sigma_{xz} = \mu(\partial u/\partial z + \partial w/\partial x) \tag{5}$$

where σ_x, σ_{xz} and σ_z are the components of the stress tensor; u and w are the components of the displacements; ρ, λ and μ are the density and elastic constants of each material. Due to the absence of viscosity in the fluid (water) its relevant field equations corresponding to eqs. (1-5) are given by

$$\partial \sigma_x^{(f)}/\partial x = \rho_f \partial^2 u^{(f)}/\partial t^2, \tag{6a}$$

$$\partial \sigma_z^{(f)}/\partial z = \rho_f \partial^2 w^{(f)}/\partial t^2 \tag{6b}$$

$$\sigma_z^{(f)} = \sigma_x^{(f)} = \lambda_f(\partial u^{(f)}/\partial x + \partial w^{(f)}/\partial z) \tag{7}$$

Equations (1-7) must be supplemented with the appropriate interfacial continuity conditions. For rigid bonding between the individual layers of the plate these are

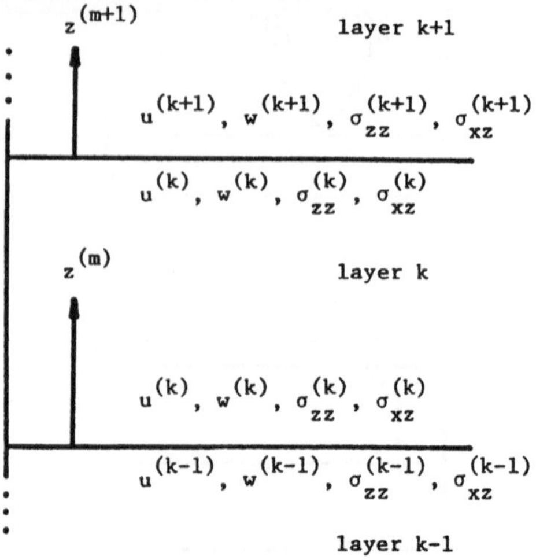

Fig. 2. Representative lamina with interfacial field variables.

$$\sigma_{xz}^{(k)} = \sigma_{xz}^{(k+1)}, \quad \sigma_z^{(k)} = \sigma_z^{(k+1)}$$

$$u^{(k)} = u^{(k+1)}, \quad w^{(k)} = w^{(k+1)} \; , \; k = 1,2,\ldots,n-1 \tag{9}$$

at $z^{(k)} = d^{(k)}$, or $z^{(k+1)} = 0$. Similarly, at the substrate's interface the rigid bonding continuity conditions are given by

$$\sigma_{xz}^{(1)} = \sigma_{xz}^{(s)}, \quad \sigma_z^{(1)} = \sigma_z^{(s)}$$

$$u^{(1)} = u^{(s)}, \quad w^{(1)} = w^{(s)} \tag{10}$$

at $z^{(1)} = 0$, or global $z = 0$. Here subscript (s) designates the substrate. Finally, at the fluid-plate interface, the appropriate bonding conditions are

$$\sigma_{xz}^{(n)} = 0, \quad \sigma_z^{(n)} = \sigma_z^{(f)}, \quad w^{(n)} = w^{(f)} \tag{11}$$

at $z^{(n)} = d^{(n)}$, or global $z = d$.

(b) Analysis

In this subsection we shall describe the propagation process in the plate by solving the field equations in each of its layers and satisfying the interfacial continuity conditions. If for each layer, k, eqs. (1) - (5) are combined into two coupled equations in u and w then a formal solution in the form

$$(u,w) = (U^*, W^*)\exp[iq(x-ct+\alpha z)] \equiv (U,W)\exp(i\alpha z) \tag{12}$$

where U^* and W^* are constant amplitudes, q is the wave number, c is the phase velocity and α is the ratio of the z and x-directions wave numbers is assumed, one gets a characteristic equation relating α to q. This equation admits four solutions and by using superposition one finally gets,

$$\begin{bmatrix} u \\ w \\ \sigma_z^* \\ \sigma_{xz}^* \end{bmatrix}_k = \begin{bmatrix} 1 & 1 & 1 & 1 \\ \alpha_1 & -\alpha_1 & -1/\alpha_2 & 1/\alpha_2 \\ D_1 & D_1 & D_2 & D_2 \\ D_3 & -D_3 & D_4 & -D_4 \end{bmatrix}_k \begin{bmatrix} U_1\exp(iq\alpha_1 z) \\ U_2\exp(-iq\alpha_1 z) \\ U_3\exp(iq\alpha_2 z) \\ U_4\exp(-iq\alpha_2 z) \end{bmatrix}_k \tag{13}$$

where

$$\alpha_1^2 = (c^2/c_L^2) - 1, \quad \alpha_2^2 = (c^2/c_T^2) - 1$$

$$D_1 = \mu[(c^2/c_T^2) - 2], \quad D_2 = -2\mu,$$

$$D_3 = 2\mu\alpha_1, \quad D_4 = (\mu/\alpha_2)[(c^2/c_T^2) - 2],$$

$$\sigma_z^* = \sigma_z/iq \quad \text{and} \quad \sigma_{xz}^*. \tag{14}$$

Since eqs. (13) and (14) hold for each layer k ($k = 1,\ldots,n$), eq. (13) can be used to relate the displacements and stresses at $z^{(k)} = 0$ to those at $z^{(k)} = d^{(k)}$. This can be done by specializing (13) to $z^{(k)} = 0$ and to $z^{(k)} = d^{(k)}$, and eliminating the common amplitude column made up of $U_1^{(k)}$, $U_2^{(k)}$, $U_3^{(k)}$ and $U_4^{(k)}$ resulting in

$$\begin{bmatrix} u^{(k)} \\ w^{(k)} \\ \sigma_z^{*(k)} \\ \sigma_{xz}^{*(k)} \end{bmatrix}_{z^{(k)}=d^{(k)}} = \begin{bmatrix} a_{11} & a_{12} & a_{13} & a_{14} \\ a_{21} & a_{22} & a_{23} & a_{24} \\ a_{31} & a_{32} & a_{33} & a_{34} \\ a_{41} & a_{42} & a_{43} & a_{44} \end{bmatrix} \begin{bmatrix} u^{(k)} \\ w^{(k)} \\ \sigma_{z}^{*(k)} \\ \sigma_{xz}^{*(k)} \end{bmatrix}_{z^{(k)}=0} \tag{15}$$

where

$$[a_{ij}]_k = \begin{bmatrix} B_1 & B_2 & B_3 & B_4 \\ \alpha_1 B_1 & -\alpha_1 B_2 & -B_3/\alpha_2 & B_4/\alpha_2 \\ D_1B_1 & D_1B_2 & D_2B_3 & D_2B_4 \\ D_3B_1 & -D_3B_2 & D_4B_3 & -D_4B_4 \end{bmatrix}_k \begin{bmatrix} 1 & 1 & 1 & 1 \\ \alpha_1 & -\alpha_1 & -1/\alpha_2 & 1/\alpha_2 \\ D_1 & D_1 & D_2 & D_2 \\ D_3 & -D_3 & D_4 & -D_4 \end{bmatrix}_k^{-1} \tag{16}$$

and

$$B_1 = \exp(iq\alpha_1 d^{(k)}), \quad B_2 = \exp(-iq\alpha_1 d^{(k)})$$

$$B_3 = \exp(iq\alpha_2 d^{(k)}), \quad B_4 = \exp(-iq\alpha_2 d^{(k)}). \tag{17}$$

By applying the above procedure for each layer and invoking the continuity relations on the top and bottom of each layer we can finally relate the displacements and stresses at the top of layer n to those at the bottom of layer 1 via the transfer matrix multiplications

$$[A_{ij}] = [a_{ij}]_n[a_{ij}]_{n-1} \ldots [a_{ij}]_1 \tag{18}$$

which can be written in the expanded form

$$\begin{bmatrix} u^{(n)} \\ w^{(n)} \\ \sigma_z{}^{*(n)} \\ \sigma_{xz}{}^{*(n)} \end{bmatrix}_{z=d} = \begin{bmatrix} A_{11} & A_{12} & A_{13} & A_{14} \\ A_{21} & A_{22} & A_{23} & A_{24} \\ A_{31} & A_{32} & A_{33} & A_{34} \\ A_{41} & A_{42} & A_{43} & A_{44} \end{bmatrix} \begin{bmatrix} u^{(1)} \\ w^{(1)} \\ \sigma_z{}^{*(1)} \\ \sigma_{xz}{}^{*(1)} \end{bmatrix}_{z=0} \tag{19}$$

Now, in order to satisfy the remaining continuity conditions (10) and (11) at the substrate-plate and the plate-fluid interfaces, respectively, we need to solve the field equations in the substrate and in the fluid. By inspection, such solutions can be deduced from the formal solution (13). First, due to the absence of shear deformation, specializing (13) to the fluid half-space yields

$$\begin{bmatrix} u^{(f)} \\ w^{(f)} \\ \sigma_z{}^{*(f)} \end{bmatrix} = \begin{bmatrix} 1 & 1 \\ \alpha_f & -\alpha_f \\ \rho_f c^2 & \rho_f c^2 \end{bmatrix} \begin{bmatrix} U_1{}^{(f)} \exp[iq\alpha_f(z-d)] \\ U_2{}^{(f)} \exp[-iq\alpha_f(z-d)] \end{bmatrix} \tag{20a}$$

where

$$\alpha_f{}^2 = (c^2/c_f{}^2) - 1, \quad U_r{}^{(f)} = U_r{}^{*(f)} \exp[iq(x-ct)], \; r = 1,2. \tag{20b}$$

with $U_1{}^{*(f)}$ is the constant amplitude of the incoming wave, $U_2{}^{*(f)}$ is that of the reflected wave and z is the global coordinate. Also, the sub and superscripts f denote quantities belonging to the fluid. Next, specializing (13) to the substrate yields

$$\begin{bmatrix} u^{(s)} \\ w^{(s)} \\ \sigma_z{}^{*(s)} \\ \sigma_{xz}{}^{*(s)} \end{bmatrix} = \begin{bmatrix} 1 & 1 & 1 & 1 \\ \alpha_1 & -\alpha_1 & -1/\alpha_2 & 1/\alpha_2 \\ D_1 & D_1 & D_2 & D_2 \\ D_3 & -D_3 & D_4 & -D_4 \end{bmatrix}_s \begin{bmatrix} U_1{}^{(s)} \exp(iq\alpha_1 z) \\ 0 \\ U_3{}^{(s)} \exp(iq\alpha_2 z) \\ 0 \end{bmatrix}_s \tag{21}$$

were the 4x4 characteristic material matrix in (21) designates the $[a_{ij}]_s$ of the substrate, and $U_1{}^{(s)}$ and $U_2{}^{(s)}$ are related to $U_1{}^{*(s)}$ and $U_2{}^{*(s)}$ in a manner similar to that of eq. (20b). Notice that in eq. (21) the reflected wave amplitudes $U_2{}^{*(s)}$ and $U_4{}^{*(s)}$ vanish since our solutions must be bounded for large values of $\|z\|$ and the substrate is considered to be an infinite half-space. Here also we choose α_1 and α_3 which insure boundedness at infinity. Again, z in (21) is the global coordinate.

By specializing (20a) and (21) to the fluid-plate interface (z=d) and plate-substrate interface (z=0) and followed by invoking the continuity conditions (10) and (11) we finally get

$$\begin{bmatrix} \alpha_f & -\alpha_f \\ \rho_f c^2 & \rho_f c^2 \\ 0 & 0 \end{bmatrix} \begin{bmatrix} U_1{}^{(f)} \\ U_{2(f)} \end{bmatrix} = \begin{bmatrix} R_{21} & R_{23} \\ R_{31} & R_{33} \\ R_{41} & R_{43} \end{bmatrix} \begin{bmatrix} U_1{}^{(s)} \\ U_3{}^{(s)} \end{bmatrix} \tag{22}$$

where $[R_{ij}] = [A_{ij}][a_{ij}]_s$. (23)

Since the incident wave amplitude $U_1^{*(f)}$ is assumed to be known, the matrix eq. (22) represents three equations for three unknowns. It can thus be solved to yield the reflection, longitudinal transmission, and shear transmission coefficients, respectively as

$$R = U_2^{(f)}/U_1^{(f)} = (G_{31} - Q_f G_{21})/(G_{31} + Q_f G_{21}) \quad (24)$$

$$T_L = U_1^{(s)}/U_1^{(f)} = (2\rho_f c^2)/(G_{31} + Q_f G_{21}) \quad (25)$$

$$T_S = U_3^{(s)}/U_1^{(f)} = -(R_{41}/R_{43})T_L \quad (26)$$

where

$$G_{21} = R_{21} - R_{23}(R_{41}/R_{43}),\ G_{31} = R_{31} - R_{33}(R_{41}/R_{43}),\ Q_f = (\rho_f c^2)/\alpha_f. \quad (27)$$

The expression (24) for the reflection coefficient contains, as a by-product, the characteristic equation for the propagation of modified (leaky) Rayleigh surface waves which propagate along the fluid-layered solid interface. The vanishing of the denominator in eq. (24), namely,

$$G_{31} + Q_f G_{21} = 0 \quad (28)$$

defines the characteristic equation for such waves. Furthermore, in the absence of the fluid, i.e., for $\rho_f = 0$, eq. (28) reduces to

$$G_{31} = 0 \quad (29)$$

which defines the characteristic equation for Rayleigh surface waves on the multilayered plate bonded to a semi-infinite solid substrate.

For given real frequency ω (or fd), the real wavenumber solutions $\xi_p = k_r$ of (29) define propagating Rayleigh surface modes. It is important to indicate that in the absence of the plate only a single real solution will exist. This will be the classical surface wave mode which propagates on a half-space. In the presence of the liquid these real wavenumbers will be perturbed rather mildly and become complex. This, of course, is confirmed by eq. (28) which in general admits the complex solutions

$$\xi_p = k_r + i\alpha. \quad (30)$$

From eq. (30) the phase velocity is given as $c_r = \omega/k_r$ and α is the energy leakage coefficient. Notice that α vanishes in the absence of the fluid and, hence, no attenuation (leaking of energy in the fluid) occurs. Hence, in the presence of the fluid these surface waves are called leaky waves. It is also known that c_r is hardly affected by the presence of the fluid.[1-3] However, as has been shown earlier[1-3] c_r is important because it is related to the lateral displacement of the reflected beam; in fact, the beam displacement parameter Δ_s is defined to be equal to $2/\alpha$.

Since we have concluded that the vanishing of G_{31} defines the propagating surface modes, then it is clear from (24) that as $G_{31} \to 0$, $R \to -1$, and we find that we have an alternative method for deducing the leaky wave propagation constant. Accordingly, the reflection coefficient at Rayleigh angle can be represented by expanding its phase factor about the incident wave vector in powers ξ and retaining the leading term[1-3]

$$R(\xi) \sim \exp[i(\xi - k_r)S'(k_r)] \tag{31}$$

where k_r is the Rayleigh wave vector, and $S'(k_r)$ is the derivative with respect to ξ of the phase of R evaluated at k_r.

NUMERICAL RESULTS

For our material menu we choose steel, copper, and chromium; properties of which are collected in Table 1.

Table 1. Material Properties

Material Type	C_L (x10^5 cm/s)	C_E (x10^5 cm/s)	ρ (g/cm^3)
Steel	5.69	3.13	7.9
Copper	4.76	2.32	8.9
Chromium	6.6	4.0	7.2

In all of our numerical calculations we use steel for the substrate. The plate's constituents, on the other hand, can be chosen from all of the menu materials. Since it is known that chromium stiffens steel and that copper loads steel,[3] we shall show that combinations of these materials (to form the plate) can either stiffen or load the steel substrate depending upon their volume fractions and ordering. Without any loss of generality the thickness d of the plate will be kept constant, and the plate's constituents (layers) will be assigned volume fractions adding to unity. Of prime importance is keeping track of the constituents order, however.

Numerical results are presented below in two different categories. In the first, we will illustrate variations of the reflection coefficient and function G_{31} with phase velocity c (or equivalently with incident angle θ since $\sin\theta = c^f/c$). This will be done in order to display the criteria for the surface mode identification. In the second category we present dispersion relations in the form of variations of phase velocities with fd, where f is the frequency and d is the layered plate thickness.

In Figs. 3a-3d, the variations of the real and imaginary parts of the reflection coefficients with phase velocity are shown at four values of fd for a copper plate rigidly bonded to the steel substrate. Also displayed on this figure are normalized values of the corresponding parameters G_{31}. These figures clearly demonstrate the surface wave identification criteria where the real value of the reflection coefficient approaches -1 which also coincides with the rapid variation (through zero) of its phase and the vanishing of G_{31}. Furthermore, at fd=0 the mode occurs at the phase velocity of 2.89 x 10^5 cm/s which is the surface wave speed of steel. This is expected since at the zero frequency limit, i.e., for very long wavelengths, the plate will be essentially "washed" out. As the frequency increases, other modes will

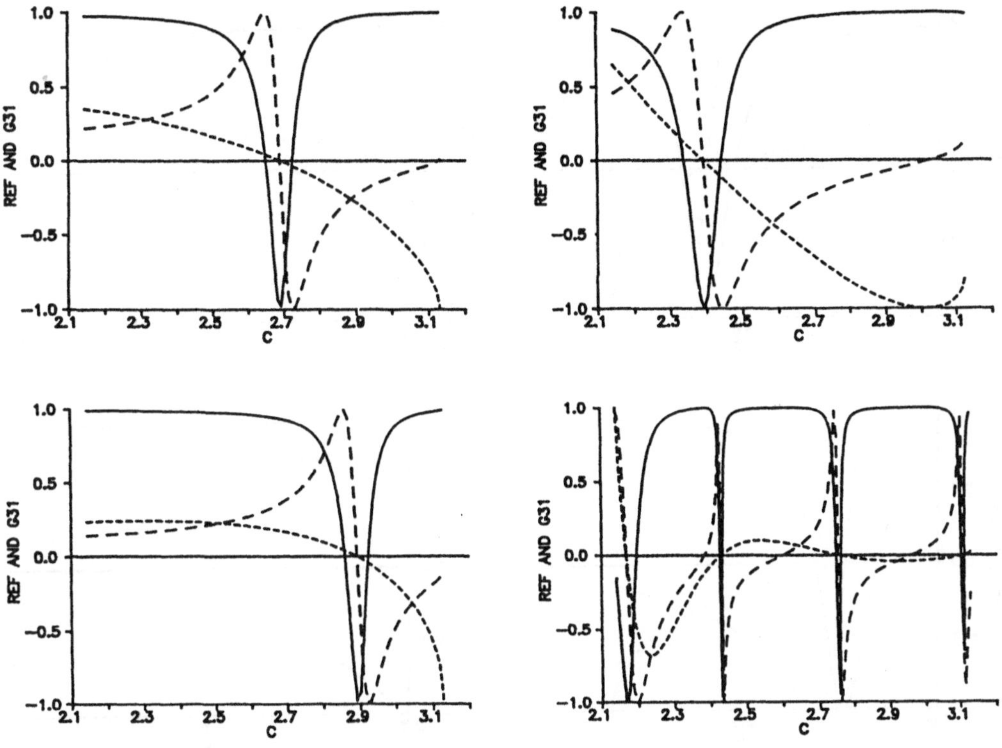

Fig. 3. Variations of the real and imaginary parts of R and normalized values of G_{31} for a copper plate rigidly bonded to a steel half-space. Solid line = real (R), long dash = imag(R), short dash = G_{31}.

appear successively; this behavior is typical of all softening (loading) materials.

In Figs. 4a-4d, similar results are presented for a chromium plate rigidly bonded to the steel substrate. Here, the behavior is entirely different from that of Figs. 3a-3d except at, obviously, the zero frequency limit. As the frequency increases, no other modes appear which is typical of stiffening materials.

Based upon the identification criteria of Figs. 3 and 4, in the two Figs. 5 and 6 dispersion curves are displayed for various plates made up of equal thickness copper and chromium layers. The order of the specific plates chosen are given by c/ch, ch/c for Fig. 5, and c/ch/c/ch/c, ch/c/ch/c/ch for Fig. 6, respectively, where c stands for copper and ch for chromium. Notice, from this group of figures, the influence of layers ordering on the propagation process. The general conclusion is that for plates whose upper layer is copper the phase velocity tends to decrease at higher values of fd and visa versa for plates with chromium top. In fact, in Fig. 7, we confirm this conclusion by presenting, for comparison, dispersion results obtained for a plate made up of a periodic array of 21 copper and chromium layers rigidly bonded to a steel substrate. The solid curve correspond to the case where a copper layer is at the top, and the broken curve corresponds to the change in ordering of the plate's layer, resulting in a chromium layer at the top.

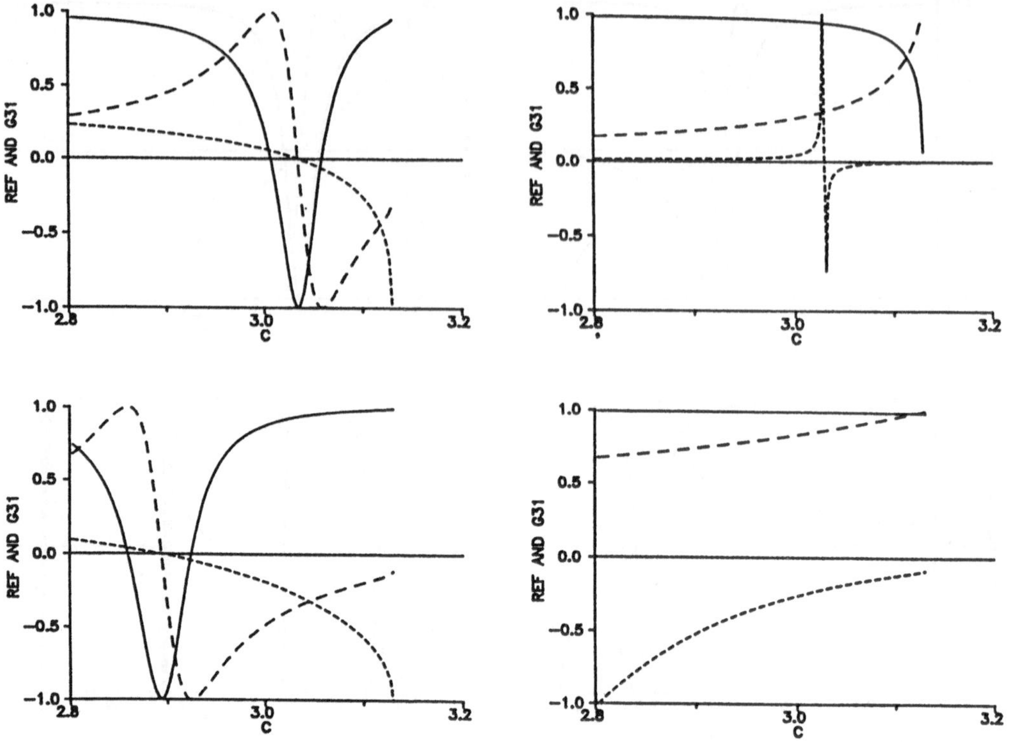

Fig. 4. Variations of the real and imaginary parts of R and normalized values of G_{31} for a chromium plate rigidly bonded to a steel half-space. Solid line = real(R), long dash = imag(R), short dash = G_{31}.

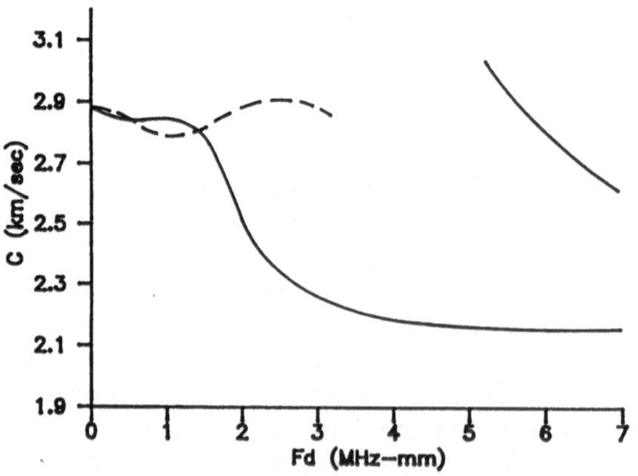

Fig. 5. Dispersion relation curves for a plate composed of two equal thickness layers of copper and chromium and bonded to a steel substrate. The solid lines are for the case when the top layer is copper and the broken line is for the case when the top layer is chromium.

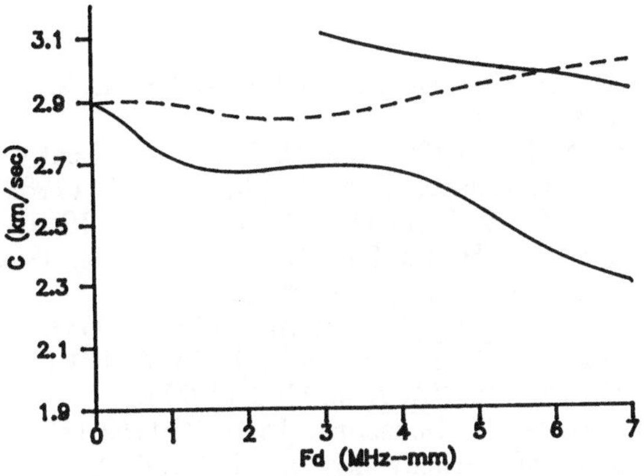

Fig. 6. Dispersion relation curves for a plate composed of a periodic array of 5 equal thickness layers of copper alternating with chromium and bonded to a steel substrate. The solid lines are for the case when the top layer is copper and the broken line is for the case when the top layer is chromium.

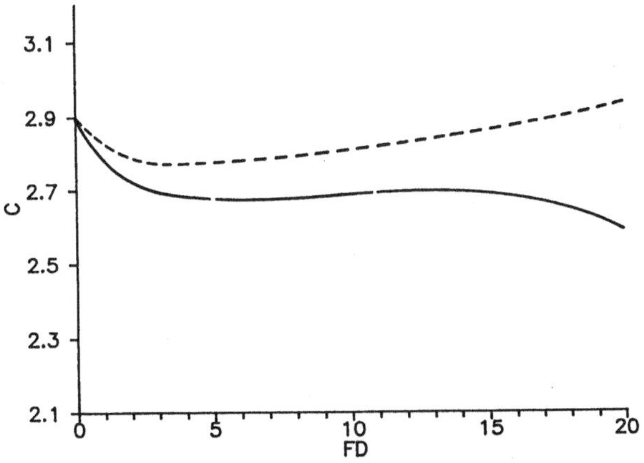

Fig. 7. Dispersion relation curves for a plate bonded to a steel substrate and composed of a periodic array of 21 equal thickness layers of copper alternating with chromium. The solid line is for the case when the top layer is copper and the dashed line is for the case when the top layer is chromium.

ACKNOWLEDGEMENT

This research has been supported by AFOSR 86-0052 grant.

REFERENCES

1. A. H. Nayfeh, D. E. Chimenti, L. Adler, and R. L. Crane, J. Applied Phys. 521:4985 (1981).
2. D. E. Chimenti, A. H. Nayfeh, and D. L. Butler, Leaky Rayleigh Waves on a Layered Halfspace, J. Appl. Phys. 53:170 (1982).
3. A. H. Nayfeh and D. E. Chimenti, JASA 75:1860 (1985).
4. D. E. Chimenti and A. H. Nayfeh, Appl. Phys. Lett. 49(a) (1986).
5. W. T. Thomson, J. Appl. Phys. 21:89 (1950).
6. N. A. Haskell, Bull. Seismol. Soc. Am. 43:17 (1953).
7. D. F. McCammon and S. T. McDaniel, JASA 77:499 (1985).
8. D. Folds and C. Loggins, JASA 62:1102 (1977).
9. P. D. Jackins and G. C. Gaunaurd, JASA 80:1762 (1986).
10. D. B. Bogy and S. M. Gracewski, Int. J. Solids Struc. 20:747 (1984).

DIFFUSE WAVES FOR MATERIALS NDE

R. L. Weaver

Dept. of Theoretical and Applied Mechanics
University of Illinois
Urbana, IL

ABSTRACT

Diffuse Field Analysis (DFA) applies the concepts of statistical room acoustics to the study of diffuse, short wavelength, reverberant ultrasound in solids. Multiply reflected and/or scattered ultrasound has lost all phase information (the coherent part can be said to have fully attenuated) but retains some information in its spectral energy density. DFA studies this spectral energy as it evolves in time and space and deduces information on the source which emitted the ultrasound and on the medium in which it reverberates. There are three major applications which have so far shown promise. These are acoustic emission source characterization, material characterization by internal friction, and microstructure characterization by incoherent transport.

This paper reviews the promise and limitations of Diffuse Field Analysis for the NDE of materials, and in particular, discusses possible applications towards the better understanding of acousto-ultrasonics.

INTRODUCTION

At times long after an ultrasonic source has acted in a finite solid body, long that is compared to an acoustic transit time across the body or between reflections, the ultrasonic field has lost most of its phase coherence. In all but the very simplest of geometries the field after many reflections is essentially incalculable by the theorist. These late time fields will give rise to received signals which appear incoherent, and empty of information. In a recent series of theoretical and experimental papers,[1-9] however, it has been shown that the incoherent spectral energy density E(f), which has dimensions of energy per volume per frequency interval, as studied by DFA, carries information about the source and about the medium. Much of this information is not recoverable from conventional ultrasonic signals.

In the next section we discuss the parameter range in which DFA applies. The past and ongoing applications are then summarized in the following section and finally, possible applications to acousto-ultrasonics (AU) are discussed.

CONSTRAINTS

DFA is a statistical approach attempting to describe the temporal and spatial evolution of acoustic energy. Phase information is discarded at the outset. Certain necessary and reasonable assumptions are made regarding the statistical partition of energy amongst different modes. This assumption of energy "equipartition," or steady state dynamic balance of power flows, is like similar assumptions in thermal physics, architectural acoustics and statistical energy analysis. The assumption is crucial to the development of the theory. In consequence there are certain constraints imposed on the experimental parameters if DFA is to apply. These conditions appear to occur in a variety of circumstances including at least some acousto-ultrasonic configurations. It is therefore suggested that DFA be employed for interpretation and guidance in AU measurements. Acousto-ultrasonics will not become a fully respected NDE technique until it has a firmer grounding in wave propagation theory.

The parameter range in which DFA is appropriate has been investigated in a series of theoretical and experimental papers. The essential requirements are found to be two: The legitimacy of a statistical treatment and the establishment of equipartition, and the collection of sufficient quantities of data (signal duration and bandwidth) for confident assessments of mean square signals in spite of the inevitable stochastic fluctuations.

More specifically, one requires sufficiently weak absorption, σ (measured in units of nepers/msec), sufficiently high frequencies, f, sufficiently large samples, sufficiently short mean free ray paths, ℓ, between random scatterings, sufficiently large bandwidths, Δf, over which one is willing and able to average the mean square signals, and sufficient signal observation time, T. The constraints are:

$$\ell\sigma/c \ll 1 \tag{1}$$

$$(NT\Delta f)^{.5} \gg 1 \tag{2}$$

$$D(f)\Delta f \gg 1 \tag{3}$$

where c is a shear wave speed (most of the energy in a diffuse field is in the form of shear waves), where D(f) is the spectral density of modes[2]

$$D(f) \simeq 8\pi f^2 V/c$$

and V is the specimen volume. ℓ is the average path length of an acoustic ray between randomizing reflections. ℓ is generally no larger than the specimen size and can be as short as typical grain diameters or fiber spacings if the scattering from internal surface is strong. N is the number of independent receivers over which received energy is averaged; N is usually equal to 1.

Constraint number one attempts to assure that the randomizing influences are rapid compared to the decay, so that as the diffuse field decays, it remains in equipartition. The second constraint follows from standard arguments concerning the accuracy of an assessment of spectral power density in a random process. Constraint number three assures that there are a large number of normal modes within the bandwidth Δf and thus that one may do good statistics within the bandwidth. That constraint can be relaxed but only at the expense of increased uncertainty in the precision of theoretical predictions and at the expense of increased spatial fluctuations.

In addition to the above constraints, it is usually necessary that the signal being studied has been allowed to suffer at least a couple of randomizing scatterings. The initial, direct, part of a signal is coherent and depends critically on the nature of the source. The usual procedure has been to wait a short amount of time before recording the data[4].

Much of the work done to data on diffuse ultrasonic fields has been carried out in aluminum plates. Typical parameters are 50 kHz $< f <$ 1 MHz, $c \leq$ 300 cm/msec, $\Delta f \leq$ 10 kHz, $\sigma \simeq$ 0.2 neper/msec, $T \simeq$ 10 msec. Specimen dimensions were typically of order 30 x 30 x 1 cm, so $\ell \simeq$ 30 cm. The above constraints are all obeyed under these conditions. If, however, σ had been substantially higher than 0.2, an observation time of 10 msec would not have been possible because the field would have decayed too quickly. In practice, it has been found that constraint number two has been the most difficult to enforce.

CURRENT APPLICATIONS

In an early theoretical paper[3], it was suggested that source character (orientation, position, duration) can be recovered by measurement of spectral energy density. This follows from the mechanical impedance, $Z(\omega)$, presented to the source being dependent on these parameters. A later study[4] substantiated the suggestion by showing detailed experimental agreement with the predictions.[3] Further work, as yet unpublished, showed that the spectra of force-time source functions and the actual acoustic field energy (in ergs per kHz) could be extracted from the diffuse field measurements even with severely limited data (T = 500 microsec, Δ = 100 kHz).

One of the more readily observed properties of a diffuse field is its temporal decay. Depending on the material, a diffuse field can suffer appreciable damping in times as short as $1/\sigma$ = hundreds of microseconds or as long as $1/\sigma$ = tens of milliseconds. σ is a measure of ultrasonic absorption, or high frequency internal friction, and is loosely related to the better known absorption component α_a of ultrasonic attenuation by $\sigma = \alpha_a c$ where c is the wave speed. The identification is not exact, however, because a diffuse field is composed of all types of waves, each propagating with its own α_a and c. Thus σ is more properly a mode type and volume weighted average of $\alpha_a c$. Even this identification is misleading, though, as σ also has contributions from surface mechanisms and from acoustic energy transmissions into supports. It is clear, however, that σ is unrelated to scattering attenuation α_s.

The decay has been found to follow an exponential law

$$\mathcal{E}(f;t) = \mathcal{E}_o(f) \exp[-2\sigma(f)t]$$

to within experimental accuracies. The phenomenon is shown in Fig. 1 for two different frequencies in a specimen of cast iron. These diffuse field measurements of σ are the first simple measurements of high frequency internal friction in polycrystalline materials. Inasmuch as internal friction is thought to be damage sensitive, measurements of $\sigma(f)$ are promising as an NDE tool.

In extremely heterogeneous media, in which ultrasound suffers from strong and densely packed scatterings (e.g., from grain boundaries, cracks, voids, fibers, delaminations and inclusions depending on the wavelengths) acoustic energy will not be coherently transported across a body. In other words, when attenuation is too high, no direct coherent signal is received. However, one might expect incoherent transport of energy. As a first approximation, one would imagine a random walk process with a mean free path $\ell \simeq 1/\alpha_s$ of the order of the inverse of the shear wave beam attenuation due to scattering. One would then expect the spectral energy density, E(f), to be governed by a heat equation as it evolves in time and space. The diffusion constant associated with this heat equation will be a measure of mean free path and thus in a sense, of scattering loss attenuation and scatterer density.

This idea has been studied[8] in aluminum plates cut with hundreds of densely distributed randomly placed slits. Broadband (0.1 to 1.0 MHz) pulses

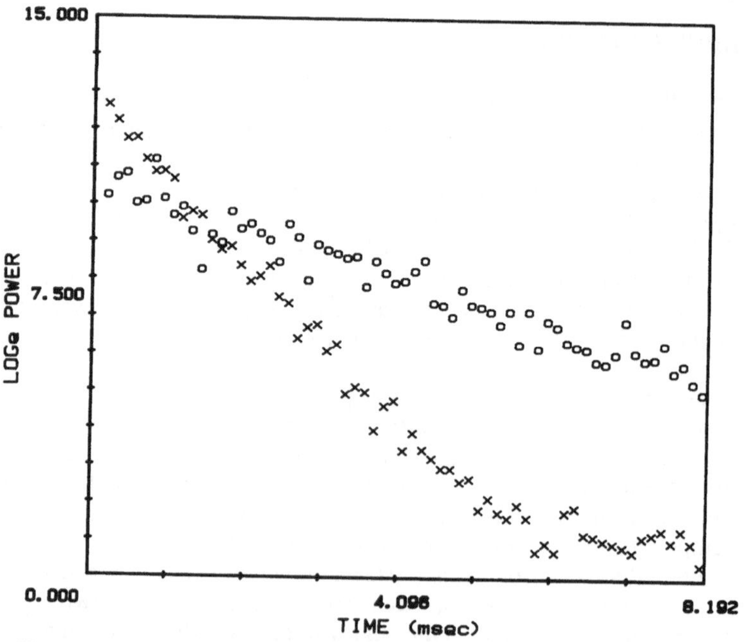

Fig. 1. The Spectral Energy Density is studied in a cast iron specimen as a function of time. The energy in two different frequency bands, centered on 262 and 638 KHz, each with a width of 85 KHz, decay exponentially in time. Note that the higher frequencies decay faster, reaching the level of the background electronic noise after only 7 msec. The specimen measured 10 x 10 x 1.2 cm. Note also the stochastic fluctuations in measured power levels.

of ultrasound were introduced into a corner of the plate and the resulting incoherent energy density $\mathcal{E}(f;x,t)$ was studied as the energy spread over the plate. A diffusion model would expect $\mathcal{E}$ to be governed by a heat equation (with a modification due to internal friction σ)

$$D\nabla^2 e - 2\sigma\mathcal{E} = \partial\mathcal{E}/\partial t \qquad (4)$$

The observed energy density $\mathcal{E}$ was readily fit to the prediction of equation (4), and a best fit diffusion constant D was found. This D was of the expected order, $D \simeq \ell c/4$, where ℓ is a typical interslit spacing of a centimeter and c the shear wave speed.

Similar behavior has been observed by Guo, Holler, and Goebbels[9] at frequencies of several megahertz in large grained steel specimens where the random scattering is presumably due to grain boundaries.

Diffuse fields distribute their energy in a specimen in space, between strain and kinetic energy, and in polarization in characteristic fashion[3,6]. In particular, the relative participation of a given point in the general stochastic motion is dependent upon local values of moduli and material density, upon the frequency, and upon the local geometry. The participation is independent of the location and type of ultrasonic source. By reciprocity, the relative mechanical impedance of that point is dependent in the same way. The predictions of DFA in this regard have been substantiated in the work on source characterization[4] showing the theoretically predicted complicated frequency dependence of that participation. The predictions have also been substantiated in some unpublished work showing that in a specimen composed of equal volumes of brass and steel, the brass held 80% of the acoustic energy. In consequence, the mean square motions in the brass were higher than in the steel. Thus, measurements of participations show promise of sensitivity to local moduli and geometry.

APPLICABILITY TO ACOUSTO-ULTRASONICS

AU, in spite in many reported correlations between damage and AU parameters like the SWF, remains an unorthodox NDE technique. This is partly due to the well known reproducibility problems and the related unexplained spatially fluctuating values of SWF's. It is mostly due, however, to the lack of any deep theoretical basis for the technique beyond a few plausibility arguments.

There would be many advantages to the successful integration of wave propagation theory and AU experimentation. These include the possible explanation and amelioration of the reproducibility-fluctuation problems, the possible deeper explanation of correlations with other NDE techniques, the prospect of theoretical guidance in the design of experiments, and the prospect that theory can suggest new parameters (other than, e.g., the SWF or M_o[5]) to extract from specimens. It appears that the SWF and M_o have been chosen in the past primarily for the ease with which these parameters can be extracted using standard equipment.

The theories developed to date have been concerned with the analysis of what is convenient to call the "direct" signal, whether that signal is studied by means of ray theory and plate surface reflections, or by means of

Lamb modes. In these approaches the material is assumed to be effectively homogeneous, and reflections from plate edges are ignored. In consequence one finds that long time signals, random fluctuations, and heterogeneity are beyond the scope of these theories.

Application of DFA theory is not a priori appropriate. In reviewing the literature one finds that many of the critical parameters, like specimen volume, frequency range, absorption ratio, and wave speeds are not reported. Nevertheless, from the published data and waveforms one can make some guesses.

Govada et al.[10] have introduced their $M_o{}^{.5}$ parameter based on the spectral power in the first 10 or 20 microseconds of (presumably direct) sub Megahertz signal. For DFA purposes, this is clearly an inadequate signal duration time T. Thus M_o cannot be understood by Diffuse Wave Theory. If, however, their signal were to endure for a longer amount of time than the several microseconds which they captured or were it to have greater bandwidth, and were this time to be sufficient to entail a good number of randomizing reflections, DFA would be appropriate.

Henneke et al.[11] show waveforms in composite plates with durations in excess of 200 μsec. However, the frequencies present in the signal after 80 μsec are all low. This feature is presumably due to a stronger σ at high frequencies. Diffuse field analysis is possibly appropriate. The specimen dimensions were of order 30 x 200 x 3 mm, and while reflections from fibers and laminae are presumably weak at these 0 to 1 MHz frequencies (wavelengths of order or greater than 3 mm) randomizing reflections will occur at rates at least of order $c/\ell \simeq (3mm/\mu sec)/30mm$, or every 10 μsec (every 70 μsec if one uses ℓ = 200mm). The presence of only 80 μsec of high bandwidth data will, however, limit the quality and reproducibility of any of their measurements. Hence one would expect small shifts in transducer positions to result in large changes in measured quantities. Diffuse field theory would predict fluctuations in measured spectral powers of a part in $(T\Delta f)^{.5}$. Taking T = 80 μsec, and Δf = 1MHz (parameters chosen by eye from the published waveforms) one expects fluctuations of 11%. Interestingly, this is equal to the level of random spatial fluctuation observed in their plots of SWF versus position.

Vary[12] shows a plot of the signal and spectrum of a typical AU configuration. Durations appear to be of order 500 μsec, bandwidths of order 1 MHz. Diffuse field analysis is clearly appropriate. The quality of the data here would permit one to artificially reduce Δf and T in order to be able to study the dependence of the signal on frequency and its evolution in time.

One would like to be able to say at this point what DFA can tell us about past AU tests, that is, what in fact, they have been measuring, and about future prospects for measuring material properties. It is difficult to be precise in regard to specific past experiments because not all the necessary experimental parameters have been reported. Nevertheless one can readily describe three different general parameter domains in which DFA makes clear predictions.

At very early times (say within 20 μsec) the signal is dominated by direct propagation. The arriving wave, if it has reflected at all, has usually done so coherently with no random influence. The signal should be treated either by ray theory or by guided Lamb wave theory depending on the frequency domain. If material damage is present in the vicinity of the source receiver, however, one can well imagine a strong and random effect on the received signal. For example, intervening damage should backscatter the Lamb waves, preventing them from coherent arrival at a receiver position on the other side of the damaged region and decreasing the power in the received waveform. DFA tells us, though, that unless the bandwidth is large, we can expect strong statistical fluctuations in whatever aspect of the signal is damage dependent. Thus measurements on this time scale will require large bandwidths (many MHz) in order to unambiguously detect the presence of small amounts of damage. This requirement could be ameliorated somewhat by performing local spatial averages as well, in essence increasing the factor N of eq. (2),

One imagines defining an AU parameter R,

$$R \equiv \frac{\int \langle P(f)\rangle_{damage}\, B(f)\, df}{\int \langle P(f)\rangle_{undamaged}\, B(f)\, df}$$

where the brackets indicate a local spatial average of power spectral density P. The power is integrated over the full bandwidth with an arbitrary but smooth frequency dependent balancing factor B(f) inserted to improve the statistics. B should be chosen so as to make the integrand of the denominator roughly constant across the frequency band. Optimal choice of B will weight all frequencies equally and reduce the expected error in R to a part in order $(N\Delta T/2)^{.5}$. The parameter R is similar to the parameter M_o found in Govada et al.[10].

At very late times, if the damping σ allows duration of the signal until such times, the energy field has become smoothly distributed across the entire specimen. At such times there are two material dependent parameters available. As discussed above, σ measures an average internal friction plus the effects of specimen supports and other attachments. It is unclear whether the volume averaged σ will be significantly affected by serious amounts of damage. A damage dependence of σ is plausible, and is indeed observed for machining in aluminum[5] and plastic strains in steels and other metals. A similar effect in composites has not yet been demonstrated. If damage had a significant effect on σ, a SWF or other AU parameter extracted from long time scale signals would depend on damage through σ, large σ leading to weak signals at late times. On such long time scales $T\Delta f$ would be large and stochastic fluctuations of a part in $(T\delta f)^{.5}$ would generally be small. It has been found, though, that σ can fluctuate due to variations in the supports and due to variations in transducer contact. Reproducible measures of σ, and in general of all long time signals, must take these possible fluctuations into account.

The other material parameter available from these long time scale signals is the relative participation of different points of the structure, strong signals indicating high participation of the receiver and/or source positions. Roughly speaking, that would indicate a low value for the local moduli, and therefore localized damage. The theory is complex, though, and not well developed, with polarization dependence in the participation, especially in anisotropic media. It is conceivable that it would be the weak signals that would correspond to localized damage.

In an intermediate time domain, after the field has become randomized but before it has spread throughout the structure, the signal strength will vary strongly in time and space as the energy diffuses according to an equation like eq. (4) with a diffusion constant, D, having spatial dependence due to different mean free paths, ℓ, in damaged and undamaged parts of the specimen. The evolution is in general considerably complicated by possible spatial dependence in σ and likely frequency dependence in D and in σ. Even in the simple case of vanishing σ and spatially constant D, the behavior of any AU parameter dependent upon signal strength, or energy, is complex. If source and receiver are separated by a distance $d > \ell$ then the energy detected at the receiver will rise from zero to a maximum on a time scale of order $\tau_{max} = d^2 \simeq 2d^2/c\ell$ and then fall again at later times as the energy spreads beyond the distance d. This time scale is of order hundreds of microseconds if ℓ is of order centimeters and if typical AU source-receiver separations d are used. More complex behavior would be observed if a non-zero σ were considered and/or spatial dependencies in D and σ. In any case, it is clear that any AU parameter extracted from a signal on this time scale will depend strongly on ℓ and σ. The sign of this dependence and what that in turn implies for damage dependence is not immediately obvious.

ACKNOWLEDGMENT

This work was supported by the National Science Foundation Solid Mechanics Program, grant number MSM-8412178.

REFERENCES

1. D. M. Egle, Diffuse Waves in Solid Media, JASA 70:476 (1981).
2. R. L. Weaver, On Diffuse Waves in Solid Media, JASA 71:1608 (1982).
3. R. L. Weaver, Diffuse Wave in Finite Plates, J. Sound Vib. 94:319 (1984).
4. R. L. Weaver, Laboratory Studies of Diffuse Waves in Plates, J. ASA 79:919 (1986).
5. R. L. Weaver, Diffuse Field Decay Rates for Material Characterization, in: "Solid Mechanics Research for QNDE", Nijhoff, Dordrecht (in press).
6. R. L. Weaver, Diffuse Elastic Waves at a Free Surface, JASA 78:131 (1985).
7. R. L. Weaver, On the Time and Geometry Independence of Elastodynamic Spectral Energy Density, JASA 80:1539 (1986).
8. R. L. Weaver, Indications of Material Character from the Behavior of Diffuse Ultrasonic Fields, in: "Proceedings of the Second International Symposium on Nondestructive Characterization of Materials," (1986).
9. C. B. Guo, P. Holler, and K. Goebbels, Scattering of Ultrasonic Waves in Anisotropic Polycrystalline Metals, Acoustica 59:112 (1985).
10. A. K. Govada, J. C. Duke, Jr., E. G. Henneke, II, and W. W. Stinchcomb, "NASA Contractor Report 174870," NASA, Cleveland (1985).
11. E. G. Henneke, II, J. C. Duke, Jr., W. W. Stinchcomb, A. Govada, and A. Lemascon, "A Study of the Stress Wave Factor Technique for the Char-acterization of Composite Materials, NASA CR-3670," NASA, Cleveland (1983).

12. A. Vary, Ultrasonic Measurement of Material Properties, in: "Research Techniques in Nondestructive Testing, 4," R. S. Sharpe, ed., Academic Press, New York (1980).

Low Frequency Flexural Wave Propagation in Laminated Composite Plates

B. Tang, E. G. Henneke II, and R. C. Stiffler

Department of Engineering Science and Mechanics
Virginia Polytechnic Institute and State University
Blacksburg, VA 24061-4899

ABSTRACT

Shear deformation and rotary inertia are included in plate theory to determine the dispersion curves for flexural waves propagating in laminated composite plates. The results of a unidirectional laminate are compared with the elasticity solutions for flexural waves traveling in transversely isotropic plates to determine the shear correction factors in the low frequency, long wavelength range. The values of the shear correction factors for the unidirectional composite laminate are in good agreement with the theoretical values calculated from static cylindrical bending. An acousto-ultrasonic technique using narrowband excitation frequencies is used to obtain experimental data for flexural waves. By measuring the phase velocities for different excitation frequencies, dispersion curves are generated. There is excellent agreement between the experimentally determined values and the theoretical results for aluminum and unidirectional composite plates. For symmetric cross-ply and quasi-isotropic laminates, the data definitely have the characteristic of a dispersion curve for flexural waves, although the agreement between analytic and experimental results is not quite as good. The results of the present work indicate that the inclusion of shear deformation and rotary inertia in plate theory improves the prediction of dispersion curves for flexural waves propagating in composite laminates and suggest that the acousto-ultrasonic technique can be used to characterize composite plates with and without damage since each material and stacking sequence gives distinct dispersion curves.

INTRODUCTION

Laminated composite plates have become increasingly popular for a wide range of applications, especially for high performance structural components. Usually, in this type of structure, the composite plates are in a complex state of stress. Reliability assurance of the structure requires the availability of nondestructive evaluation techniques. Appropriate nondestructive testing and evaluation of composite plates are needed to evaluate and predict the mechanical properties of the material in the as-received condition and during service loading. Wave propagation in isotropic and anisotropic plates has been the subject of numerous investigations for many years.[1-6] Recently, wave propagation in composite plates has drawn some attention.[7-9] These plate (or Lamb) waves can be utilized as a nondestructive means to characterize laminated composite plates. A simple method of detecting low frequency plate waves has been proposed by Stiffler and Henneke.[10-12] In low frequency plate waves, there are two distinct types of harmonic motion. These are called symmetric or extensional waves and antisymmetric or flexural waves. Stiffler and Henneke employ the classical plate theory in their analysis of flexural wave propagation. The classical plate theory is based on the Kirchhoff hypothesis, which assumes that normals to the midplane before deformation remain straight and normal to the midplane after deformation, and hence neglects transverse shear deformation effects. For homogeneous isotropic thin plates, the effects of the transverse shear deformation are negligible. However, these effects are significant in the case of laminated composite plates due to the relatively low transverse shear modulus. Thus, the effects of transverse shear deformation should be included in the formulation. The present study deals with an analysis of the low frequency flexural wave propagation using a shear deformation theory which incorporates the effects of transverse shear deformation and rotary inertia. Elasticity solutions for the flexural wave propagation in transversely isotropic plates are also presented.

FLEXURAL WAVE PROPAGATION IN LAMINATED COMPOSITE PLATES

The classical plate theory is a special case of the shear deformation theory. In the classical plate theory, the shear moduli are taken to be very large so that the transverse shear deformation can be neglected. A variety of shear deformation theories have been proposed to date. The analysis presented here is based on the work of Yang, Norris and Stavsky [13] who extended Mindlin's theory for homogeneous plates[14] to laminates consisting of an arbitrary number of anisotropic layers.

Considering a laminated composite plate of thickness h , the origin of the global coordinate system is located at the midplane, with the z axis being normal to the midplane, and the x and y axes in the midplane of the plate. The approach of Yang, Norris and Stavsky, which takes into account the effects of transverse shear deformation and rotary inertia, assumes the following displacement field:

$$
\begin{aligned}
u &= u_o(x,y,t) + z\psi_x(x,y,t) \\
v &= v_o(x,y,t) + z\psi_y(x,y,t) \qquad (1) \\
w &= w(x,y,t)
\end{aligned}
$$

where u, v, w are the displacement components in the x, y and z directions, u_o and v_o are the midplane displacement components, and ψ_x and ψ_y are the rotation components along the x and y axes, respectively. From the strain-displacement relations, we have

$$\varepsilon_x = \partial u_o/\partial x + z\partial\psi_x/\partial x$$

$$\varepsilon_y = \partial v_o/\partial y + z\,\partial\psi_y/\partial y$$

$$\varepsilon_z = 0 \tag{2}$$

$$\gamma_{xy} = \partial u_o/\partial y + \partial v_o/\partial x + z(\partial\psi_x/\partial y + \partial\psi_y/\partial x)$$

$$\gamma_{xz} = \psi_x + \partial w/\partial x$$

$$\gamma_{yz} = \psi_y + \partial w/\partial y$$

The stress-strain relations for any layer are given by

$$\begin{bmatrix} \sigma_x \\ \sigma_y \\ \tau_{yz} \\ \tau_{xz} \\ \tau_{xy} \end{bmatrix} = \begin{bmatrix} Q_{11} & Q_{12} & 0 & 0 & Q_{16} \\ Q_{12} & Q_{22} & 0 & 0 & Q_{26} \\ 0 & 0 & Q_{44} & Q_{45} & 0 \\ 0 & 0 & Q_{45} & Q_{55} & 0 \\ Q_{16} & Q_{26} & 0 & 0 & Q_{66} \end{bmatrix} \begin{bmatrix} \varepsilon_x \\ \varepsilon_y \\ \gamma_{yz} \\ \gamma_{xz} \\ \gamma_{xy} \end{bmatrix} \tag{3}$$

where Q_{ij} for i,j = 1,2,6 are plane-stress reduced stiffnesses, and Q_{ij} for i,j = 4,5 are transverse shear stiffnesses. Defining the force and moment resultants per unit length as

$$(N_x, N_y, N_{xy}) = \int_{-h/2}^{h/2} (\sigma_x, \sigma_y, \tau_{xy})dz$$

$$(Q_x, Q_y) = \int_{-h/2}^{h/2} (\tau_{xz}, \tau_{yz})dz \tag{4}$$

$$(M_x, M_y, M_{xy}) = \int_{-h/2}^{h/2} (\sigma_x, \sigma_y, \tau_{xy})zdz$$

where h is the thickness of the plate, we have

$$\begin{bmatrix} N_x \\ N_y \\ Q_y \\ Q_x \\ N_{xy} \\ M_x \\ M_y \\ M_{xy} \end{bmatrix} = \begin{bmatrix} A_{11} & A_{12} & 0 & 0 & A_{16} & B_{11} & B_{12} & B_{16} \\ A_{12} & A_{22} & 0 & 0 & A_{26} & B_{12} & B_{22} & B_{26} \\ 0 & 0 & A_{44} & A_{45} & 0 & 0 & 0 & 0 \\ 0 & 0 & A_{45} & A_{55} & 0 & 0 & 0 & 0 \\ A_{16} & A_{26} & 0 & 0 & A_{66} & B_{16} & B_{26} & B_{66} \\ B_{11} & B_{12} & 0 & 0 & B_{16} & D_{11} & D_{12} & D_{16} \\ B_{12} & B_{22} & 0 & 0 & B_{26} & D_{12} & D_{22} & D_{26} \\ B_{16} & B_{26} & 0 & 0 & B_{66} & D_{16} & D_{26} & D_{66} \end{bmatrix} \begin{bmatrix} \partial u_o/\partial x \\ \partial v_o/\partial y \\ \partial w/\partial y + \psi_y \\ \partial w/\partial x + \psi_x \\ \partial u_o/\partial y + \partial v_o/\partial x \\ \partial\psi_x/\partial x \\ \partial\psi_y/\partial y \\ \partial\psi_x/\partial y + \partial\psi_y/\partial x \end{bmatrix} \tag{5}$$

where the laminate stiffnesses are given by

$$(A_{ij}, B_{ij}, D_{ij}) = \int_{-h/2}^{h/2} (Q_{ij})_k(1,z,z^2)dz \qquad i,j = 1,2,6$$

and $A_{ij} = k_i k_j \int_{-h/2}^{h/2} (Q_{ij})_k dz \qquad i,j = 4,5.$

The shear correction factors $k_i k_j$ are included to account for the fact that the transverse shear strain distributions are not uniform across the thickness of the plate. Neglecting body forces, the equations of motion are

$$\partial N_x/\partial x + \partial N_{xy}/\partial y = \rho^* \partial^2 u_o/\partial t^2 + R\partial^2 \psi_x/\partial t^2$$

$$\partial N_{xy}/\partial x + \partial N_y/\partial y = \rho^* \partial^2 v_o/\partial t^2 + R\partial^2 \psi_y/\partial t^2$$

$$\partial Q_x/\partial x + \partial Q_y/\partial y = \rho^* \partial^2 w/\partial t^2 \tag{6}$$

$$\partial M_x/\partial x + \partial M_{xy}/\partial y - Q_x = R\partial^2 u_o/\partial t^2 + I\partial^2 \psi_x/\partial t^2$$

$$\partial M_{xy}/\partial x + \partial M_{yy}/\partial y - Q_y = R\partial^2 v_o/\partial t^2 + I\partial^2 \psi_y/\partial t^2$$

where $(\rho^*,R,I) = \int_{-h/2}^{h/2} \rho(1,z,z^2)dz$ and ρ is the mass density.

Substituting eqs. (5) into (6), we obtain the equations of motion in terms of the displacements and the rotations as

$$A_{11}\partial^2 u_o/\partial x^2 + 2A_{16}\partial^2 u_o/\partial x\partial y + A_{66}\partial^2 u_o/\partial y^2 + A_{16}\partial^2 v_o/\partial x^2$$
$$+ (A_{12} + A_{66})\partial^2 v_o/\partial x\partial y + A_{26}\partial^2 v_o/\partial y^2 + B_{11}\partial^2 \psi_x/\partial x^2 + 2B_{16}\partial^2 \psi_x/\partial x\partial y$$
$$+ B_{66}\partial^2 \psi_x/\partial y^2 + B_{16}\partial^2 \psi_y/\partial x^2 + (B_{12} + B_{66})\partial^2 \psi_y/\partial x\partial y + B_{26}\partial^2 \psi_y/\partial y^2$$
$$= \rho^* \partial^2 u_o/\partial t^2 + R\partial^2 \psi_x/\partial t^2$$

$$A_{16}\partial^2 u_o/\partial x^2 + (A_{12} + A_{66})\partial^2 u_o/\partial x\partial y + A_{26}\partial^2 u_o/\partial y^2 + A_{66}\partial^2 v_o/\partial x^2$$
$$+ 2A_{26}\partial^2 v_o/\partial x\partial y + A_{22}\partial^2 v_o/\partial y^2 + B_{16}\partial^2 \psi_x/\partial x^2 + (B_{12} + B_{66})\partial^2 \psi_x/\partial x\partial y$$
$$+ B_{26}\partial^2 \psi_x/\partial y^2 + B_{66}\partial^2 \psi_y/\partial x^2 + 2B_{26}\partial^2 \psi_y/\partial x\partial y + B_{22}\partial^2 \psi_y/\partial y^2$$
$$= \rho^* \partial^2 v_o/\partial t^2 + R\partial^2 \psi_y/\partial t^2$$

$$A_{55}(\partial \psi_x/\partial x + \partial^2 w/\partial x^2) + A_{45}(\partial \psi_x/\partial y + \partial \psi_y/\partial x + 2\partial^2 w/\partial x\partial y)$$
$$+ A_{44}(\partial \psi_y/\partial y + \partial^2 w/\partial y^2) = \rho^* \partial^2 w/\partial t^2 \tag{7}$$

$$B_{11}\partial^2 u_o/\partial x^2 + 2B_{16}\partial^2 u_o/\partial x\partial y + B_{66}\partial^2 u_o/\partial y^2 + B_{16}\partial^2 v_o/\partial x^2$$
$$+ (B_{12} + B_{66})\partial^2 v_o/\partial x\partial y + B_{26}\partial^2 v_o/\partial y^2 + D_{11}\partial^2 \psi_x/\partial x^2 + 2D_{16}\partial^2 \psi_x/\partial x\partial y$$
$$+ D_{66}\partial^2 \psi_x/\partial y^2 + D_{16}\partial^2 \psi_y/\partial x^2 + (D_{12} + D_{66})\partial^2 \psi_y/\partial x\partial y + D_{26}\partial^2 \psi_y/\partial y^2$$
$$- A_{55}(\psi_x + \partial w/\partial x) - A_{45}(\psi_y + \partial w/\partial y) = R\partial^2 u_o/\partial t^2 + I\partial^2 \psi_x/\partial t^2$$

$$B_{16}\partial^2 u_o/\partial x^2 + (B_{12} + B_{66})\partial^2 u_o/\partial x\partial y + B_{26}\partial^2 u_o/\partial y^2 + B_{66}\partial^2 v_o/\partial x^2$$
$$+ 2B_{26}\partial^2 v_o/\partial x\partial y + B_{22}\partial^2 v_o/\partial y^2 + D_{16}\partial^2 \psi_x/\partial x^2 + (D_{12} + D_{66})\partial^2 \psi_x/\partial x\partial y$$

$$+ D_{26}\,\partial^2\psi_x/\partial y^2 + D_{66}\,\partial^2\psi_y/\partial x^2 + 2D_{26}\,\partial^2\psi_y/\partial x\,\partial y + D_{22}\,\partial^2\psi_y/\partial y^2$$

$$- A_{45}(\psi_x + \partial w/\partial x) - A_{44}(\psi_y + \partial w/\partial y) = R\,\partial^2 v_o/\partial t^2 + I\,\partial^2\psi_y/\partial t^2$$

For symmetric laminates, the coupling stiffnesses B_{ij}'s and the coupling normal-rotary inertia coefficient R are identically zero. The equations of motion decouple into two sets of equations governing the in-plane and the transverse motions respectively:

$$A_{11}\,\partial^2 u_o/\partial x^2 + 2A_{16}\,\partial^2 u_o/\partial x\,\partial y + A_{66}\,\partial^2 u_o/\partial y^2 + A_{16}\,\partial^2 v_o/\partial x^2$$

$$+ (A_{12} + A_{66})\,\partial^2 v_o/\partial x\,\partial y + A_{26}\,\partial^2 v_o/\partial y^2 = \rho^*\,\partial^2 u_o/\partial t^2$$

$$A_{16}\,\partial^2 u_o/\partial x^2 + (A_{12} + A_{66})\,\partial^2 u_o/\partial x\,\partial y + A_{26}\,\partial^2 u_o/\partial y^2 + A_{66}\,\partial^2 v_o/\partial x^2 \qquad (8)$$

$$+ 2A_{26}\,\partial^2 v_o/\partial x\,\partial y + A_{22}\,\partial^2 v_o/\partial y^2 = \rho^*\,\partial^2 v_o/\partial t^2$$

and

$$A_{55}(\partial\psi_x/\partial x + \partial^2 w/\partial x^2) + A_{45}(\partial\psi_x/\partial y + \partial\psi_y/\partial x + 2\,\partial^2 w/\partial x\,\partial y)$$

$$+ A_{44}(\partial\psi_y/\partial y + \partial^2 w/\partial y^2) = \rho^*\,\partial^2 w/\partial t^2 \qquad (9)$$

$$D_{11}\,\partial^2\psi_x/\partial x^2 + 2D_{16}\,\partial^2\psi_x/\partial x\,\partial y + D_{66}\,\partial^2\psi_x/\partial y^2 + D_{16}\,\partial^2\psi_y/\partial x^2$$

$$+ (D_{12} + D_{66})\,\partial^2\psi_y/\partial x\,\partial y + D_{26}\,\partial^2\psi_y/\partial y^2 - A_{55}(\psi_x + \partial w/\partial x)$$

$$- A_{45}(\psi_y + \partial w/\partial y) = I\,\partial^2\psi_x/\partial t^2$$

$$D_{16}\,\partial^2\psi_x/\partial x^2 + (D_{12} + D_{66})\,\partial^2\psi_x/\partial x\,\partial y + D_{26}\,\partial^2\psi_x/\partial y^2 + D_{66}\,\partial^2\psi_y/\partial x^2$$

$$+ 2D_{26}\,\partial^2\psi_y/\partial x\,\partial y + D_{22}\,\partial^2\psi_y/\partial y^2 - A_{45}(\psi_x + \partial w/\partial x) - A_{44}(\psi_y + \partial w/\partial y)$$

$$= I\,\partial^2\psi_y/\partial t^2$$

For wave propagation, we consider plane waves of the type

$$w = W\,\exp\{i[k(l_1 x + l_2 y) - \omega t]\}$$

$$\psi_x = \Psi_x\,\exp\{i[k(l_1 x + l_2 y) - \omega t]\} \qquad (10)$$

$$\psi_y = \Psi_y\,\exp\{i[k(l_1 x + l_2 y) - \omega t]\}$$

where k is the wave number, l_1 and l_2 are the direction cosines of the wave vector in the x and y directions, respectively, omega is the circular frequency, and W, Ψ_x and Ψ_y are the amplitudes of the plane harmonic waves. Substituting eq. (10) into eq. (9), the determinant of this resulting set of equations gives the characteristic equation for flexural wave propagation. For symmetric quasi-isotropic laminates, in addition to $B_{ij} = 0$ and $R = 0$, we have $A_{16} = A_{26} = A_{45} = 0$ and $D_{16} = D_{26}$. The characteristic equation for flexural wave propagation is

$$\begin{bmatrix} D_{11}k^2l_1^2 + 2D_{16}k^2l_1l_2 + D_{66}k^2l_2^2 + A_{55} - i\omega^2 & D_{16}k^2 + (D_{12} + D_{66})k^2l_1l_2 & iA_{55}kl_1 \\ D_{16}k^2 + (D_{12} + D_{66})k^2l_1l_2 & D_{66}k^2l_1^2 + 2D_{16}k^2l_1l_2 + D_{22}k^2l_2^2 + A_{44} - I\omega^2 & iA_{44}kl_2 \\ iA_{55}kl_1 & iA_{44}kl_2 & -A_{55}k^2l_1^2 - A_{44}k^2l_2^2 \end{bmatrix} = 0 \quad (11)$$

If $l_1 = 1$ and $l_2 = 0$ (i.e. waves propagating in the x direction), eq. (11) gives

$$(D_{11}k^2 + A_{55} - I\omega^2)(D_{66}k^2 + A_{44} - I\omega^2)(A_{55}k^2 - \rho^*\omega^2 2) - (D_{16}k^2)^2(A_{55}k^2 - \rho^*\omega^2) - (A_{55}k)^2(D_{66}k^2 + A_{44} - I\omega^2) = 0 \quad (12)$$

If $l_1 = 0$ and $l_2 = 1$, the characteristic equation is

$$(D_{22}k^2 + A_{44} - I\omega^2)(D_{66}k^2 + A_{55} - I\omega^2)(A_{44}k^2 - \rho^*\omega^2) - (D_{16}k^2)^2(A_{44}k^2 - \rho^*\omega^2) - (A_{44}k)^2(D_{66}k^2 + A_{55} - I\omega^2) = 0 \quad (13)$$

For symmetric cross-ply laminates, if $l_1 = 1$ and $l_2 = 0$, we obtain

$$(D_{11}k^2 + A_{55} - I\omega^2)(A_{55}k^2 - \rho^*\omega^2) - A_{55}^2k^2 = 0 \quad (14)$$

and if $l_1 = 0$ and $l_2 = 1$, we have

$$(D_{22}k^2 + A_{44} - I\omega^2)(A_{44}k^2 - \rho^*\omega^2) - A_{44}^2k^2 = 0 \quad (15)$$

In each situation, the characteristic equation has more than one root. However, only one root approaches zero circular frequency as the wave number approaches zero, and this is the root corresponding to the lowest antisymmetric branch of the frequency spectrum for plate waves. It is apparent that the phase velocity omega/k of the flexural wave is dispersive since it is a function of the wave number. Equations (12-15) are the dispersion relations for flexural waves traveling in symmetric quasi-isotropic and symmetric cross-ply laminates. A similar approach has been investigated by Sun and Tan.[15] However, shear correction factors are not included in their model.

FLEXURAL WAVES IN TRANSVERSELY ISOTROPIC PLATES

Considering an infinite plate of transversely isotropic material with the boundary surfaces at $z = \pm h/2$ and the axis of symmetry in the x direction, the stress-strain relations for the transversely isotropic material are given by

$$\begin{bmatrix} \sigma_x \\ \sigma_y \\ \sigma_z \\ \tau_{yz} \\ \tau_{xz} \\ \tau_{xy} \end{bmatrix} = \begin{bmatrix} C_{11} & C_{12} & C_{12} & 0 & 0 & 0 \\ C_{12} & C_{22} & C_{23} & 0 & 0 & 0 \\ C_{12} & C_{23} & C_{22} & 0 & 0 & 0 \\ 0 & 0 & 0 & (C_{22} - C_{23})/2 & 0 & 0 \\ 0 & 0 & 0 & 0 & C_{55} & 0 \\ 0 & 0 & 0 & 0 & 0 & C_{55} \end{bmatrix} \begin{bmatrix} \varepsilon_x \\ \varepsilon_y \\ \varepsilon_z \\ \gamma_{yz} \\ \gamma_{xz} \\ \gamma_{xy} \end{bmatrix} \quad (16)$$

Also, the strain-displacement relations are

$$
\begin{aligned}
\varepsilon_x &= \partial u/\partial x \\
\varepsilon_y &= \partial v/\partial x \\
\varepsilon_z &= \partial w/\partial z \\
\gamma_{xy} &= \partial u/\partial x + \partial v/\partial x \\
\gamma_{xz} &= \partial u/\partial z + \partial w/\partial x \\
\gamma_{yz} &= \partial v/\partial z + \partial w/\partial x
\end{aligned}
\tag{17}
$$

where u, v, w are the displacement components in the x, y and z directions, respectively. Our problem is to solve the equations of motion

$$
\begin{aligned}
\partial\sigma_x/\partial x + \partial\tau_{xy}/\partial x + \partial\tau_{xz}/\partial z &= \rho\,\partial^2u/\partial t^2 \\
\partial\tau_{xy}/\partial x + \partial\sigma_y/\partial x + \partial\tau_{yz}/\partial z &= \rho\,\partial^2v/\partial t^2 \\
\partial\tau_{xz}/\partial x + \partial\tau_{yz}/\partial x + \partial\sigma_z/\partial z &= \rho\,\partial^2w/\partial t^2
\end{aligned}
\tag{18}
$$

subjected to the traction-free boundary conditions

$$\sigma_z = \tau_{xz} = \tau_{yz} = 0 \qquad \text{at} \qquad z = \pm h/2 \tag{19}$$

Green [16] suggested that for waves propagating in the plane of the plate in a direction making an angle with the x axis the displacements have the form

$$
\begin{aligned}
u &= U(z)\sin[k(l_1x + l_2y) - \omega t] \\
v &= V(z)\sin[k(l_1x + l_2y) - \omega t] \\
w &= W(z)\cos[k(l_1x + l_2y) - \omega t]
\end{aligned}
\tag{20}
$$

For flexural waves, the lateral displacement w must be an even function of z, while the in-plane displacements u and v must be odd in z. Substituting eq. (20) into eqs. (17) and (16), we obtain

$$
\begin{aligned}
\sigma_x &= [C_{11}kl_1U + C_{12}(kl_2V + \partial W/\partial z)]\cos[k(l_1x + l_2y) - \omega t] \\
\sigma_y &= [C_{12}kl_1U + C_{22}kl_2V + C_{23}\partial W/\partial z]\cos[k(l_1x + l_2y) - \omega t] \\
\sigma_z &= [C_{12}kl_1U + C_{23}kl_2V + C_{22}\partial W/\partial z]\cos[k(l_1x + l_2y) - \omega t] \\
\tau_{xy} &= C_{55}k(l_2U + l_1V)\cos[k(l_1x + l_2y) - \omega t] \\
\tau_{xz} &= C_{55}(\partial U/\partial z - kl_1W)\sin[k(l_1x + l_2y) - \omega t] \\
\tau_{yz} &= [(C_{22} - C_{23})/2](\partial V/\partial z - kl_2W)\sin[k(l_1x + l_2y) - \omega t]
\end{aligned}
\tag{21}
$$

Introducing the stresses, eq. (21), into eq. (18), the equations of motion become

$$-C_{55}\,\partial^2U/\partial z^2 + k^2(C_{11}l_1^2 + C_{55}l_2^2 - \rho v^2)U + (C_{12} + C_{55})k^2l_1l_2V + (C_{12} + C_{55})kl_1\,\partial W/\partial z = 0$$

$$(C_{12} + C_{55})k^2l_1l_2U - [(C_{22} - C_{23})/2]\partial^2V/\partial z^2 + k^2(C_{55}l_1^2 + C_{22}l_2^2 - \rho v^2)V + \{[(C_{22} - C_{23})/2] + C_{23}\}kl_2\,\partial W/\partial z = 0 \qquad (22)$$

$$(C_{12} + C_{55})kl_1\,\partial U/\partial z + \{[(C_{22} - C_{23})/2] + C_{23}\}kl_2\,\partial V/\partial z + C_{22}\,\partial^2W/\partial z^2 - k^2(C_{55}l_1^2 + [(C_{22} - C_{23})/2]l_2^2 - \rho v^2)W = 0$$

where v, which is equivalent to ω/k, is the phase velocity. From the solution of the isotropic plate problem [1-5] and the requirements that W(z) must be even, and U(z) and V(z) must be odd for flexural waves propagating in the plate, we let

$$U = A \sinh kpz$$

$$V = B \sinh kpz \qquad (23)$$

$$W = C \cosh kpz$$

Substituting these expressions into the equations of motion, (22), and setting the determinant of the resulting system of three homogeneous equations to zero for nontrivial solutions, the outcome takes the form

$$\begin{bmatrix} C_{11}l_1^2 + C_{55}l_2^2 & (C_{12} + C_{55})l_1l_2 & (C_{12} + C_{55})pl_1 - C_{55}p^2 - \rho v^2 \\ (C_{12} + C_{55})l_1l_2 & C_{55}l_1^2 + C_{22}l_2^2 - [(C_{22} - C_{23})/2]p^2 - \rho v^2 & \{[(C_{22} - C_{23})/2] + C_{23}\}pl_2 \\ (C_{12} + C_{55})pl_1 & \{[(C_{22} - C_{23})/2] + C_{23}\}pl_2 & -C_{55}l_1^2 - [(C_{22} - C_{23})/2]l_2^2 + C_{22}p^2 + \rho v^2 \end{bmatrix} = 0 \qquad (24)$$

If $l_1 = 1$ and $l_2 = 0$ (i.e. waves propagating in the x direction), V is uncoupled from U and W in eq. (22). Therefore, eq. (24) gives

$$(C_{11} - C_{55}p^2 - \rho v^2)(C_{55} - C_{22}p^2 - \rho v^2) + (C_{12} + C_{55})^2p^2 = 0 \qquad (25)$$

for nontrivial solutions of U and W. Consequently, the general solution of eq. (22) is

$$U = A_1\sinh kp_1z + A_2\sinh kp_2z$$

$$W = [-(C_{11} - C_{55}p_1^2 - \rho v^2)/(C_{12} + C_{55})][A_1/p_1]\cosh kp_1z - [(C_{11} - C_{55}p_2^2 - \rho v^2)/(C_{12} + C_{55})][A_2/p_2]\cosh kp_2z \qquad (26)$$

where A_1 and A_2 are arbitrary constants, and p_1^2 and p_2^2 are the roots of eq. (25). With the boundary conditions, eq. (19), we arrive at

$$A_1[C_{12}(C_{12} + C_{55}) - C_{22}(C_{11} - C_{55}p_1^2 - \rho v^2)]\sinh kp_1(h/2)$$

$$+ A_2[C_{12}(C_{12} + C_{55}) - C_{22}(C_{11} - C_{55}p_2^2 - \rho v^2)]\sinh kp_2(h/2) = 0 \qquad (27)$$

$$A_1[C_{11} + C_{12}p_1^2 - \rho v^2](\cosh kp_1(h/2)/p_1)$$

$$+ A_2(C_{11} + C_{12}p_2^2 - \rho v^2)(\cosh kp_2(h/2)/p_2) = 0$$

Finally, in order to have a solution to our problem, eq. (27) yields the condition

$$(C_{11} + C_{12}p_2^2 - \rho v^2)[C_{12}(C_{12} + C_{55}) - C_{22}(C_{11} - C_{55}p_1^2 - \rho v^2)]P_1 \qquad (28)$$

$$- (C_{11} + C_{12}p_1^2 - \rho v^2)[C_{12}(C_{12} + C_{55}) - C_{22}(C_{11} - C_{55}p_2^2 - \rho v^2)]P_2 = 0$$

where $P_1 = p_1\tanh kp_1(h/2)$ and $P_2 = p_2\tanh kp_2(h/2)$

Thus, the phase velocity v is a function of kh and eq. (28) is the dispersion relation for flexural waves traveling in the x direction. Similarly, the condition for flexural waves traveling in the y direction of an infinite transversely isotropic plate having traction-free boundary surfaces is

$$[(C_{22} - C_{23}) - \rho v^2]^2\tanh kp_1(h/2) = (C_{22} - C_{23})^2 p_1 p_2 \tanh kp_2(h/2) \qquad (29)$$

with p_1^2 and p_2^2 satisfying the equation

$$(C_{22} - C_{22}p^2 - \rho v^2)\{[(C_{22} - C_{23})/2] - [(C_{22} - C_{23})/2]p^2 - \rho v^2\} = 0 \qquad (30)$$

EXPERIMENTAL PROCEDURE

An acousto-ultrasonic technique suggested by Stiffler and Henneke [10-12] was used to obtain experimental data for dispersion curves. Low frequency plate waves, usually in the range of 10 kHz to 1 MHz, were generated in our experiments. A summary of the experimental procedure is given here. Figure 1 is a schematic diagram of the experimental set-up for phase velocity measurements in a plate. A piezoelectric transducer, which acted as an acoustic generator, was excited by a gated sine wave. The gated sine wave provided an excitation signal with a very narrow frequency band. A second piezoelectric transducer was used as a receiver. The position of the receiver was advanced a known distance, and the change in time for a particular phase point was noted on a digitizing oscilloscope. Dividing the displacement by the time yielded the phase velocity. Then, the wavenumber was calculated from the phase velocity for the frequency at which the measurement was obtained. Next, the frequency f and the wave number k were each multiplied by the thickness of the plate and a dispersion plot was then made of fh versus kh. The phase velocity for any fh is related to the ratio fh/kh, and the group velocity is related to the slope of the tangent to the dispersion curve.

It was noted that as the frequency of the excitation signal was increased, there existed a frequency at which an abrupt change in the phase velocity took place. Further investigation suggested that this occurred due to a change in modes from the flexural wave propagation at low frequencies to the extensional wave propagation at high frequencies. This transitional

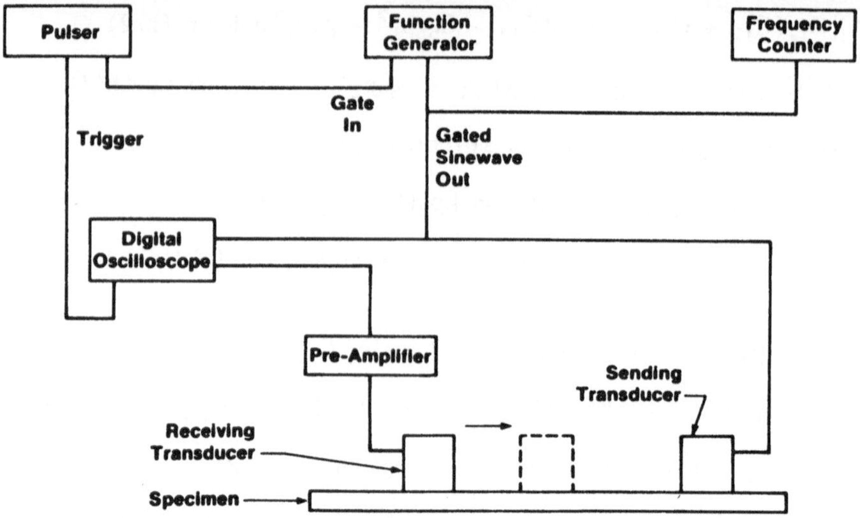

Fig. 1. Schematic diagram of the experimental set-up for phase velocity measurements in a plate.

frequency is related to the center frequency of the sending transducer and the thickness of the plate. Also, from observing the received signals, it was noted that the signals at low frequencies were dispersive while the signals at high frequencies were not. A similar method has been studied by Liu.[17] Liu used a shear wave transducer to generate flexural, extensional and SH modes in graphite/aluminum composite plates, and used an electromagnetic acoustic transducer (EMAT), which was not sensitive to the SH modes, to detect the signals.

DATA AND NUMERICAL RESULTS

The previous sections have provided the elasticity solutions for flexural waves propagating in transversely isotropic plates and the approximate solutions obtained by shear deformation theory for flexural waves traveling in composite laminates. We will restrict our discussion to the long wavelength range since all our experimental data were taken with wavelengths which are much larger than the thickness of the plate; thus the interfaces between plies can be neglected. An AS-4(Gr)/Pr 288 $[0]_8$ graphite/epoxy composite plate supplied by General Electric can be treated approximately as a transversely isotropic plate with the axis of symmetry in the fiber direction. The lamina properties of AS-4(Gr)/Pr 288 graphite/epoxy are $E_{11} = 18.30 \times 10^6$ lb/in^2 (1.26×10^2 GPa), $E_{22} = 1.30 \times 10^6$ lb/in^2 (8.96GPa), $G_{12} = G_{13} = 0.66 \times 10^6$ lb/in^2 (4.55GPa), $G_{23} = 0.56 \times 10^6$ lb/in^2 (3.86GPa), $\nu_{12} = \nu_{13} = 0.32$ and $\rho = 1.487 \times 10^{-4}$ lbf-sec^2/in^4 (1,589kg/m^3). In Fig. 2, the lowest branch of the dispersion curves from the elasticity solution (ES) for flexural waves traveling in the fiber direction of the $[0]_8$8 graphite/epoxy laminate is plotted along with the numerical solutions obtained by the shear deformation theory with different values of shear correction factor $k_5{}^2$ for the $[0]_8$ laminate. The dispersion curves are normalized with respect to the thickness of the composite plate, thus the curves are independent of the ply thickness. The elasticity solution (ES) for waves propagating perpendicular to the fiber

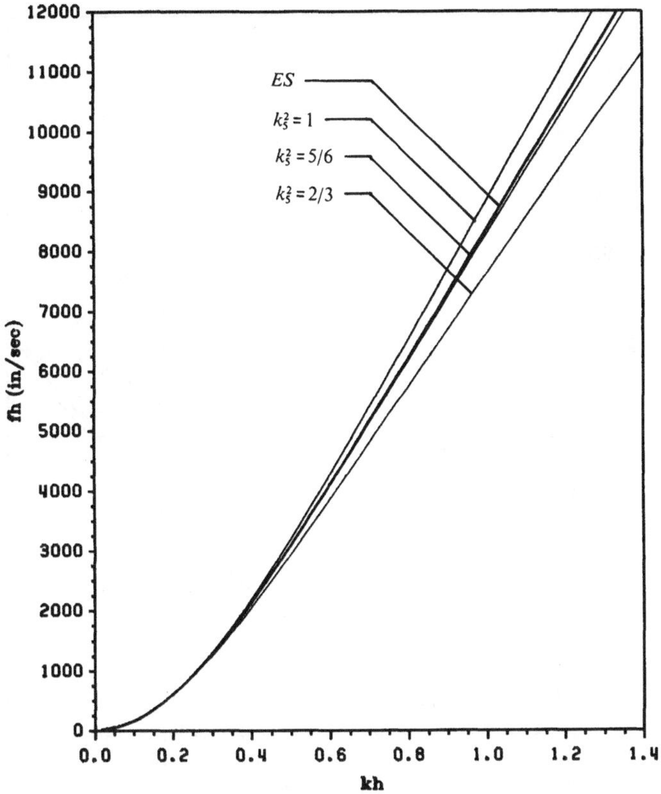

Fig. 2. Theoretical dispersion curves for low frequency flexural waves propagating along the fiber direction of the $[0]_8$ graphite/epoxy laminate. (1in/sec = 0.0254m/sec)

direction and the shear deformation theory solutions with different values of k_4^2 for the $[0]_8$ laminate are showed in Fig. 3. Judging from these two figures, it is acceptable to assume that $k_4^2 = k_5^2 = 5/6$ for this unidirectional graphite/epoxy composite laminate in the low frequency, long wavelength range. Experimental data are compared with the results obtained by the classical plate theory (CPT),[10-12] the elasticity solution (ES) and the shear deformation theory (SDT) with $k_4^2 = k_5^2 = 5/6$ in Figs. 4 and 5 for waves propagating parallel and perpendicular to the fiber direction, respectively.

In order to apply the shear deformation theory to multiply laminated composite plates, the selection of appropriate values for the shear correction factors is necessary. Chow[18] and Whitney,[19] using a procedure which follows the approach of Reissner[20] for homogeneous isotropic plates, have determined k_4^2 and k_5^2 for cross-ply laminates under static cylindrical bending. In general, k_4^2 and k_5^2 have different values for symmetric laminates. The same procedure can be applied to symmetric quasi-isotropic laminates. It has been shown that the factors calculated from static cylindrical bending yield a good approximation to the exact solution for cross-ply laminates subjected to static or buckling loading.[19,21] Using this approach, the values of the shear correction factors are found to be 5/6 for homogeneous isotropic plates. This is the classic value determined by Reissner.[20]

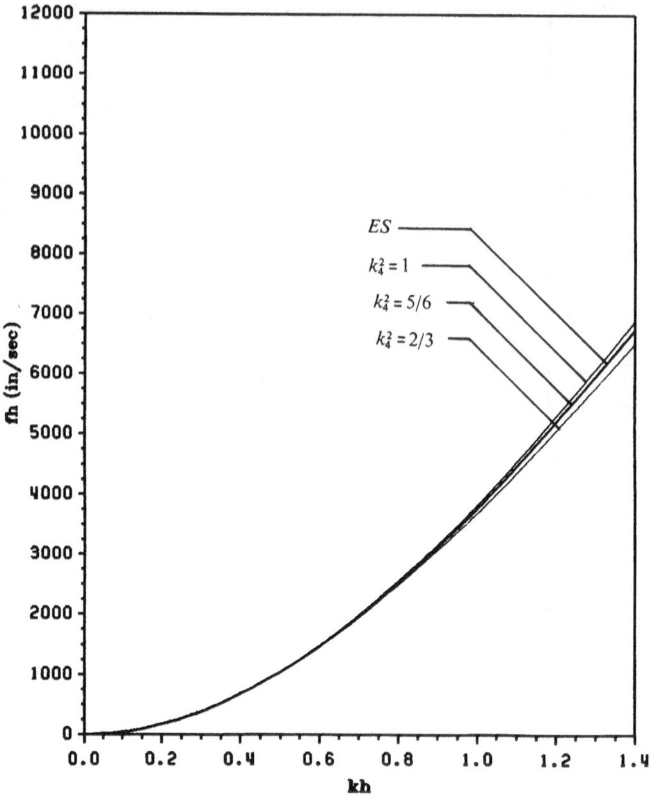

Fig. 3. Theoretical dispersion curves for low frequency flexural waves propagating perpendicular to the fiber direction of the $[0]_8$ graphite/epoxy laminate. (1in/sec = 0.0254m/sec)

For unidirectional composite laminates, the same value of 5/6 is obtained for the shear correction factors. Thus, this is consistent with our results for flexural waves propagating in the $[0]_8$ composite plate. For a $[0/90]_{2s}$ cross-ply graphite/epoxy laminate, we obtain $k_5^2 = 0.8607$ and $k_4^2 = 0.7133$, and for a $[0/45/90/-45]_s$ quasi-isotropic laminate, we have $k_5^2 = 0.8837$ and $k_4^2 = 0.6462$. The results of the classical plate theory (CPT)[10-12] and the shear deformation theory (SDT) with the shear correction factors calculated from static cylindrical bending together with the experimental data are plotted in Figs. 6 and 7 for the $[0/90]_{2s}$ cross-ply laminate. Figures 8 and 9 show the results for the symmetric quasi-isotropic laminate. However, for these laminates the experimental data for flexural waves traveling in the x direction are below our predicted curves obtained by the shear deformation theory, while the data for the y direction are above our predicted dispersion curves. Figure 10 shows the results for a 6061-T4 aluminum plate. It is suspected that the shear correction factors for static cylindrical bending are not precisely applicable to the cross-ply and quasi-isotropic laminates and cause the predicted dispersion curves to deviate from the data since the transverse shear strain distributions in these cases are more complicated. Using a higher order shear deformation theory may improve the results. With regard to the validity of cylindrical bending in our experimental method, there is excellent correlation between the experimental data and the theoretical results in the aluminum plate and the unidirectional laminate, which

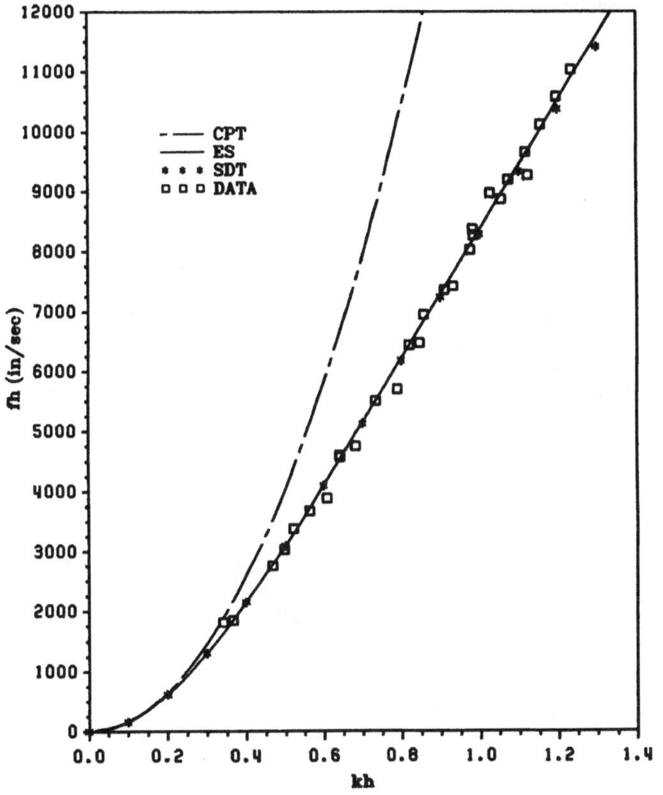

Fig. 4. Dispersion curves for low frequency flexural waves propagating along the fiber direction of the $[0]_8$ graphite/epoxy laminate. (1in/sec = 0.0254m/sec)

suggests that the cylindrical bending assumption is appropriate for these cases. Other factors, such as the combined effects of residual thermal stresses and the bimodulus character of composite materials, may also contribute to the deviation between the theory and the data. Further investigation is needed to identify the source of the deviation.

It is apparent that the effects of transverse shear deformation and rotary inertia are significant in laminated composite plates while the effects are relatively small in aluminum plates. For thin aluminum plates, in addition to the small thickness h, if the region of interest is in the low frequency, long wavelength range, the classical plate theory can be used instead of the shear deformation theory without much discrepancies. Surprisingly, in Fig. 5, the inclusion of transverse shear deformation and rotary inertia in the plate theory has relatively little effect on the results of waves propagating perpendicular to the fiber direction in the $[0]_8$ graphite/epoxy composite plate. That is, there is little deviation from classical plate theory. This is a special situation for laminated composite plates because the ratio between the transverse shear modulus and the Young's modulus perpendicular to the fiber direction for this transversely isotropic plate and the similar ratio for the isotropic aluminum plate are of the same order of magnitude. Thus, the effects of transverse shear deformation and rotary inertia are small as in the case of aluminum plate in Fig. 10.

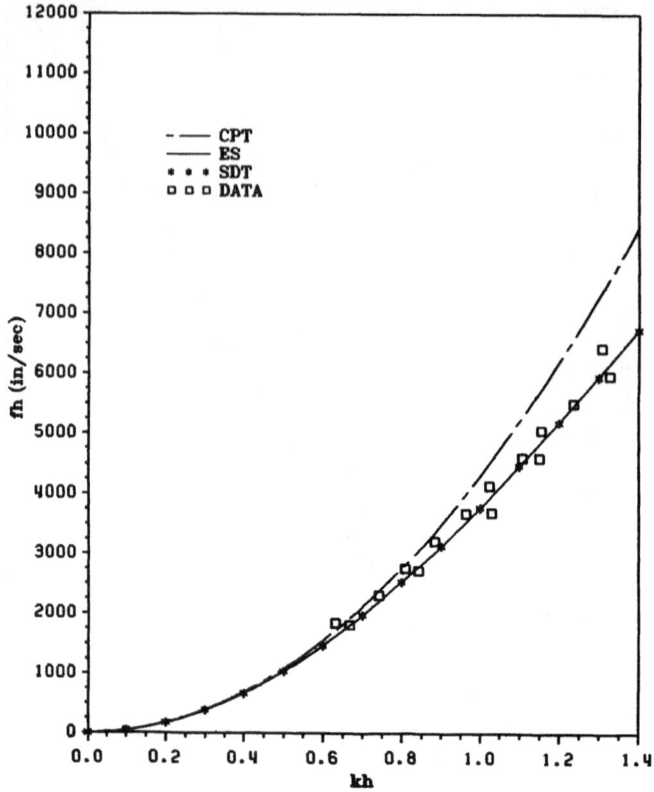

Fig. 5. Dispersion curves for low frequency flexural waves propagating perpendicular to the fiber direction of the $[0]_8$ graphite/epoxy laminate. (1in/sec = 0.0254m/sec)

Finally, the characteristic equation derived from the shear deformation theory has three distinct roots, however, only one approaches zero circular frequency as the wave number approaches zero. This is the root corresponding to the lowest antisymmetric branch of the frequency spectrum for plate waves and is the one shown in the figures. In the case of waves propagating parallel or perpendicular to the fiber direction in the $[0]_8$ graphite/epoxy composite plate, this lowest root from the shear deformation theory with a static bending shear correction factor begins to deviate from the lowest branch of the elasticity solution as kh becomes large. A different value of shear correction factor is needed to correct the deviation. In addition, one of the two remaining roots correctly predicts the cut-off frequency of the second lowest antisymmetric branch. Similar results have been reported by Mindlin[2,14] for isotropic plates. The displacement field in the Yang, Norris and Stavsky approach is a one term approximation of the power series expansion of the displacement field in the elasticity approach for transversely isotropic plates. Using a higher order approximation may improve the results for transversely isotropic plates as well as the results for composite laminates.

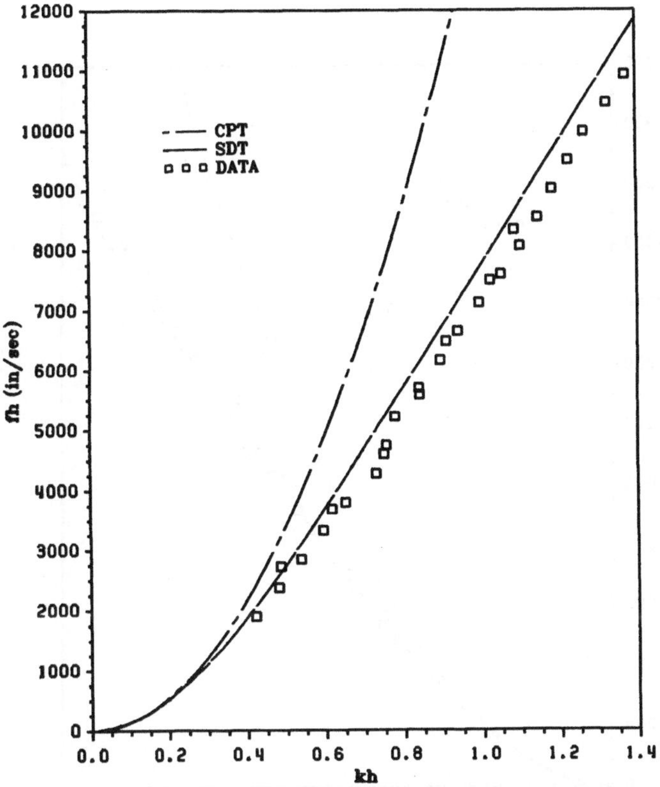

Fig. 6. Dispersion curves for low frequency flexural waves propagating along the x axis of the $[0/90]_{2s}$ graphite/epoxy laminate. (1in/sec = 0.0254m/sec)

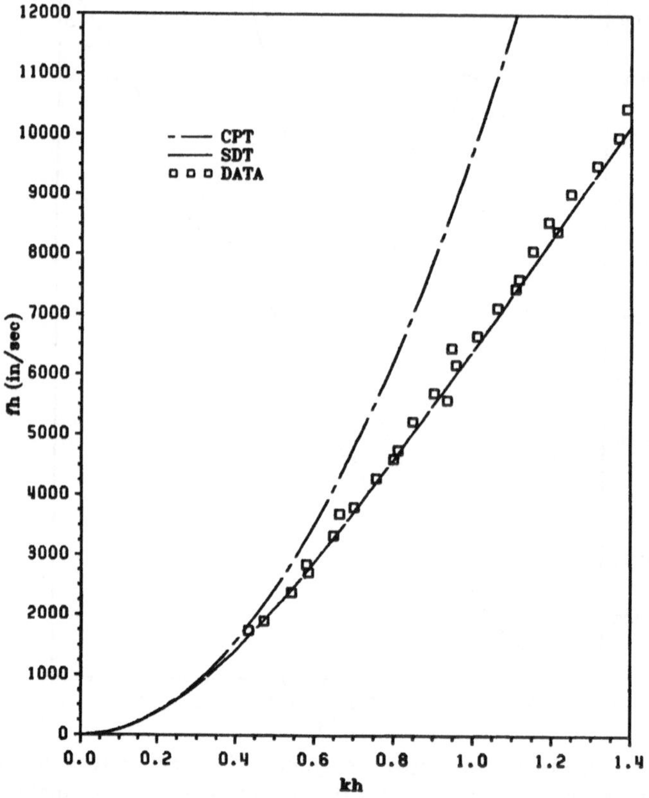

Fig. 7. Dispersion curves for low frequency flexural waves propagating along the y axis of the $[0/90]_2$ graphite/epoxy laminate. (1in/sec = 0.0254m/sec)

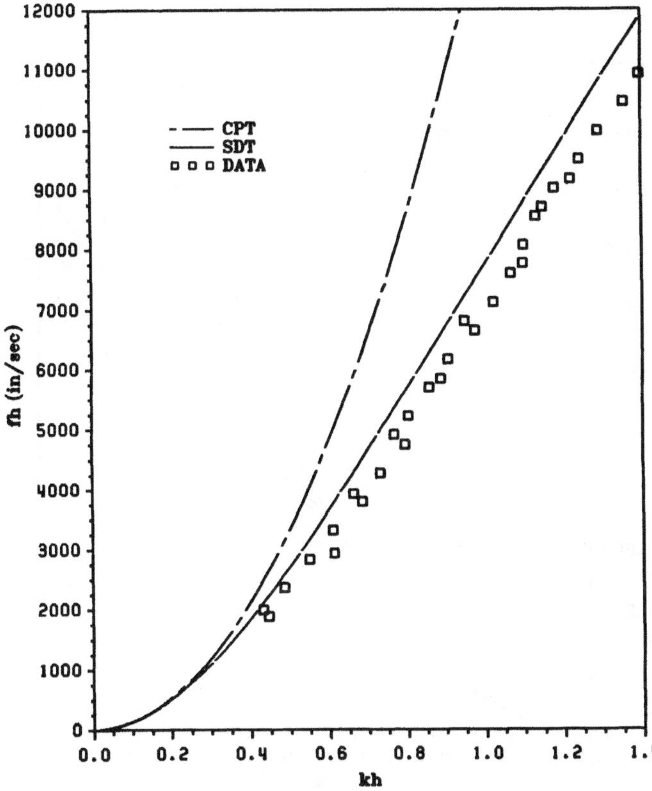

Fig. 8. Dispersion curves for low frequency flexural waves propagating along the x axis of the $[0/45/90/-45]_s$ graphite/epoxy laminate. (1in/sec = 0.0254m/sec)

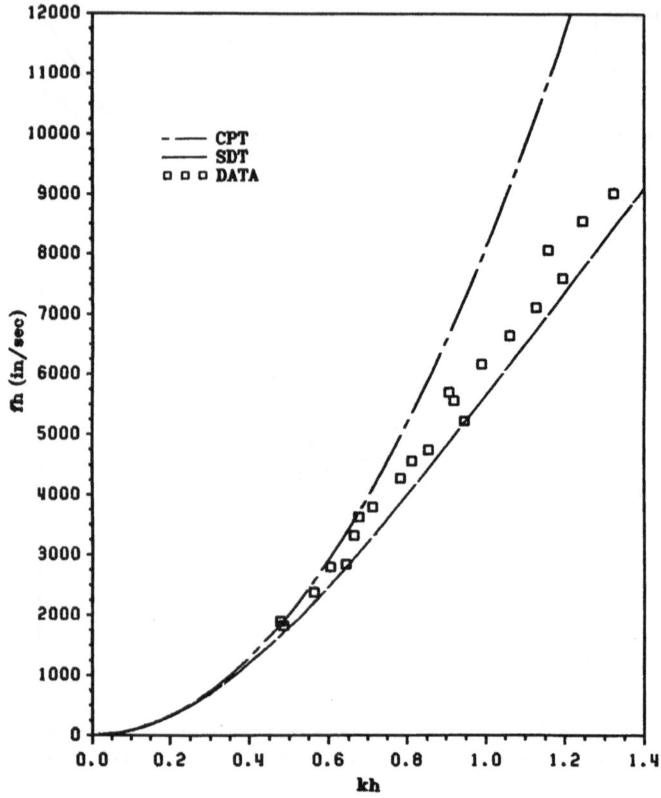

Fig. 9. Dispersion curves for low frequency flexural waves propagating along the y axis of the $[0/45/90/-45]_s$ graphite/epoxy laminate. (1in/sec = 0.0254m/sec)

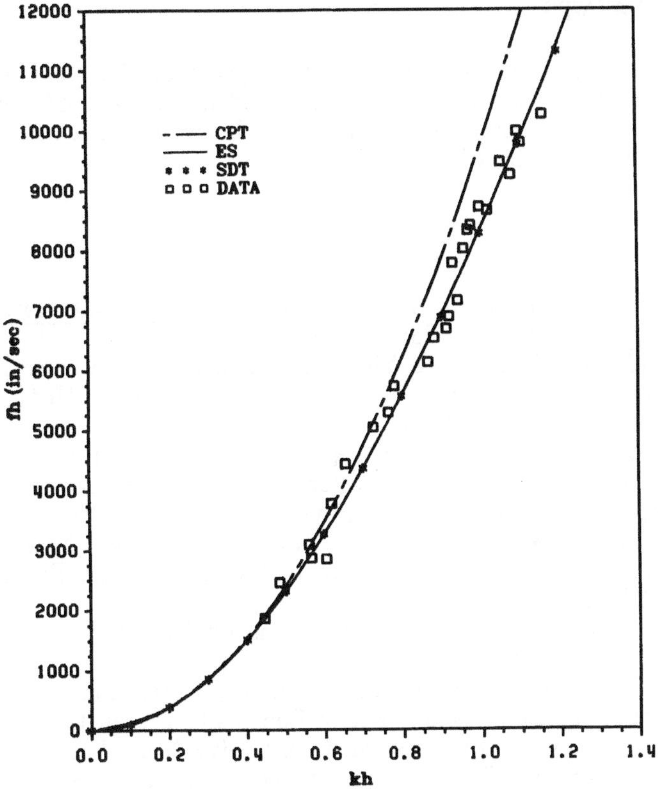

Fig. 10. Dispersion curves for low frequency flexural wave propagation in a 6061-T4 aluminum plate. (1in/sec = 0.0254m/sec)

CONCLUSIONS

The inclusion of shear deformation and rotary inertia in classical plate theory drastically improves the theoretical dispersion curves for flexural waves propagating in composite laminates, but has relatively little effect on isotropic aluminum plates which are modeled quite well at long wavelengths by classical plate theory. The shear deformation theory can be used to estimate the long wavelength portion of the dispersion curves for laminated composite plates. The shear correction factors for static cylindrical bending are applicable to the problem of low frequency, long wavelength flexural wave propagation in isotropic plates and unidirectional laminates. Using a higher order approximation of the displacement field may improve our results and the prediction of the higher order modes for transversely isotropic plates and composite laminates. Experimental data for the dispersion curves are obtained by using an acousto-ultrasonic technique. There is excellent correlation between the experimental data and the theoretical results in the aluminum plate and the unidirectional laminate. The data for the symmetric cross-ply and the symmetric quasi-isotropic laminates definitely have the characteristic of a dispersion curve for flexural waves. This method of measuring low frequency, long wavelength flexural wave phase velocity is reliable and reproducible. It can be used to characterize composite plates since each material and stacking sequence gives distinct dispersion curves.

ACKNOWLEDGMENTS

The authors gratefully acknowledge the support of the General Electric Co. under Contract No. 200-14AC-14G-21243 and the support of Institute of Materials Science and Engineering, Center for Innovative Technology under Contract No. MAT-86-0135-07

REFERENCES

1. H. Lamb, On Waves in an Elastic Plate, Proc of the Royal Soc of London, Series A, 93:114 (1917).
2. R. D. Mindlin, Waves and Vibrations in Isotropic, Elastic Plates, in: "Structural Mechanics," J. N. Goodier and N. J. Hoff, eds., Pergamon Press, New York (1960).
3. M. Redwood, "Mechanical Waveguides," Pergamon Press, New York (1960).
4. T. R. Meeker and A. H. Meitzler, Guided Wave Propagation in Elongated Cylinders and Plates, in: "Physical Acoustics: Principles and Methods," W. P. Mason, ed., Academic Press, New York (1964).
5. I. A. Viktorov, "Rayleigh and Lamb Waves," Plenum Press, New York (1967).
6. H. Ekstein, High Frequency Vibrations of Thin Crystal Plates, Phys Rev. 68:11 (1945).
7. W. R. Rose, S. I. Rokhlin, and L. Adler, Evaluation of Anisotropic Properties of Graphite/Epoxy Composite Plates Using Lamb Waves, in: "Review of Progress in Quantitative NDE, Vol. 6B," D. O. Thompson and D. E. Chimenti, eds., Plenum Press, New York (1987).
8. D. E. Chimenti and A. H. Nayfeh, Leaky Lamb Waves in Fibrous Composite Laminates, J of Appl Phys. 58:4531 (1985).
9. W. T. Yost and J. H. Cantrell, Surface Generation and Detection of Coupled Fiber-Matrix Mode Acoustic Wave Propagation in Fiber-Reinforced Composites, in: "Review of Progress in Quantitative NDE, Vol 5B," D. O. Thompson and D. E. Chimenti, eds., Plenum Press, New York (1986).
10. R. C. Stiffler and E. G. Henneke, II, "The Application of Low Frequency Acoustic Waves for Determining the Extensional and Flexural Stiffnesses in Composite Plates, Interim Report to General Electric Co., Contract No. 14-G-45-480," Virginia Tech, Blacksburg (1985).
11. J. C. Duke, Jr., E. G. Henneke, II, and W. W. Stinchcomb, "Ultrasonic Stress Wave Characterization of Composite Materials, NASA CR-3976," NASA, Cleveland (1986).
12. R. C. Stiffler, "Wave Propagation in Composite Plates," Ph.D. Dissertation, College of Engineering, Virginia Polytechnic Institute and State University, Blacksburg, (1986).
13. P. C. Yang, C. H. Norris, and Y. Stavsky, Elastic Wave Propagation in Heterogeneous Plates, Intl J of Solids and Struc. 2:665 (1966).
14. R. D. Mindlin, Influence of Rotatory Inertia and Shear on Flexural Motions of Isotropic, Elastic Plates, J of Appl Mech. 18:31 (1951).
15. C. T. Sun and T. M. Tan, Wave Propagation in a Graphite/Epoxy Laminate, J of the Astro Sc. 32:269 (1984).
16. W. A. Green, Bending Waves in Strongly Anisotropic Elastic Plates, Quarterly J of Mech and Appl Math. 35:485 (1982).
17. J. M. Liu, The Frequency Dependence of Ultrasonic Wave Propagation in

Metal-Matrix Composite Plates, in: "Proceedings of the 15th Symposium on NDE," D. W. Moore and G. A. Matzkanin, eds., Southwest Research Institute, San Antonio (1985).

18. T. S. Chow, On the Propagation of Flexural Waves in an Orthotropic Laminated Plate and Its Response to an Impulsive Load, J of Comp Matls. 5:306 (1971).
19. J. M. Whitney, Shear Correction Factors for Orthotropic Laminates Under Static Load, J of Appl Mech. 40:302 (1973).
20. E. Reissner, The Effect of Transverse Shear Deformation on the Bending of Elastic Plates, J of Appl Mech. 12:69 (1945).
21. A. K. Noor, Stability of Multilayered Composite Plates, Fibre Sc and Tech. 8:81 (1975).

PRELIMINARY EVALUATION OF NON-CONTACT ACOUSTO-ULTRASONIC DISPLACEMENT FIELDS IN POLYMERIC MATRIX COMPOSITE

James H. Williams, Jr.

Professor of Applied Mechanics
Massachusetts Institute of Technology
Cambridge, MA 02139

Peter Liao

Member of Technical Staff
AT&T Bell Laboratory
Whippany, NJ

ABSTRACT

A unidirectional fiberglass epoxy composite is modelled as a homogeneous transversely isotropic continuum plate medium. Acousto-ultrasonic non-contact input-output characterization by tracing the stress waves (SH, P and SV waves) in the continuum is studied as if a transmitting and a receiving transducer were located on the same face of the plate but without direct contact.

The output displacements associated with each type of stress wave in the fiberglass epoxy composite are approximated by an asymptotic solution for an infinite transversely isotropic medium subjected to a harmonic point load. The polar diagrams for the amplitudes due to each type of stress wave in the plate are given.

INTRODUCTION

Acousto-ultrasonics (AU) is a term devised by Alex Vary and his co-workers to denote an NDE technique that synthesizes some aspects of acoustic emission methodology with the ultrasonic generation of stress waves. AU's quantitative counting measures (such as those common in acoustic emission) have been correlated with variations in overt flaw states (such as crack size) as well as with distributed flaw states (such as void content or degree of resin cure). Although the emphasis thus far in the brief history of AU has been on detecting and mapping distributed variations of mechanical properties, the potential for quantitative material characterization via AU is very broad.

Given the requirements of AU for quantitative assessments via wave propagation analyses, it is clear that an understanding of wave propagation in the materials and structures to be evaluated is fundamental to progress in AU research and reliable implementation. Further, there has recently developed an increased interest in the use of non-contact generation and detection transducers for AU in order to obviate the adulteration of AU signals by transducers in mechanical contact with evaluated structures. This study represents the initial phase of a larger effort to evaluate the AU displacement fields in various composite materials which would be available for detection by non-contact transducers.

Fiber reinforced composites are attractive materials for a broad range of applications because of their high specific mechanical properties. It has been assumed by several authors that a unidirectional fiberglass epoxy composite, as shown in Fig. 1, may be modelled as a homogeneous transversely isotropic continuum. Acousto-ultrasonic non-contact input-output characterizations of a unidirectional fiberglass plate specimen are studied by tracing three types of stress waves (SH, P and SV waves) in the plate specimen. The output displacements, associated with each type of stress wave in the fiberglass epoxy composite plate specimen, which may be detected by a non-contact receiving transducer are approximated by an asymptotic solution for an infinite transversely isotropic medium subjected to a harmonic point load. The amplitudes of the output displacements are plotted for each type of traced stress wave. This study should enhance the quantitative understanding of non-contact acousto-ultrasonic nondestructive evaluation of fiber reinforced composites.

ACOUSTO-ULTRASONIC NON-CONTACT INPUT-OUTPUT CHARACTERIZATION OF FIBERGLASS EPOXY COMPOSITE PLATE SPECIMEN

Materials and Test Configuration

It has been assumed[1-4] that a unidirectional fiber composite such as shown in Fig. 1 may be modelled as a homogeneous transversely isotropic continuum. For the axes shown in Fig. 1 and the test configuration shown in Fig. 2, the

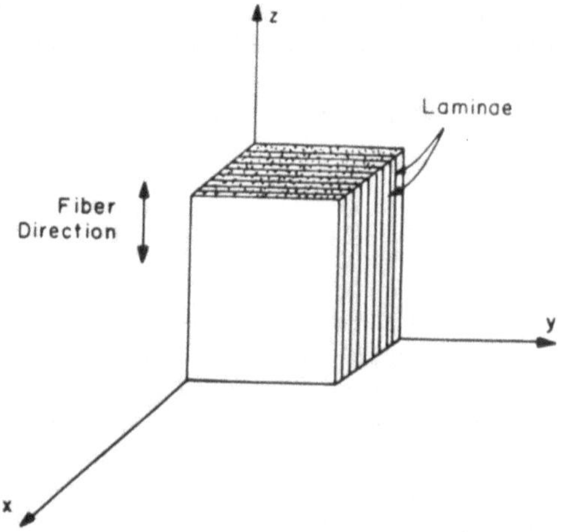

Fig. 1. Fiber reinforced polymeric matrix composite modelled as transversely isotropic medium.

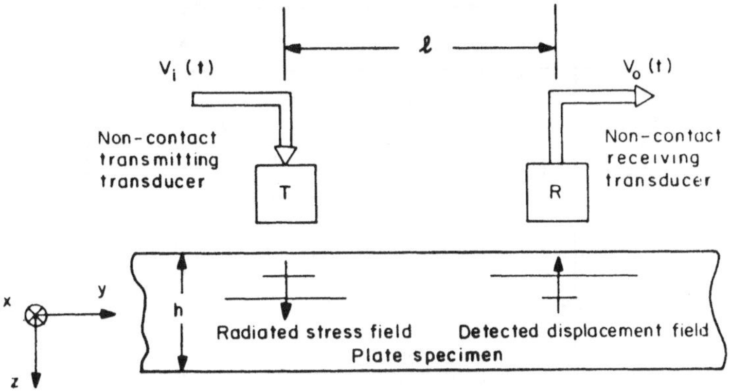

Fig. 2. Schematic of non-contact acousto-ultrasonic test configuration.

isotropic plane of its equivalent continuum lies in the midplane of the plate. A Cartesian coordinate system (x,y,z) is chosen so that the x - y plane is the isotropic plane; thus, the upper and the lower surfaces are at $z = h/2$ and $z = -h/2$, respectively, where h is the plate thickness. The properties of the equivalent continuum model of the fiberglass epoxy composite plate to be considered are[3]

$$\begin{aligned} h &= 0.1\text{m} \\ C_{11} &= 10.581 \times 10^9 \text{N/m}^2 \\ C_{12} &= 4.098 \times 10^9 \text{N/m}^2 \\ C_{13} &= 4.679 \times 10^9 \text{N/m}^2 \\ C_{33} &= 40.741 \times 10^9 \text{N/m}^2 \\ C_{44} &= 4.422 \times 10^9 \text{N/m}^2 \\ \rho &= 1850 \text{kg/m}^3 \end{aligned} \tag{1}$$

The C_{ij} are constitutive coefficients of the stiffness matrix and ρ is the mass density.

A transmitting transducer and a receiving transducer are located on the same face of the plate specimen without direct contact, as shown in Fig. 2. The plate specimen shown in Fig. 2 is considered as a plate of thickness h and of infinite planar (x-y) extent. The input electrical voltage to the transmitting transducer is $V_i(t)$ and the output electrical voltage from the receiving transducer is $V_o(t)$ where t represents time. The transmitting transducer converts an input electrical voltage into a stress, whereas the receiving transducer converts a displacement associated with stress waves arriving at it into an output voltage. Three types of stress waves (SH, P and SV waves) are traced in the plate. Each type of stress wave which is generated by the transmitting transducer located above point O experiences multiple reflections at each face of the plate, and then reaches the receiving transducer located above point M, as shown in Fig. 3. Since the isotropic plane lies in the midplane and is parallel to both the top and the bottom faces where the multiple reflections occur, the angle of the reflected stress wave is equal to the angle of incidence of the incident stress wave for each reflection at each face of the plate.[1,5-7] Accordingly, except for the approximate reflections coefficients, the stress wave traveling from the point O to the point M may be considered as a wave propagating to a semi-infinite transversely isotropic medium, and traveling to point M' as if there were no bottom face, as shown in Fig. 3.

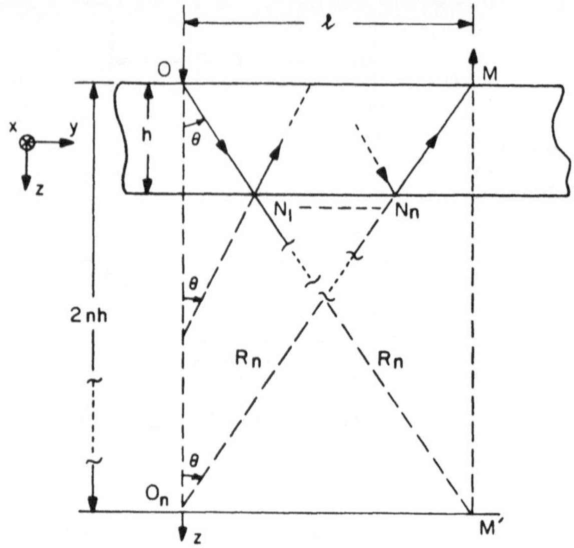

Fig. 3. Path of each type of stress wave which arrives at point M after n reflections from bottom boundary.

Output Displacements Associated with SH Waves in Plate

The output displacements, radiated by the non-contact transmitting transducer and detected by the non-contact receiving transducer above point M, are assumed to be equivalent to the displacements at point M' associated with each type of stress wave propagating in a semi-infinite transversely isotropic medium as if there were no bottom boundary (except of the cumulative effect of the reflection coefficient), as shown in Fig. 3. The displacements at point M' are approximated by the far-field asymptotic solution for large R_n of an infinite transversely isotropic medium subjected to a harmonic point load.[5]

Since the SH wave traced in the plate specimen shown in Fig. 3 is traveling in the y-z plane, it follows that only the displacement component along the x-axis, u, is detectable at the point M'. The amplitude of the u-displacement at point M' is $U_{M'}$, and is given as[5]

$$U_{M'} = [(C_{44}\rho)^{0.5} s_z^* X_o]/[2\pi R_n(C_{11} - C_{12})] \tag{2}$$

where X_o is the point load generated by the non-contact transmitting transducer along the x-direction; and s_z^* is the z-component of the point on the slowness surface of the SH wave where the normal to the slowness surface is parallel to the direction OM'. R_n can be obtained from the geometry in Fig. 3 as

$$R_n = \ell/\sin\theta \tag{3}$$

where ℓ is the separation distance between the input O and the output M, and

$$\theta = \tan^{-1}(\ell/2nh) \tag{4}$$

where n is the number of reflections at the bottom face of the plate specimen experienced by the SH wave in traveling from the input O to the output M; as shown in Fig. 3.

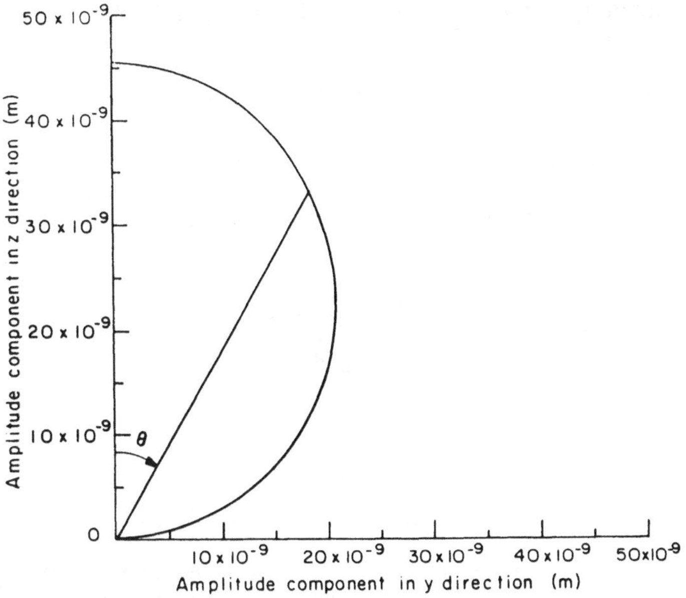

Fig. 4. Polar diagram for u-displacement amplitude for points along arc $y^2 + z^2 = 1$, due to SH waves in the fiberglass epoxy composite.

The amplitude of the u-displacement evaluated along the arc $y^2 + z^2 = 1$ in the positive y - z quadrant due to an applied point load of unit amplitude associated with the SH wave in the fiberglass epoxy composite is obtained by substituting eq. (1) into eq. (2), and then by setting $X_o = 1$ and $R_n = 1$. The numerical result is shown in Fig. 4. It has been shown in eq. (4) that the angle θ in Fig. 3 is determined by the separation distance ℓ and the number of reflections n. For a given number of reflections n and a given separation distance n, the amplitude of the u-displacement at the point M' can be determined graphically by measuring the distance between the origin and the intersection of the ray oriented at an angle of θ, determined by the given n and the given ℓ, with respect to the z-axis of the polar diagram, as shown in Fig. 4. The output displacement at the point M in Fig. 3 is thus determined by taking into account the cumulative effect of the reflection coefficient.

Output Displacements Associated with P Waves in Plate

Since the P wave traced in the plate specimen shown in Fig. 3 is traveling in the y - z plane, it follows that only the displacement components along the y and z axes, u and w, respectively, are detectable at the point M'. The amplitude of the y-component displacement due to the applied point load acting along the y-direction. D_v^Y is given as[6]

$$D_v^Y = f_1(s_y^*, s_z^*)(Y_o/R_n) \qquad (5)$$

where Y_o is the applied point load acting along the y-direction; and s_y^* and s_z^* are the y-component and the z-component, respectively, of the point on the slowness surface of the P wave where the normal is parallel to the direction OM'.

$$f_1(s_y^*, s_z^*) = (\lambda_n/2\pi)\|(C_{44}/\rho)s_y^{*2} + (C_{33}/\nabla)s_z^{*2} - 1\| \qquad (6)$$

where the symbol $\|\|$ denotes "the magnitude of"; and λ_n is the amplitude coefficient and is given by

$$\lambda_n = \{(H^2{,}_{s_x} + H^2{,}_{s_y} + H^2{,}_{s_z})/\|K_n\|\}^{0.5} \quad (7)$$

where s_x, s_y and s_z are the components of the slowness vector of the P wave along the x, y, and z axes, respectively; and "," denotes partial differentiation with respect to the slowness vector components which follow. H is defined by

$$H(s_x,s_y,s_z) = [(C_{44}/\rho)s_z^2 + (C_{11}/\rho)(s_x^2 + s_y^2) - 1] \\ [(C_{44}/\rho)(s_x^2 + s_y^2) + (C_{33}/\rho)s_z^2 - 1] \\ - [(C_{44} + C_{13})/\rho]^2 s_z^2(s_x^2 + s_y^2) \quad (8)$$

and K_n is given by

$$K_n = \sum [H^2{,}_{s_z}(H{,}_{s_x s_x}H{,}_{s_y s_y} - H^2{,}_{s_x s_y}) \\ + 2H{,}_{s_x}H{,}_{s_y}(H{,}_{s_x s_z}H{,}_{s_y s_z} - H{,}_{s_x s_y}H{,}_{s_z s_z})] \quad (9)$$

where $\sum$ denotes the sum with respect to cyclic permutation of s_x, s_y and s_z. Similarly, the amplitude of the z-component displacement, D_w^Y is given by[6]

$$D_w^Y = f_2(s_y^*,s_z^*)/(Y_o/R_n) \quad (10)$$

where $f_2(s_y^*,s_z^*) = (\lambda_n/2\pi)[(C_{13} + C_{44})/\rho]s_y^*s_z^*$. The amplitude of the y-component displacement due to the applied point load acting along the z-direction, D_v^Z is given as[6]

$$D_v^Z = f_2(s_y^*,s_z^*)/(Z_o/R_n) \quad (11)$$

where Z_o is the point load acting along the z-direction. Similarly, the amplitude of the z-component displacement due to the applied point load acting along the z-direction, D_w^Z is given as[6]

$$D_w^Z = f_3(s_y^*,s_z^*)/(Z_o/R_n) \quad (12)$$

where $f_3(s_y^*,s_z^*) = (\lambda_n/2\pi)\|(C_{11}/\rho)s_y^{*2} + (C_{33}/\rho)s_z^{*2} - 1\|$.

The amplitudes of the v-displacement and the w-displacement components evaluated along the $y^2 + z^2 = 1$ in the positive y - z quadrant due to an applied point load of unit amplitude associated with the P wave in the fiberglass epoxy composite are obtained by substituting eq. (1) into eqs. (5), (10), (11) and (12), and then by setting $Y_o = Z_o = 1$ and $R_n = 1$. The numerical results are shown in Figs. 5-8. It is interesting to note that in the asymptotic solution, whereas the stress is a function of frequency, the displacement is independent of frequency.

Output Displacements Associated with SV Waves in Plate

Since the SV wave traced in the plate specimen shown in Fig. 3 is traveling in the y - z plane, it follows that only the displacement components along the y and z axes, v and w, are detectable at the point M'. The amplitude of the y-component displacement due to the applied point load acting along the y-direction, D_v^Y is given as[7]

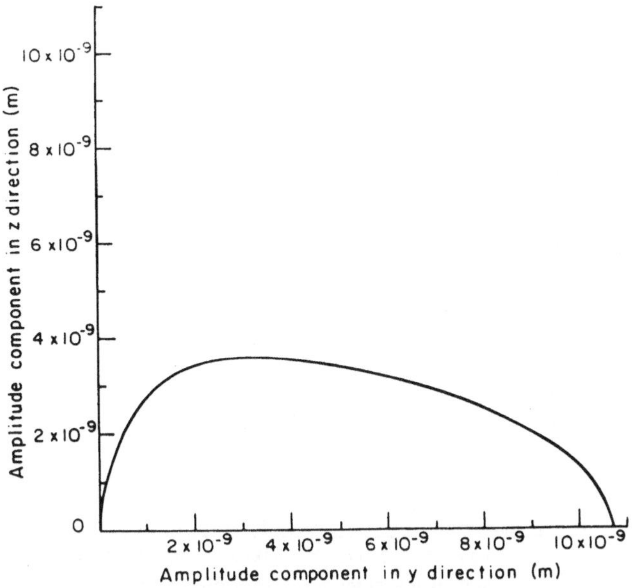

Fig. 5. Polar diagram for v-displacement amplitude associated with P wave in fiberglass epoxy composite for points along the arc $y^2 + z^2 = 1$ due to applied point force of unit amplitude acting along y-direction.

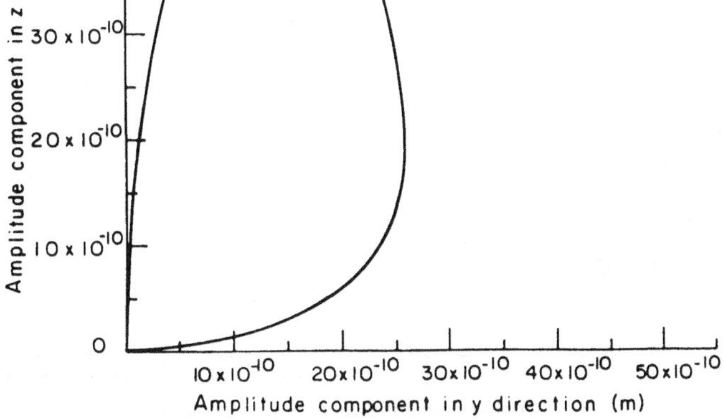

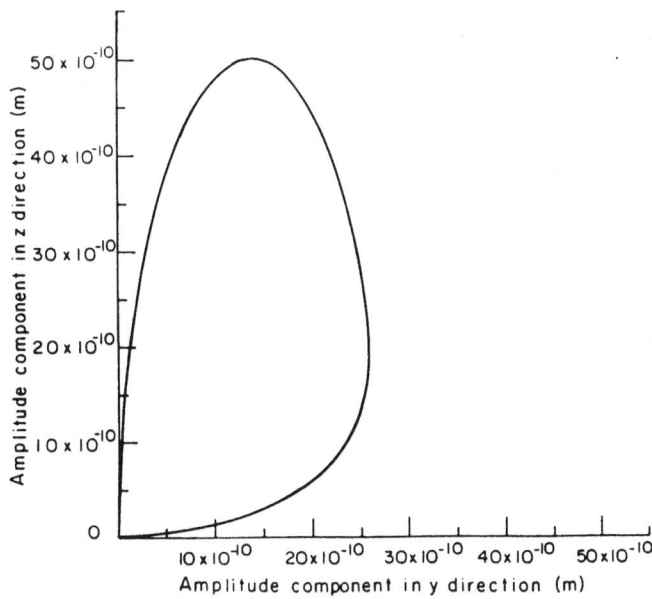

Fig. 6. Polar diagram for w-displacement amplitude associated with P wave in fiberglass epoxy composite for points along the arc $y^2 + z^2 = 1$ due to applied point force of unit amplitude acting along y-direction.

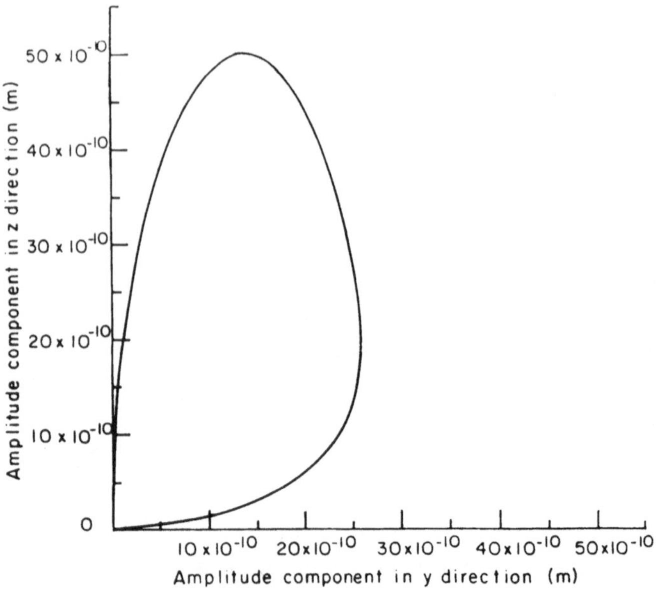

Fig. 7. Polar diagram for v-displacement amplitude associated with P wave in fiberglass epoxy composite for points along the arc $y^2 + z^2 = 1$ due to applied point force of unit amplitude acting along z-direction.

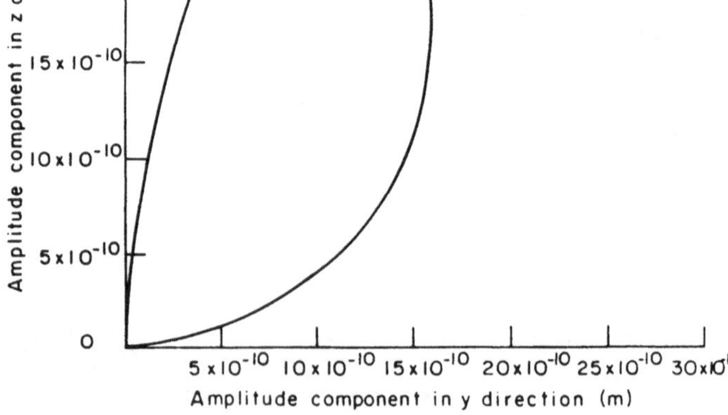

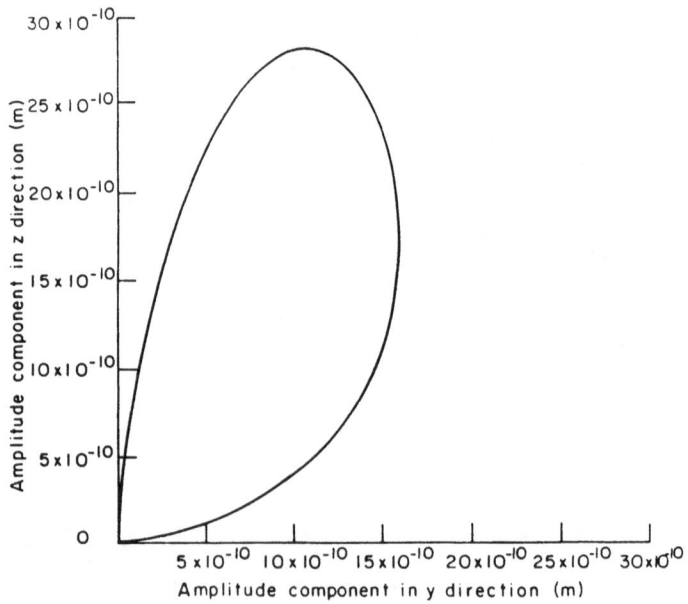

Fig. 8. Polar diagram for w-displacement amplitude associated with P wave in fiberglass epoxy composite for points along the arc $y^2 + z^2 = 1$ due to applied point force of unit amplitude acting along z-direction.

$$D_v^Y = f_1(s_{yn},s_{zn})(Y_o/R_n) \tag{13}$$

where s_{yn} and s_{zn} are the y-components and the z-components of the points on the slowness surface of the SV wave where the normal is parallel to the direction OM'; and

$$f_1(s_{yn},s_{zn}) = (1/2\pi)\left\| \sum_{n=1}^{N} A_n\lambda_n[(C_{44}/\rho)s_{yn}^2 + (C_{33}/\rho)s_{zn}^2 - 1] \exp\{i\omega(s_{yn}y + s_{zn}z)\}\right\| \tag{14}$$

where N is the total number of points on the slowness surface of the SV wave where the normal is parallel to the direction OM'; ω denotes radian frequency; and A_n is the phase constant and is determined as follows: $A_n = 1$ if $K_n > 0$ or $A_n = i$ if $K_n < 0$. The existence of the exponential terms accounts for the cuspidal behavior of the SV wave. Similarly, the amplitude of the z-component displacement due to the applied point load acting along the y-direction, D_w^Y, is given as[7]

$$D_w^Y = f_2(s_{yn},s_{zn})(Y_o/R_n) \tag{15}$$

where

$$f_2(s_{yn},s_{zn}) = (1/2\pi)\left\| \sum_{n=1}^{N} A_n\lambda_n[(C_{44} + C_{13})/\rho]s_{yn}s_{zn} \exp\{i\omega(s_{yn}y + s_{zn}z)\}\right\|.$$

The amplitude of the y-component displacement due to the applied point load acting along the z-direction, D_v^Z, is given as[7]

$$D_v^Z = f_2(s_{yn},s_{zn})(Z_o/R_n) \tag{16}$$

And, the amplitude of the z-component displacement due to the applied point load acting along the z-direction, D_w^Z, is given as[7]

$$D_w^Z = f_3(s_{yn},s_{zn})(Z_o/R_n) \tag{17}$$

where

$$f_3(s_{yn},s_{zn}) = (1/2\pi)\left\| \sum_{n=1}^{N} A_n\lambda_n[(C_{11}/\rho)s_{yn}^2 + (C_{33}/\rho)s_{zn}^2 - 1] \exp\{i\omega(s_{yn}y + s_{zn}z)\}\right\|$$

The amplitudes of the v-displacement component and the w-displacement component evaluated along the arc $y^2 + z^2 = 1$ in the positive y - z quadrant due to an applied point load of unit amplitude associated with the SV waves in the fiberglass epoxy composites are obtained by substituting eq. (1) into eqs. (13), (15), (16) and (17), and then by setting $Y_o = Z_o$ and $R_n = 1$. The numerical results are shown in Figs. 9-12.

ACKNOWLEDGEMENT

The Materials and Structures Division (Alex Vary, project monitor), NASA Lewis Research Center, Cleveland, Ohio, is gratefully acknowledged for its support of this research.

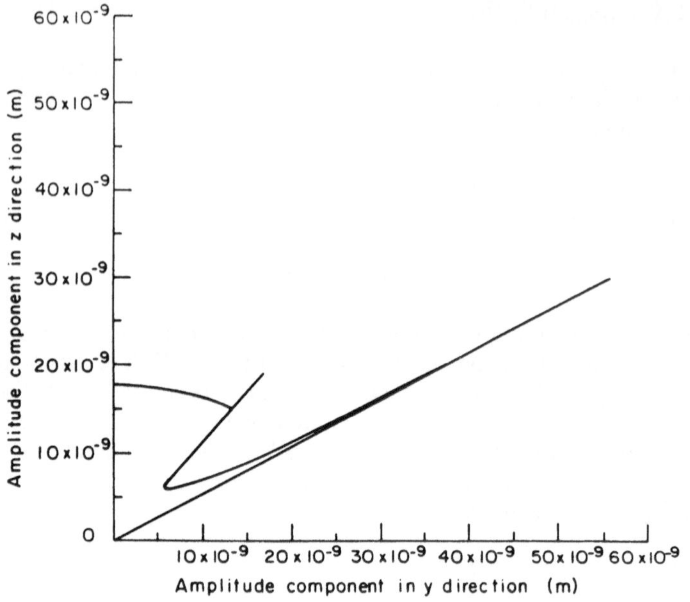

Fig. 9. Polar diagram for v-displacement amplitude associated with SV wave in fiberglass epoxy composite for points along the arc $y^2 + z^2 = 1$ due to applied point force of unit amplitude acting along y-direction.

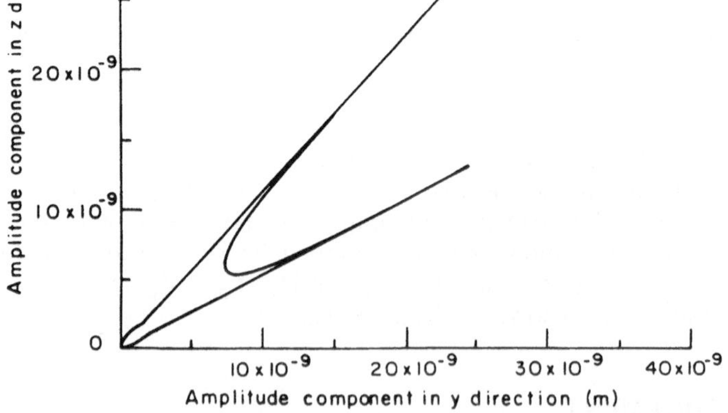

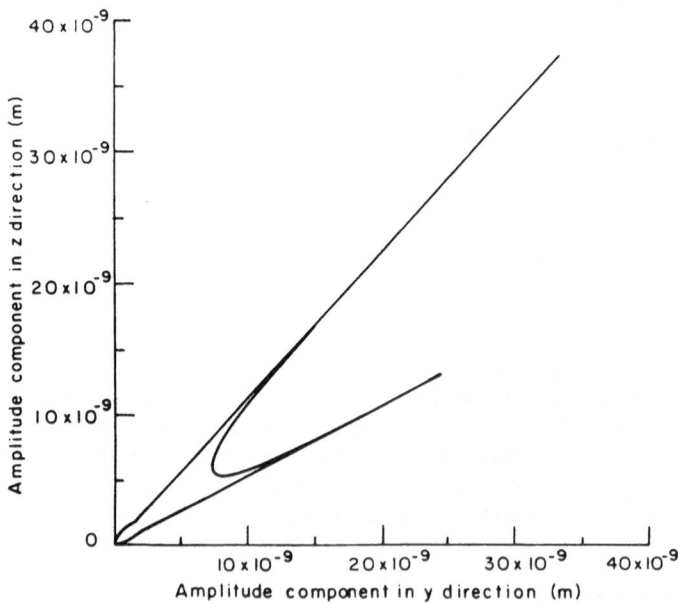

Fig. 10. Polar diagram for w-displacement amplitude associated with SV wave in fiberglass epoxy composite for points along the arc $y^2 + z^2 = 1$ due to applied point force of unit amplitude along y-direction.

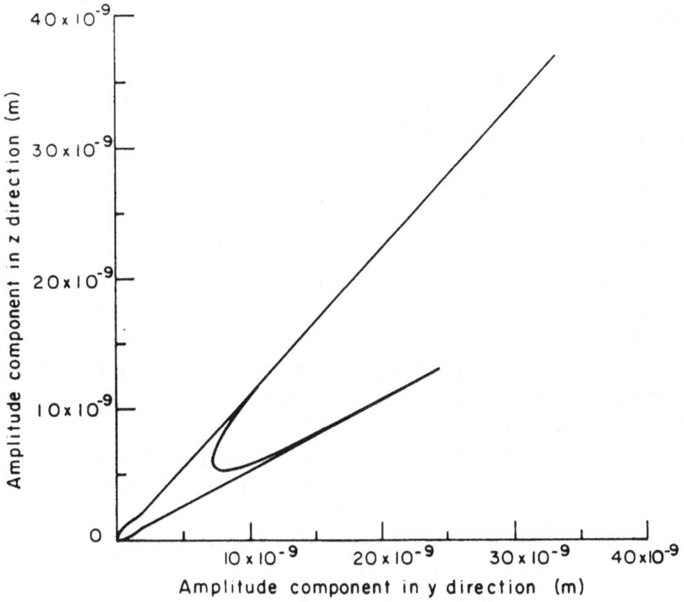

Fig. 11. Polar diagram for v-displacement amplitude associated with SV wave in fiberglass epoxy composite for points along the arc $y^2 + z^2 = 1$ due to applied point force of unit amplitude acting along z-direction.

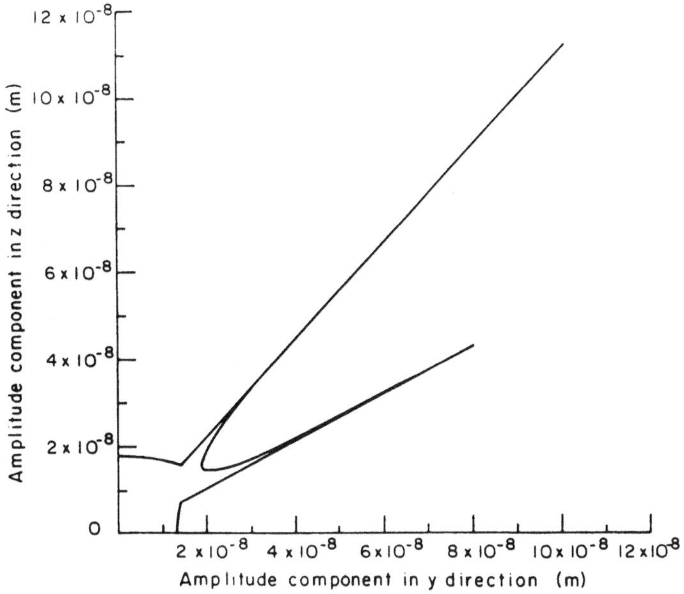

Fig. 12. Polar diagram for w-displacement amplitude associated with SV wave in fiberglass epoxy composite for points along the arc $y^2 + z^2 = 1$ due to applied point force of unit amplitude acting along z-direction.

REFERENCES

1. E. G. Henneke, "Reflection-Refraction of a Stress Wave at a Plane Boundary Between Anisotropic Media," JASA 51:210 (1972).
2. R. D. Kriz and H. M. Ledbetter, Elastic Representation Surfaces of Unidirectional Graphite/Epoxy Composites, in: "Recent Advances in Composites in the United States and Japan, ASTM STP 864," J. R. Vinson and M. Taya, eds., American Society for Testing Materials, Philadelphia (1985).
3. E. R. C. Marques and J. H. Williams, "Stress Waves in Transversely Isotropic Media, CR- 3977," NASA, Cleveland (1986).
4. J. H. Williams, Jr., E. R. C. Marques, and S. S. Lee, "Wave Propagation in Anisotropic Medium Due to an Oscillatory Point Source with Application to Unidirectional Composites, CR-4001," NASA, Cleveland (1986).
5. J. H. Williams, Jr. and T. P. Liao, "Acousto-Ultrasonic Input-Output Characterization of Unidirectional Fiber Composite Plate by SH Waves," NASA Contractor Report (in press).
6. T. P. Liao and J. H. Williams, Jr., "Acousto-Ultrasonic Input-Output Characterization of Unidirectional Fiber Composite Plate by P Waves," NASA Contractor Report (in preparation).
7. T. P. Liao and J. H. Williams, Jr., "Acousto-Ultrasonic Input-Output Characterization of Unidirectional Fiber Composite Plate by SV Waves," NASA Contractor Report (in preparation).

UTILIZATION OF OBLIQUE INCIDENCE IN ACOUSTO-ULTRASONICS

A. Pilarski, J. L. Rose, K. Balasubramaniam, and J. Da-Le

Department of Mechanical Engineering and Mechanics
Drexel University
Philadelphia, PA

ABSTRACT

The acousto-ultrasonic test technique has proven itself as a very valuable nondestructive evaluation procedure. The technique, however, is severely limited by using normal beam incidence of two transducers side by side. The current theoretical explanation of the wave propagation phenomena in ultrasonic NDE is limited. The purpose of this paper is to explore the full potential of the acousto-ultrasonic technique by way of including oblique incidence, and some theoretical explanation of wave propagation in the structure. By doing this, greater benefit of acousto-ultrasonic nondestructive evaluation can be realized.

Several sample problems that could fit into the category of acousto-ultrasonic evaluation are presented and outlined in the paper; the goal being to demonstrate the versatility of the acousto-ultrasonic technique. Four sample problems are covered: An adhesive bond inspection problem involving a single interface considering variations in bond rigidity varying from a smooth to a welded condition. A second problem treats a three layered structure with various boundary conditions. The third problem is associated with composite material inspection examining porosity level and fiber volume fraction variations. The fourth problem considers surface wave techniques in composite material inspection, a modification of which can indeed be considered as an extension of the acousto-ultrasonic test procedure.

INTRODUCTION

Even though the term acousto-ultrasonics is used in this paper, it could be replaced by the term "utilization of oblique incidence" (UOIT) in ultrasonic nondestructive evaluation. We envision the test procedure as simply an extension of a basic ultrasonic test procedure. The basic elements of the so-called acousto-ultrasonic test procedure are reviewed in Fig. 1. All sorts of wave motion are possible when using oblique incidence, including plate waves, interface waves and surface waves. In Fig. 1a is shown an early acousto-ultrasonic test procedure where two normal beam probes were placed very

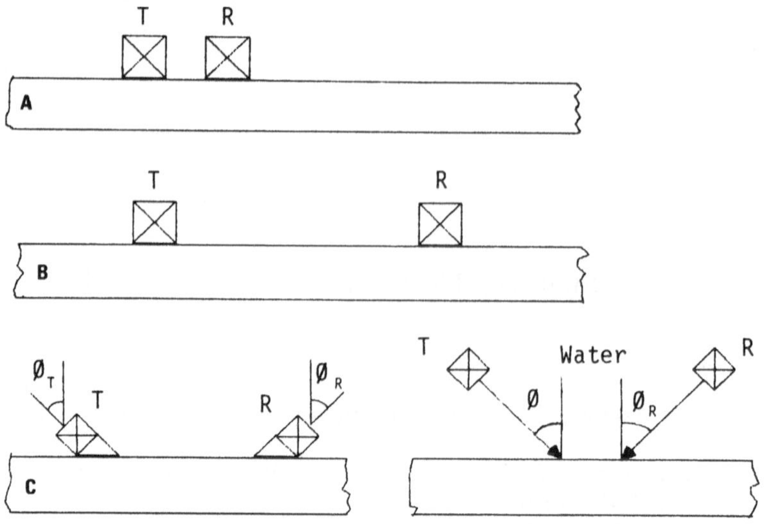

Fig. 1. Acousto-Ultrasonic Test Possibilities.

close together in an attempt to measure the variation in amplitude as energy propagated through the structure; this led to a measurement of a stress wave factor.[1] Obviously, the frequency used and the distance between the probes becomes an important variable, but the early results in acousto-ultrasonic testing did not emphasize the physics and mechanics of wave propagation. The technique was used primarily for observing an amplitude change or change in the signature of the ultrasonic wave form. A reduction in the amplitude detected by a receiver, obviously indicated some change in the material, or perhaps the presence of some defect. More recently though, as shown in Fig. 1b, the acousto-ultrasonic test procedure employed transducers that were a larger distance apart. Using a test technique like this allowed plate waves to be generated in the structure giving them some time to develop as the mechanical disturbance traveled from one transducer to another. A plate wave could indeed be generated, if the wavelength with respect to the thickness of the structure was of the same order of magnitude. Note that in this particular test procedure, there was limited control of the mode of wave propagation in the structure. In the work that is presented here, new directions in acousto-ultrasonic inspection are possible as illustrated in Fig. 1c. In this case we are utilizing oblique incidence to produce waves in the structure. By proper selection of incidence angle and frequency it becomes possible to control the mode of propagation in the plate, hence providing a source with maximum energy transmission from one probe to another. It is this particular angle beam approach, either in contact or immersion mode, that is of interest in this paper.. In concentrating on the numerous variables associated with the physics and mechanics of plate and surface wave generation in the structure, tremendous sensitivity to material characterization and defect classification becomes possible.

The technique of collecting ultrasonic information using a normal beam pulse-echo mode provides data in only one direction. In the case of the longitudinal wave, discontinuities can be detected by applying a vibration approximately normal to the surface of a defect. Additional sensitivity can often be obtained by employing transverse waves where particle motion is perpendicular to the wave propagation direction. This could be particularly useful in composite material and layered media inspection where many interfaces must be examined. Composite material inspection is different than the more conventional material inspection for two principal reasons: the anisotropic nature and the multi-material global failure possibilities. When inspecting areas with interfacial weaknesses in adhesively bonded structures, we may also need the tangential vibration with respect to the interface in order to obtain increased sensitivity to the imperfections. To obtain complete information, therefore, it seems appropriate to utilize oblique incidence ultrasonic beams to produce guided waves, which will then enable us to penetrate the test object from different directions and also with various ultrasonic wave type. The information collected will then represent a multi-dimensional view of the material state and also will provide us with better results on material integrity and defect analysis. Some of the data collection procedures presented in Rose et. al.[2] are used to accomplish this multi-dimensional material analysis. Explanations of the physical models and the various aspects of wave propagation mechanics that are useful in acousto-ultrasonic inspection of composite materials will also be reviewed.

BACKGROUND

Basic elements of wave mechanics useful in ultrasonic NDE can be found in the literature. For example, the formulas for the reflection (or transmission) coefficients for the oblique incidence of a plane, homogeneous wave (longitudinal or transverse) onto a solid-solid interface (both materials isotropic and ideal elastic) with respect to the potential functions or for the amplitudes of the displacements (particle velocities), stresses (acoustic pressures), or intensities are given in many fundamental monographs on acoustics (e.g. Graff[3] and Miklowitz[4]). Derivations of the formulas are based on an assumption of the continuity of displacements and stresses, both normal and tangential to the interface, the so-called "welded" boundary conditions. In the literature, there are available some angular characteristics of the reflection/transmission coefficients for a few material combinations. Generally, the reflection coefficient can be a complex quantity, described by its modulus and a phase angle different than 0 or 180 degrees. Such a case corresponds to the existence of nonhomogeneous waves propagating along the interface with leakage into one of the connecting media. Usage in UOIT of only Snell's law is not sufficient. It does not give us sufficient information concerning the efficiency of mode generation. The influence of inelasticity on the reflection factor can be investigated by including into the derivation the wavenumber in complex form with the imaginary part related to the attenuation of each kind of ultrasonic wave. This influence, however, especially for two materials with a large difference in acoustic impedance, is almost negligible.

Let's now consider the interfacial weakness detection problem. To select the appropriate type of wave (longitudinal or transverse) and appropriate angle of incidence for adhesion weakness detection, one must determine the angular characteristics for both the normal and tangential reflection coefficient for smooth boundary conditions or for an interface with finite values of rigidities.[5,6] The first type of boundary condition is related to the case with vanishing tangential stresses, which we can imagine as a case with a very thin, inviscid, intermediate layer. The second type seems to describe all possible real cases of connection between ideal bonds with infinitely large rigidities and complete disbonds with rigidities equal to zero. Comparing angular characteristics for welded and smooth boundary conditions, one can find the best angle of incidence for the detection of adhesion imperfections.[7] The frequency dependence of the reflection coefficient from an interface with a finite value of rigidity is an additional feature that can be used in discriminating between cohesive and adhesive weaknesses in an adhesively bonded structure.[7]

Concerning anisotropic influences, the issue of reflectivity from an interface between two anisotropic materials was described in several papers (e.g. Henneke[8]) and recently in a unified approach for general anisotropic media by Rohlin et. al.[9] Even in the welded case described, the situation is more complex than for isotropic materials, because there exists possibly three (not two) reflected or transmitted waves. The direction of propagation of these quasi-longitudinal or quasi-transverse waves differs from the direction of energy flow in general. Also, the determination of reflected and refracted angles is more complex, because a solution of the Christoffel equation is required in order to obtain the slowness surfaces for both materials.[9] The condition that the energy flow away from the interface in determining the correct reflected and transmitted waves must be considered. Critical angles for these waves occurs when the energy flow directions are parallel to the interface. This is important in the utilization of the Critical Angle Reflectivity Technique (CART) for anisotropic material characterization. Generally, we can not evaluate directly the elastic constants with Snell's law by measuring the critical angle, since the phase velocity in this widely known relationship for isotropic materials differs in both magnitude and direction from the group velocity for anisotropic materials.

CART is based on a registration of the angular characteristics of the reflection coefficient of longitudinal waves incident onto an interface between a liquid and a solid. If the thickness of the solid material exceeds by several times the wavelength, the structure can be treated as a half-space, therefore simplifying the boundary conditions; normal stresses and displacements are continuous. If one includes attenuation losses in a solid due to a bounded beam, the angular characteristic of the reflection factor produces a pronounced minimum for the third critical angle. This angle corresponds to the generation of surface waves and is chosen as the best one for applying CART, especially for metals characterization. For some plastics or composite materials, a more pronounced effect in the first critical angle appears, which corresponds to a longitudinal or quasi-longitudinal wave. In this case the inequality of the phase and group velocities is valid.[10]

If the thickness (or total thickness) of the layer(s) is comparable with a wavelength, the angular characteristics of the reflection factor appears to have a larger number of minima (at least two), which corresponds tc total transmission. For almost all metals, the minimums also correspond to the individual modes of the plate waves. In the case of materials with densities closer to the density of the liquid, however, such coincidence does not occur, especially for modes of lower order.[11,12] Angles of total reflection (minimum reflectivity) are not the same as those showing a maximum of backscattering. This phenomenon of backward propagating leaky Rayleigh or Lamb waves has been described (e.g. Billy et. al.[13]) and utilized for a one probe variation of CART. When coincidence occurs, the angular characteristics of the reflection or transmission factors are sufficient for producing curves of the dispersion[14] for either single or multilayered plate. These relationships can be derived utilizing Thompson's approach[11] in determining the recurrence matrix. For a single layer or multilayered isotropic plate immersed in a liquid, results are available (e.g. Jackins and Gaunaurd[15]). The results for an anisotropic single layer with hexagonal symmetry are also useful and are described here.

Another approach in establishing the dispersion curves for surface or plate waves in layered structures with different boundary conditions is obtained by solving the characteristic equation for such a medium.[16,17] In this way, without a very time consuming analysis of the structure of the individual modes, one can select the best mode for a particular task, for example, detection of delamination, weakness in cohesive or adhesive strength etc.. An example of such an indirect analysis for a three layered structure,[18] is also presented here. Sometimes the roots of the characteristic equation are complex quantities.[17] Such solutions correspond to leaky waves, which transfer the energy into the surrounding liquid or into an acoustically "softer" solid medium.

Phase velocity determination for surface waves propagating along a free surface of an anisotropic material requires a simultaneous solution of the Christoffel equation for a material of a relevant class of symmetry and boundary conditions for a free surface.[19] The solution must satisfy the requirement of vanishing motion at large distances from the surface. If the direction of propagation is parallel to one of the axes of symmetry, the solution for anisotropic materials of orthorhombic or higher symmetry is simplified into a bicubic equation with respect to the surface phase velocity, that is equal to the group velocity since the direction of propagation is the same as the energy flow. The examples of such a solution for composite materials[20] are also given here. Once the velocities of the surface waves are known, it becomes possible to select an appropriate angle of incidence and reception. Unfortunately, due to a possible smaller value of surface velocity compared to the velocities of longitudinal waves in most liquids, such an approach of surface wave generation has to be replaced by some alternate technique. Some of these are less effective, as e.g. point or line sources, or inconvenient in NDE practice, as interdigital transducers etc.

METHODOLOGY

The UOIT procedure can be useful in classifying a large number of material states and possible defects. Before proceeding with the methodology associated with UOIT, let's consider some material and defect possibilities. Defects that occur in real structures can generally be divided into two groups: macro- and micro-defects. Macro-defects include planar or volumetric defects like cracks, voids, delaminations, debonds, etc. To be characterized fully, such defects require very precise location, size evaluation and shape description. Only then, can the fatigue life of the structure with such defects be estimated, using knowledge from cohesive and/or adhesive fracture mechanics combined with the loading characteristics. The second group, micro-defects, is represented not only by defects with very small size, as in the case of ceramic materials, but also by microcracks and other small changes in the structures, causing material degradation and interfacial weakness due to such loading as mechanical or hydrothermal fatigue, corrosion, etc.. These fabrication and in-service generated micro-defects are in general globally distributed and as such require material characterization analysis rather than defect detection. For example in the case of porosity, overall porosity level is usually evaluated and not the individual pore size.

The advantages of utilizing UOIT are numerous: First of all, we can generate ultrasonic waves of many types using longitudinal waves impinging at various angles of incidence from a reference medium with known material properties, as well as with an immersion or contact technique. For example, in the case of immersion inspection, UOIT is the only way that we can generate waves with tangential components of vibrations with respect to the interfaces between connected materials. Changing the angle of incidence from the reference medium, due to changes of refraction angles for longitudinal or transverse waves, we can easily manipulate the angle of impingement onto the inspection area. The phenomena of mode conversion enables us to generate such bulk waves as longitudinal or transverse waves, and also guided waves. These waves are "guided" by boundaries of the material being inspected. If the thickness of the test material (or total thickness in the case of multilayered structure) exceeds by many times the wavelength of the applied waves, we can generate surface waves of Rayleigh type. Plate waves of Lamb type are generated in plates with thickness (or total thickness) that is comparable to a wavelength. Such waves supply us with global information regarding the possible failure state of a larger area of an inspected structure. In addition, by changing the frequency of the incident wave, different depths of material can be penetrated and hence examined by using surface waves with penetration being approximately equal to one wavelength. Different vibration characteristics can also be obtained by selecting appropriate plate wave modes. The direction of propagation of such waves can be changed easily by using UOIT, a convenient approach, especially in the inspection of anisotropic materials. The reception of bulk or guided waves in UOIT also requires an understanding of both wave generation and multi-material coupled wave propagation characteristics. For surface and plate waves, reception can take place as a result of leakage accompanying the wave propagation along a interface between two materials; see Bar-Cohen and Crane[21] on the use of leaky waves.

When using oblique incidence, there are possibilities of utilizing ultrasonic reflectivity techniques. A technique, based on determining the angular characteristic of the reflection coefficient, most often for a liquid-solid interface, is appropriate for material property characterization with an assumption of the homogeneity of the material throughout. If we use an ultrasonic pulse of time duration comparable to the thickness of a layer, we can then evaluate the material properties relevant to the whole thickness of a plate. Sometimes, in the case of anisotropic materials with a plane of symmetry off angle to the plane of incidence, we can not directly evaluate the elastic constants through critical angle determination,[8,9] but a registration of the changes in angular position can provide us with some information concerning the overall state of the layered or composite structure.

Application of UOIT is more complex compared to traditional normal beam evaluation. A full understanding of the relevant physical phenomena is necessary. To choose the appropriate testing parameters (transducer frequency, angle of incidence, etc.) and to provide a quantitative analysis of the obtained results, one should be familiar with the reflection/transmission characteristics at each particular interface in the overall structure. A complete understanding of the wave propagation in a semi-space or in a layered isotropic or anisotropic medium is essential in order to effectively use UOIT and the many different kinds of waves that can be produced for ultrasonic inspection purposes.

This short survey of theoretical material on reflectivity related to oblique incidence and guided wave propagation provides us with information and knowledge that can be useful in applying the acousto-ultrasonic technique. The advantages resulting from the application of oblique incidence are numerous.

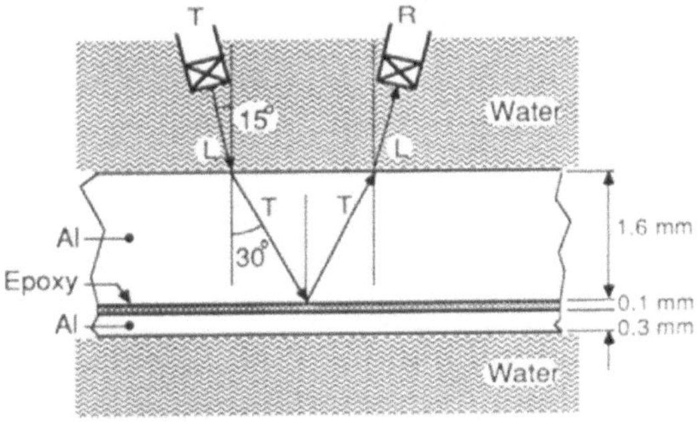

Fig. 2. Oblique Incidence Immersion Technique for producing a transverse wave input to a three layered plate bond model.

SAMPLE RESULTS

Let us now consider some sample problems on the utilization of oblique incidence and guided waves, primarily on aluminium to aluminium adhesively bonded structures and on graphite fiber unidirectionally reinforced epoxy composite materials.

a) Transverse wave for interfacial weakness determination in adhesively bonded structures: Let's consider a bonding case as shown in Fig. 2. Pilarski and Rose[7] have shown that increased sensitivity in using transverse waves compared to longitudinal waves can be obtained. This is demonstrated by examining the reflection factor for welded versus smooth boundary conditions (Fig. 3); real situations occur somewhere between these two. By comparison, we have decided that the transverse waves with a particular angle of incidence of around 30 degree would be the best for our purposes (Fig. 3). As a consequence of the above consideration, the configuration of the experimental set up, as shown in Fig. 2., was selected. In the geometrical positioning of both transducers, the displacement (shifting) of the ultrasonic beam on the Al-epoxy interface, for a frequency of 10 MHz was less than 2 mm and was hence disregarded. The frequency dependence (Fig. 4.) of the reflection factor for transverse waves incident with an angle of 30 degrees onto the imperfect interface, that corresponds in our model to a particular finite value of bond rigidity, will be the basis for a discrimination of the levels of interfacial weakness.

b) Bond integrity employing a three-layered plate adhesive bond model: Plate wave considerations for the ultrasonic inspection of adhesively bonded structures have led us to the solution of the characteristic equation for a three layered plate.[18] Fig. 5 shows the dispersion curves for the first six modes of the plate waves in the three layered bonded structure model. Also shown are two other family of curves. Thicker adhesive (0.3 mm instead of 0.1 mm) and adhesive property degradation (density and elastic constants less 15 %) can occur in cohesively weakened structures. We can see that the influence due to such imperfections or degradation will manifest itself by a decrease in velocity or an increase in the critical angle of the mode selected. Additionally, marked points on Fig. 5 are results of some experimental measurements for three frequencies 0.5, 1.0 and 2.25 MHz. The differences are a result of a small difference in density between aluminium and water. Similar calculations were performed for different kinds of boundary conditions with three combinations:

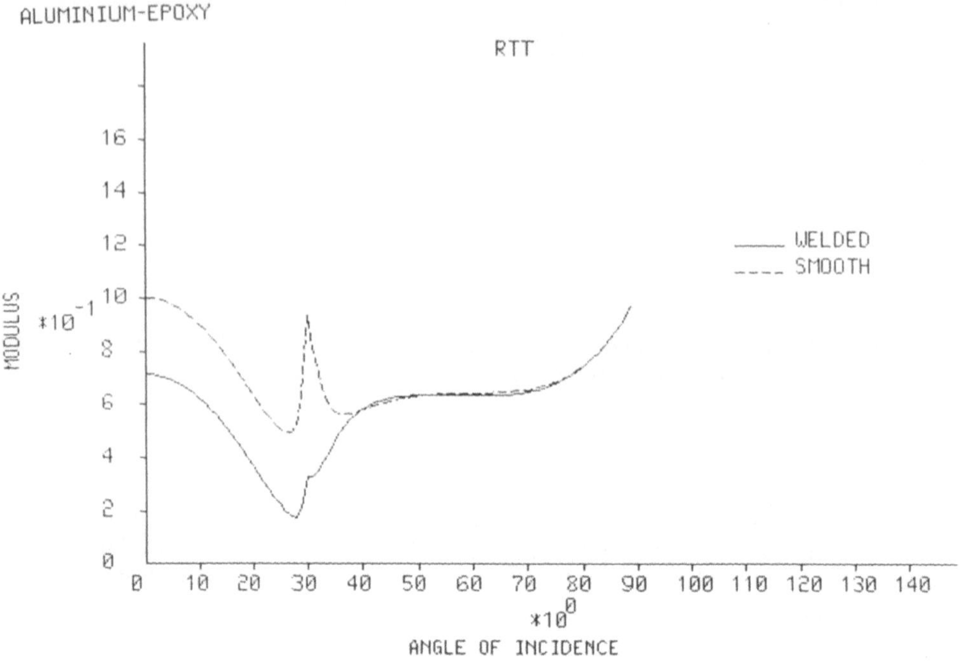

Fig. 3. The angular characteristics of the modulus of the reflection coefficient for an aluminium/epoxy resin interface with welded and smooth boundary conditions in the case of transverse wave incidence.

welded on first interface, smooth on the second, then smooth-welded and smooth-smooth for the second and third cases respectively. By comparison with the case of "welded-welded", the optimal frequencies and angles of generation of the individual modes for poor adhesion were found.[18]

c) Angular characteristics of reflection factor for porosity evaluation in composite materials: Utilizing the UOIT for the detection of porosity in composite layers has led us to the determination of the angular characteristics of reflection factor. Fig. 6 and Fig. 7 show characteristics for a graphite/epoxy unidirectional composite layer immersed in water for a constant frequency-thickness product equal to 2.0 MHz-mm and a plane of incidence parallel to the fibers. These curves were calculated for different sets of elastic constants and densities due to different degrees of porosity of the composite and also different fiber volume fractions. We can see that the changes in porosity can be evaluated by obtaining measurements of the shift in the angles of minimum reflectivity. Simultaneously, changes in fiber fraction can of course complicate the situation. In this case, similar measurements in another plane of incidence with respect to the fiber direction are suggested.

d) Utilization of surface waves for composite material characterization: A surface wave approach in composite material[20] is presented in Fig. 8. The measurements of the velocities of the surface waves, propagating in directions both parallel and perpendicular with respect to the fibers, can provide us with a simultaneous evaluation of changes in both fiber volume fraction and degree of porosity. It also seems appropriate for elastic constant determination in such anisotropic materials that measurements of surface wave velocities in a

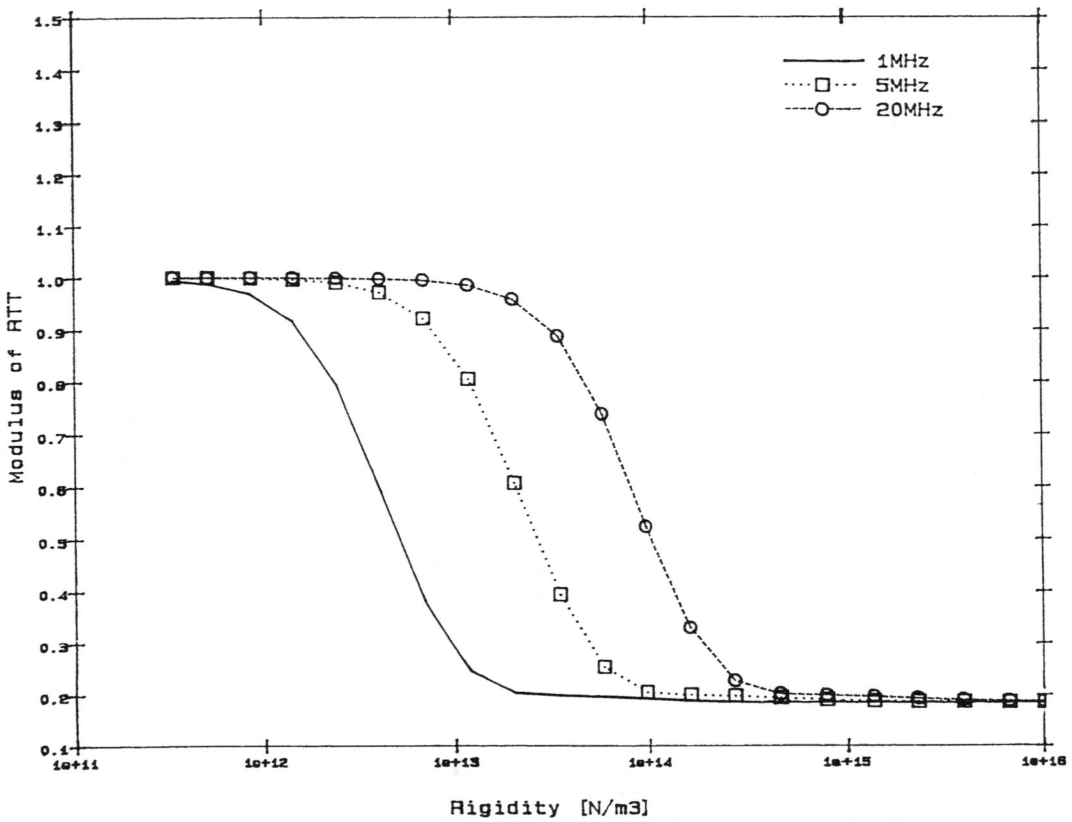

Fig. 4. The frequency influence on the modulus of the reflection coefficient, RTT, vs. the bond rigidities (KN=2KT) for an aluminium-epoxy interface, and an angle of incidence of transverse waves equal to 30 degrees.

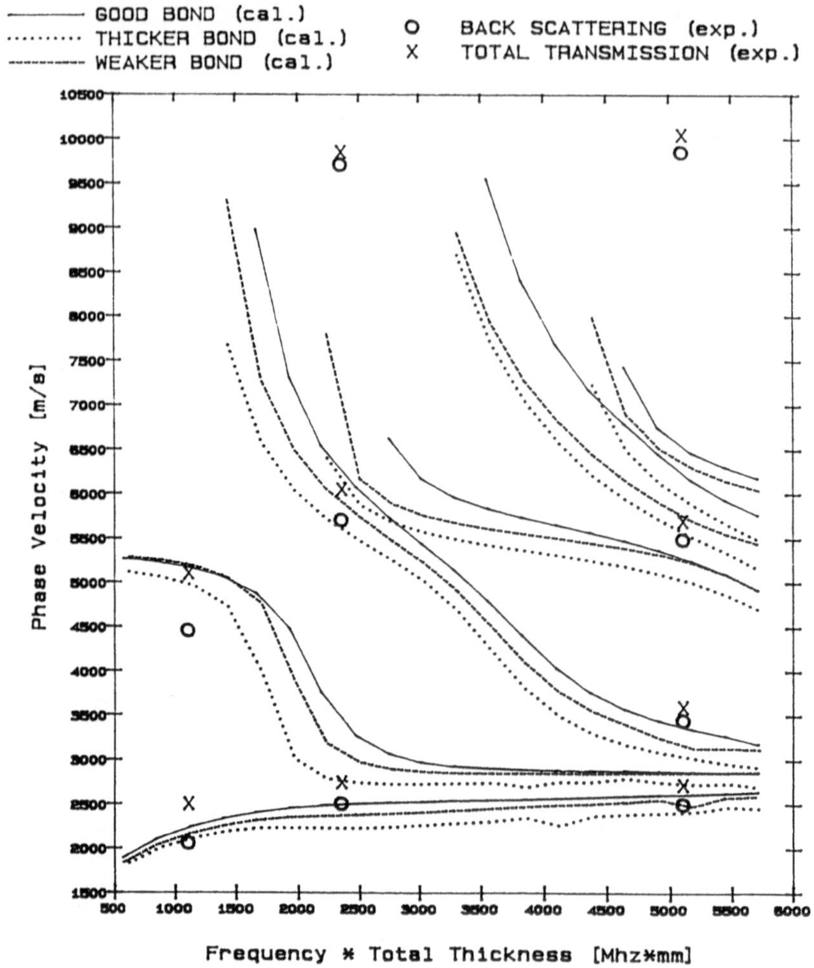

Fig. 5. Dispersion curves for the first six plate modes in the three layered plate bond model.

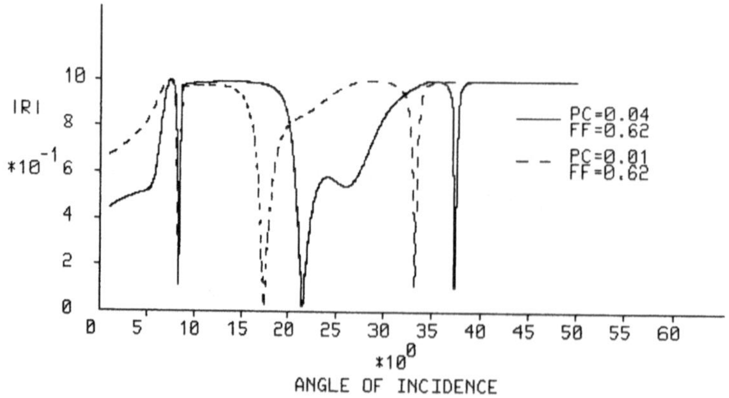

Fig. 6. Angular characteristics of the modulus of the reflection coefficient for porosity changes (PC in %) in a graphite/epoxy unidirectional composite material. (constant volume fiber fraction (FF)).

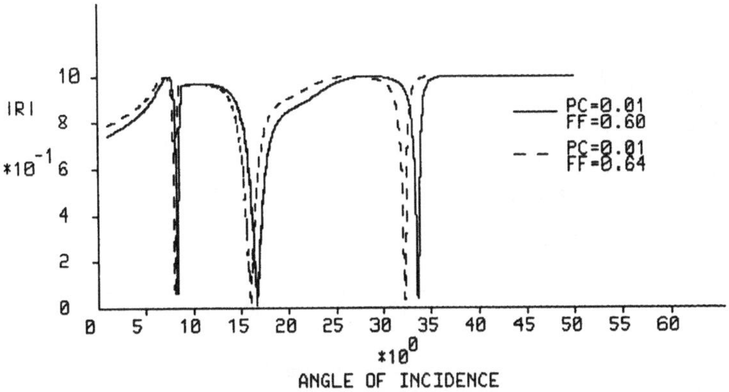

Fig. 7. Angular characteristics of the modulus of the reflection coefficient for volume fiber fraction changes (FF in %) in a graphite/epoxy unidirectional composite material.

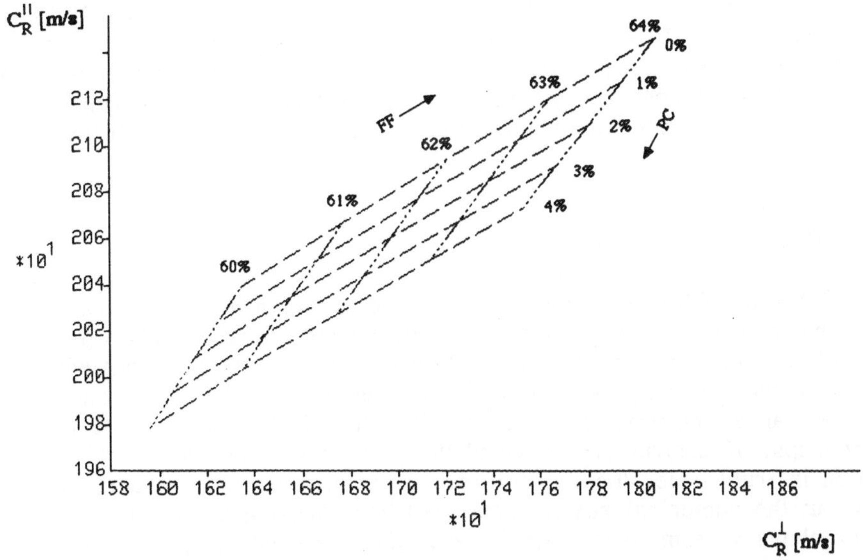

Fig. 8. Changes in velocities of surface waves propagating in parallel and perpendicular directions with respect to fibers caused by changes in porosity and fiber fraction for graphite/epoxy. PC - constant porosity.

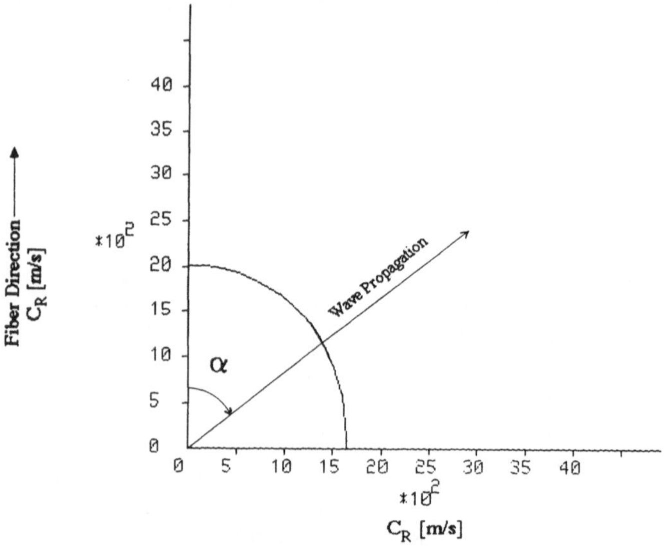

Fig. 9. The velocity surface for surface waves propagating in graphite/ epoxy composite material with 62% fiber fraction.

direction off an axis of symmetry would be useful. (The changes of the phase velocity of the surface waves in a graphite/epoxy composite with changes of angle of propagation with respect to the fibers are shown in the form of a velocity surface in Fig. 9.) For each direction of propagation, a skew angle can be easy established, the skew angle being defined as the angle between the wave vector and vector of the energy flux. For this particular case, the skew angle does not exceed 20 degrees. For applications where there is access to one side of the specimen only, additional measurements of the velocities of one longitudinal and two transverse waves seems promising for elastic constant determination for composite materials with either orthorhombic or transversely isotropic symmetry.

CONCLUSIONS

The Ultrasonic Oblique Incidence Technique presents us with a number of variables that can make acousto-ultrasonic measurements extremely valuable in all sorts of material characterization and defect analysis work. Considerations of transverse wave propagation, critical angle analysis, and surface and plate wave modes can all be used in producing "Feature Map" images, material state or defect maps, of a structure provided the physics and mechanics of wave propagation is truly understood. Additional experimental work is certainly required, but the numerical results presented here show great promise for solving problems in composite material and adhesive bonding applications.

ACKNOWLEDGEMENTS

This study was funded in part under the Ship and Submarine Materials Block sponsored by the Office of Naval Technology. Mr. Ivan L. Caplan, David W. Taylor Naval Ship Research and Development Center, was the Block Manager. The contract was administered by the Naval Research Laboratory; Dr. Irv Wolock and Henry H. Chaskelis were technical monitors.

REFERENCES

1. A. Vary, Acousto-Ultrasonic Characterization of Fiber Reinforced Composites, Mat. Eval. 40:650 (1982).
2. J. L. Rose, J.B.Nestleroth, and K.Balasubramanian, Novelty of Feature Mapping in NDT, (in progress)
3. K. F. Graff, "Wave motion in elastic solids," Ohio State University Press, Columbus (1975).
4. J. Miklowitz, "Elastic waves and waveguides," North-Holland Publishing Company, New York (1972).
5. M. Schoenberg, Elastic waves behavior across linear slip interface, J. Acoust. Soc. Am. 68:1516 (1980).
6. A. Pilarski, The coefficient of reflection of ultrasonic waves from an adhesive bond interface, Archives of Acoustics 8:41 (1983).
7. A. Pilarski and J. L. Rose, A Transverse Wave Ultrasonic Oblique Incidence Technique for Interfacial Weakness Detection in Adhesive Bonds, submitted to J. Appl. Phys..
8. E. G. Henneke, Reflection-Refraction of a Stress Wave at a Plane Boundary between Anisotropic Media, J. Acoust. Soc. Am. 51:210 (1972).
9. S. I. Rohlin, T. K. Bolland, and L. Adler, Reflection and refraction of elastic waves on a plane interface between two generally anisotropic media, J. Acoust. Soc. Am. 79:906 (1986).
10. E. G. Henneke, II and G. L. Jones, Critical angle for reflection at a liquid-solid interface in single crystals, J. Acoust. Soc. Am. 59:204 (1976).
11. L. M. Brekhovskikh, "Waves in Layered Media," Academic Press, New York (1960).
12. D. E. Chimenti and A. H. Nayfeh, Anomalous ultrasonic dispersion in fluid-coupled, fibrous composite plates, Appl. Phys. Lett. 49:492 (1986).
13. M. Billy, L. Adler, and G.Quentin, Measurements of backscattered leaky Lamb waves in plates, J. Acoust. Soc. Am. 75:998 (1984).
14. J. L. Rose and A. Pilarski, Surface and Plate Waves in Layered Structures, submitted to Mat. Eval.
15. P. D. Jackins and G. C. Gaunaurd, Resonance acoustic scattering from stacks of bonded elastic plates, J. Acoust. Soc. Am. 80:1762 (1986).
16. A. Pilarski, Ultrasonic Evaluation of the Adhesion Degree in Layered Joints, Mat. Eval. 43:765 (1985).
17. A. Pilarski, Ultrasonic Wave Propagation in a Layered Medium under Different Boundary Conditions, Archives of Acoustics 7:61 (1982).
18. J. L. Rose and A. Pilarski, Plate wave propagation in a three layered structure, (in progress).
19. G. W. Farnell, Properties of Elastic Surface Waves, in:" Physical Acoustics, Vol. 6," W. P. Mason and R. N. Thurston, eds., Academic, New York (1970)
20. A. Pilarski and J. L. Rose, Utilization of surface waves for composite materials characterization, (in progress).
21. Y. Bar-Cohen and R. L. Crane, Acoustic backscattering imaging of subcritical flaws in composites, Mat. Eval. 40:970 (1982).

ULTRASONIC VELOCITY STUDIES OF COMPOSITE AND HETEROGENEOUS MATERIALS

Subhendu K. Datta

Department of Mechanical Engineering and CIRES
University of Colorado
Boulder, Colorado 80309

Hassel M. Ledbetter

Fracture and Deformation Division
Institute for Materials Science and Engineering
National Bureau of Standards
Boulder, Colorado 80303

Arvind H. Shah

Department of Civil Engineering
University of Manitoba
Winnipeg, Canada R3T2N2

ABSTRACT

Ultrasonic measurements of wave-propagation characteristics in composite and heterogeneous materials provide an excellent means to study their mechanical properties. In recent years we have studied, both theoretically and experimentally, characteristics of elastic-wave propagation in particle-reinforced composites and heterogeneous materials as well as in homogeneous and laminated fiber-reinforced composites. Comparison of theoretical predictions with obervations of wave velocities has shown good agreement and has provided a way to evaluate microstructural dependence of mechanical properties of these materials. Modeling predictions coupled with observations can also be used to obtain mechanical properties of the reinforcing phase, which are sometimes not easily obtained. In this paper we present results of some of these recent studies.

We also present results of our study of changes in phase velocities and attenuation caused by interface layers between the reinforcing phase and the matrix. We show that this third phase measurably modifies the dispersion behavior. This should lead to effective characterization of interface layer properties by ultrasonic methods.

INTRODUCTION

Determination of effective elastic moduli and damping properties of a heterogeneous or composite material by using elastic waves (propagating and standing) is very effective. Several theoretical studies[1-13] show that for long wavelengths one can predict the effective wave speeds of plane-longitudinal and plane-shear waves through such a material. At long wavelengths, wave speeds thus calculated are nondispersive; they provide values for the static effective elastic properties. References to other studies can be found in those cited.

We present results of some of our recent studies of phase velocity and attenuation of plane longitudinal and shear waves propagating in a medium with microstructure. Microstructures studied were either inclusions or fibers. In the first case, we examined the effect of inclusion shape, volume fraction, and elastic properties on wave speeds. We studied inclusions with their interface layers separating them from the matrix medium. For fiber-reinforced materials, we studied continuous aligned fibers. In this case, the medium behaves anisotropically because of the alignment of the fibers.

The theoretical model for these microstructural studies used a wave-scattering approach and predicted the macroscopic isotropic elastic properties for the case of random orientation of inclusions and for spherical inclusions. For aligned fiber-reinforced materials the model gives the anisotropic elastic properties. The scattering approach led also to an estimation of attenuation and dispersion of waves via the optical theorem.

The scattering approach applies to a macroscopically homogeneous medium, infinite in extent. For a bounded medium, such as laminated-plate structures, we developed a hybrid numerical technique to analyze dispersion of guided waves. In this paper we present some of our computational results showing the effect on dispersement characteristics caused by interface soft layers between stiff layers in a plate.

The experimental methods[10,14,15] consisted of a pulse-echo technique and a resonance method. These were chosen to provide the advantages of small specimens and low inaccuracy. For details of the experimental techniques, the reader is referred to the references cited.

THEORY

Scattering by a Single Inclusion

Consider a single elastic ellipsoidal inclusion with material properties λ', μ', ρ' embedded in an elastic matrix with properties λ, μ, ρ. Assume that the inclusion is separated from the matrix by a thin layer of elastic material with properties λ_1, μ_1, ρ_1, which are variable through the thickness. Here λ, μ are the Lame's constants and ρ is the density. The geometry of the problem is shown in Fig. 1.

We need to find the field scattered by this inclusion when it is excited by an incident elastic wave. A low-frequency approximate solution to this problem was presented earlier[16] for the case of no intermediate layer between the inclusion and the matrix. When the time factor $e^{-i\omega t}$ is omitted and the exciting field is expressed by

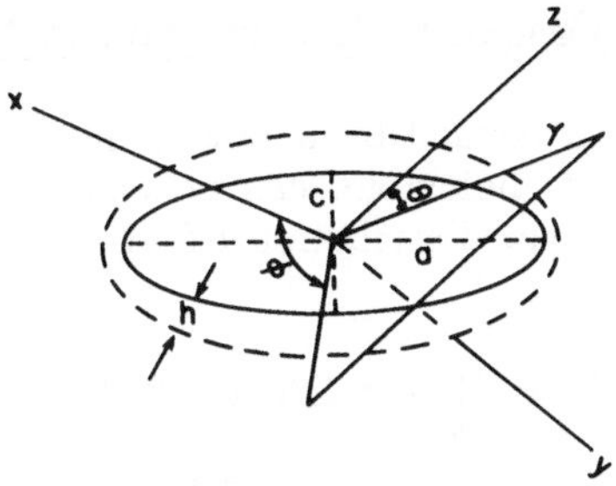

Fig. 1. An oblate spheroidal inclusion with interface layer.

$$\underline{u}^{(E)} = \sum_{n=0}^{\infty} \sum_{m=-n}^{n} \left[a_{mn}\underline{L}_{mn}^{(1)}(r,\theta,\phi) + \gamma b_{mn}\underline{N}_{mn}^{(1)}(r,\theta,\phi) + \gamma c_{mn}\underline{M}_{mn}^{(1)}(r,\theta,\phi)\right] \quad (1)$$

then the scattered field is

$$\underline{u}^{(s)} = \sum_{\nu=0}^{2} \sum_{\mu=-\nu}^{\nu} \left[A_{\mu\nu}\underline{L}_{\mu\nu}^{(3)}(r,\theta,\phi) + \gamma B_{\mu\nu}\underline{N}_{\mu\nu}(r,\theta,\phi)\right] + O(\varepsilon^4) \quad (2)$$

Here $\underline{L}_{mn}^{(1)}$, $\underline{N}_{mn}^{(1)}$, and $\underline{M}_{mn}^{(1)}$ are spherical vector wave functions[17] that are finite at $r = 0$ and $\underline{L}_{mn}$, $\underline{N}_{mn}$ are those that satisfy the radiation condition as $r \to \infty$. The constants $A_{\mu\nu}$ and $B_{\mu\nu}$ are given by[13]

$$A_{\mu\nu} = (iv_o\varepsilon^3/4\pi c^3)\sum_{v} \sum_{u=-v} \hat{T}_{\mu\nu}^{uv} \left[a_{uv} + \delta(\nu)b_{uv}\right],$$

$$B_{\mu\nu} = (iv_o\varepsilon^3/4\pi c^3)\sum_{v} \sum_{u=-v} \Delta(\nu)\hat{T}_{\mu\nu}^{uv} \left[a_{uv} + \delta(\nu)b_{uv}\right], \quad (3)$$

where v_o is the volume of the ellipsoid and

$\varepsilon = k_1C_1,\ k_1 = \omega/C_1,\ \gamma = C_1/C_2$;

$$\delta(\nu) = \begin{cases} 3\gamma^2, & \nu=0,2 \\ 2\gamma, & \nu=1 \end{cases} ;$$

$$\Delta(\nu) = \begin{cases} \gamma^2, & \nu=1 \\ \gamma^3/2, & \nu=0,2 \end{cases} .$$

C_1 and C_2 are the longitudinal and shear wave speeds in the matrix medium.

Expressions for $\hat{T}_{\mu\nu}^{uv}$, can be found elsewhere.[9]

Once the scattered field caused by a single inclusion is known, multiple scattering from a number of inclusions can be easily calculated. In particular, for a random homogeneous distribution of ellipsosidal inclusions

$$C^2{}_1/C_1{}^{*2} = \{(1+9c^*P_1)(1-3c^*P_o)[1+(3/2)c^*P_2(2+3\tau^2)]\}/\{\ \Omega\ \}, \qquad (4)$$

where $\{\ \Omega\ \} = \{1-15c^*P_2(1+3c^*P_o)+(3/2)c^*P_2(2+3\tau^2)\}$.

$$C^2{}_2/C_2{}^{*2} = (1+9c^*P_1)[1+(3/2)c^*P_2(2+3\tau^2)]/[1+(3/4)c^*P_2(4-9\tau^2)). \qquad (5)$$

Here $c^* = (4/3)\pi abcn_o$, n_o being the number density of inclusions. The constants P_o and P_2 depend on the geometrical properties of the inclusions as well as on the mechanical properties $(\lambda,\mu,\lambda',\mu')$ of the matrix and the inclusions. P_1 is simply given by

$$P_1 = (\rho'/\rho-1)/9.$$

In deriving eqs. (4) and (5) it has been assumed that the inclusions are similar in shape, size and physical properties.

When there is an intermediate layer between the inclusion and the matrix, then the coefficients $A_{\mu\nu}$ and $B_{\mu\nu}$ cannot be obtained in exact form; they have to be determined numerically for general elliposoidal inclusions. For spherical inclusions, on the other hand, exact expressions for $A_{\mu\nu}$ and $B_{\mu\nu}$, valid at arbitrary frequencies, can be obtained if the intermediate layer has constant properties. Also, for spherical inclusions with thin interface layers having variable properties, $A_{\mu\nu}$ and $B_{\mu\nu}$ can be obtained[18] approximately if $h/\lambda \ll 1$, $h/a \ll 1$. Here h is the thickness of the interface layer, a the radius of the spherical inclusions, and λ the wavelength of the incident wave. Even then, the problem of determining phase velocities of longitudinal and shear waves in the composite material at finite frequencies is complicated.[8] The problem is simpler if the volume concentration of inclusions is small.

Consider a spherical inclusion of radius a with an interface layer (Fig. 2). Let the incident field be given by

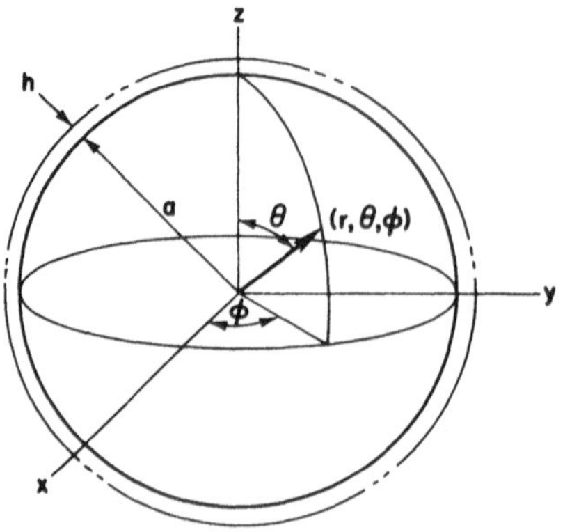

Fig. 2. A spherical inclusion with interface layer.

$$\underline{u}^{(i)} = \exp(ik_1 z)\underline{e}_z + \exp(ik_2 z)\underline{e}_x \tag{6}$$

where $k_2 = \omega/C_2$. The scattered field is then given by

$$\underline{u}^{(s)} = \underline{u}^p + \underline{u}^s \tag{7}$$

where superscripts p and s refer to longitudinal-wave and shear-wave components, respectively. Now it will be assumed that within the interface layer the elastic coefficients λ_1 and ν_1 vary with the distance from the center of the inclusions as

$$\begin{aligned} &\lambda_1(r) + 2\nu_1(r) = (\lambda_1' + 2\nu_1')f(r), \quad a<r<a+h, \\ &\nu_1(r) = \nu_1' g(r), \; a<r<a+h, \end{aligned} \tag{8}$$

where f(r) and g(r) are general integrable functions of r. If it is further assumed that $h/a \ll 1$ and $h/\lambda \ll 1$, then it can be shown that the displacement components satisfy the approximate boundary conditions on r = a.

$$\begin{aligned} &u^{(s)}{}_r + u^{(i)}{}_r - u^{(t)}{}_r = [hK_1/(\lambda'_1 + 2\mu'_1)]\tau^{(t)}{}_{rr} \\ &u^{(s)}{}_\theta + u^{(i)}{}_\theta - u^{(t)}{}_\theta = [hK_2/\mu'_1]\tau^{(t)}{}_{r\theta} \\ &u^{(s)}{}_\phi + u^{(i)}{}_\phi - u^{(t)}{}_\phi = [hK_2/\mu_1']\tau^{(t)}{}_{r\phi}, \end{aligned} \tag{9}$$

where

$$K_1 = \int_0^1 dx/f(a+hx)\;, \quad K_2 = \int_0^1 dx/g(a+hx).$$

Superscript (t) refers to the field quantities within the inclusion and τ_{ij} is the stress tensor. To this order of approximation the traction components τ_{rr}, $\tau_{r\theta}$ and $\tau_{r\phi}$ are continuous at r = a. These simplified boundary conditions allow the single scattering problem to be solved exactly.[18]

When $r\to\infty$, one obtains from eq. (7) the far-field behavior of $\underline{u}^{(s)}$ when the incident wave is given by a plane-longitudinal wave [the first term on the right-hand side of eq. (6)]. Then it can be shown that

$$\underline{u}^p \sim g^p(\theta)[\exp(ik_1 r)/r]\underline{e}_r, \quad \underline{u}^s \sim h^p(\theta)[\exp(ik_2 r)/r]\underline{e}_\theta \tag{10}$$

For the incident plane-shear wave [the second term on the right hand side of Eq. (6)], one finds

$$\begin{aligned} &\underline{u}^p \sim g^p(\theta,\phi)[\exp(ik_1 r)/r]\underline{e}_r, \\ &\underline{u}^s \sim h^s{}_1(\theta,\phi)[\exp(ik_2 r)/r]\underline{e}_\theta + h^s{}_2(\theta\div\phi)[\exp(ik_2 r)/r]\underline{e}_\phi \end{aligned} \tag{11}$$

The expressions for the amplitude functions g^p, h^p, and so on, can be obtained from our earlier study.[18] Using eqs. (10) and (11) and the forward-scattering theorem,[19] we then obtain the equations for the effective wave numbers in the composite medium:

$$k_1{}^{*2}/k_1{}^2 = 1 + [4\pi/k_1{}^2]n_o g^p(0) \tag{12}$$

for the longitudinal wave and

$$k_2^{*2}/k_2 = 1 + [4\pi/k_2^2]n_o\, h_1^s(0,0) \tag{13}$$

for the shear wave. Since $g^p(0)$ and $h_1^s(0,0)$ are complex, effective waves will be both dispersive and attenuative.

Effective Properties of a Fiber-Reinforced Composite

The analysis presented above for inclusions can be applied also to a medium reinforced by aligned continuous fibers. In that case, one can derive equations similar to (4) and (5) for longitudinal and shear waves propagating perpendicular to the fibers. By taking the x_3 axis along the fibers and assuming them to be transversely isotropic about this axis, it was shown[20] that for SH waves polarized along the fibers the effective wave number, β^*, is

$$\beta^{*2}/k_2^2 = [\rho^*(1-c^*(m-1)/(m+1))]/[\rho(1+c^*(m-1)/(m+1))] \, , \tag{14}$$

where ρ^* is the effective density and $m = C_{44}/\mu$. For longitudinal and shear waves polarized in a plane perpendicular to the fibers we get

$$k_1^{*2}/k_1^2 = \rho^*(1+c^*P_o)[1+c^*P_2(1+\gamma^2)])/[\rho(1-c^*P_2(1-\gamma^2)-2c^{*2}P_oP_2)] \, , \tag{15}$$

$$k_2^{*2}/k_2^2 = (\rho^*/\rho)/\Big[1+(2c^*(C_{66}-\mu)(\lambda+2\mu))/(2\mu(\lambda+2\mu)+ (1-c^*)(\lambda+3\mu)(C_{66}-\mu))\Big]. \tag{16}$$

Note that C_{11}, C_{33}, C_{44}, C_{66}, and C_{13} are the five independent elastic constants characterizing the fibers and $\rho^* = \rho[1+c^*(\rho'/\rho-1)]$. P_o and P_2 are defined by the relationships

$$P_o = -[K'_T - (\lambda+\mu)]/(K'_T+\mu), \quad P_2 = -[\mu(C_{66}-\mu)]/[C_{66}(\lambda+3\mu)+\mu(\lambda+\mu)] \tag{17}$$

where K'_T is the plane-strain bulk modulus of the fibers. The remaining two elastic moduli of the composite are obtained from relationships derived by Hill.[21]

Dispersion of Guided Waves in a Laminated Plate with Interface Layers

In composite materials, interfaces between the different constitutents play an important role in determining their mechanical behavior. As discussed above, the effect of interfacial layers between inclusions and matrix medium on composite properties can be evaluated approximately by using a scattering approach. For laminated-composite plate structures, the effect on the dispersion characteristics caused by interfacial layers between the laminae can be analyzed in detail. The changes in dispersion characteristics should provide a way to evaluate the interface properties ultrasonically.

In this section we briefly outline a theoretical technique to study guided-wave propagation in a laminated plate with interface layers. For simplicity of analysis, we consider only isotropic laminae and isotropic interface layers. Equations governing guided-wave propagation in such a plate can be solved exactly. However, if the laminae are anisotropic, then an exact analysis is extremely complicated. To avoid the difficulties associated with an exact analysis, we developed a stiffness method in which each lamina (and interface layer) is divided into several sublayers. Polynomial interpolation functions for through-thickness variation of displacements in

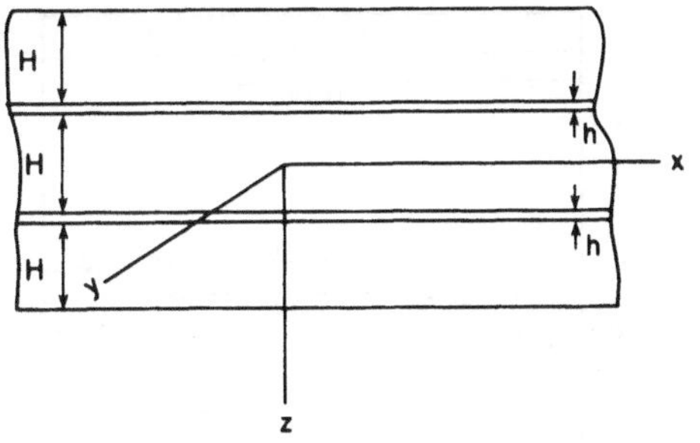

Fig. 3. A layered plate with thin interface layers.

each layer are assumed. The interpolation functions involve a discrete number of generalized coordinates that are functions of the in-plane coordinates x and y (Fig. 3) and time, t. The generalized coordinates are the displacements and tractions at the interfaces between the adjoining sublayers, thus ensuring continuity of these quantities at the interfaces. By applying Hamilton's principle the dispersion equation is obtained as a standard algebraic eigenvalue problem whose solutions yield the dispersion relations and the variation of stresses and displacements through the thickness of the plate. This technique was used earlier[22] to study wave propagation in an infinite periodically laminated medium. The application of the technique to a sandwich plate was discussed in detail elsewhere.[23] Here, for brevity, we present only the results for a particular sandwich plate.

RESULTS AND DISCUSSION

Cast-Iron Elastic Constants

Graphite-particle shape affects cast iron's properties, both physical and mechanical. Several studies[24-27] dealt with the various properties of cast iron as they depend on graphite particle shape. But all these studies, experimental and theoretical, dealt only with limiting shapes: sphere, rod, and disc. They failed to deal with arbitrary aspect ratio, c/a, of the particles. Using eqs. (4) and (5), we calculated Young's modulus, E^*, of cast-iron for various values of c/a. The results are shown in Fig. 4 along with the experimental observations reported by various investigators.[24-26] In this figure, calculated results are for two different volume fractions--10 and 12 percent--a range that contains most of the studied cast irons.

The two upper, nearly horizontal, curves correspond to graphite's upper third-order elastic-constant bounds. The two lower curves correspond to the lower bounds. From monocrystal elastic constants and equations by Kroner and Koch,[28] for graphite, Wawra, et al.[29] calculated third-order elastic-constant bounds. They found the following effective quasi-isotropic elastic constants: $E' = 4.17(1.34)$ GPa; $\mu' = 1.41(0.45)$ GPa. Values outside parentheses denote upper bounds; those inside denote lower. For the matrix phase we took the constants for alpha iron: $E = 206$ GPa, $\mu = 80.0$ GPa.

Corresponding to observation, our model predicts a strong dependence of Young's modulus on aspect ratio. Near the spherical limit ($c/a = 1$), E^* varies slowly with c/a. Near the oblate-disc limit ($c/a = 0$), E^* varies

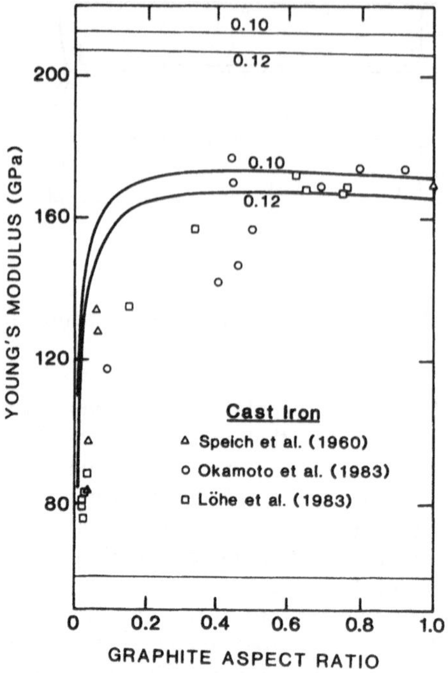

Fig. 4. For cast iron, Young's modulus versus graphite-particle aspect ratio. Symbols represent measurements. Curves represent model-calculation results for two volume fractions: 0.10 and 0.12. The Upper, nearly horizontal, curves represent graphite's upper third-order-bound (Kroner-bound) elastic constants. Lower curves represent graphite's lower third-order bound.

rapidly with c/a. An interesting result is that graphite's lower-bound quasi-isotropic elastic constants fit observation so well.

Elastic Constants of Graphite-Aluminum Composite

Figure 5 shows the microstructure of graphite-fiber-reinforced aluminum obtained from a commerical source. By Archimedes' method, we found the mass density of the composite to be 2.013 g/cm^3. For a fiber volume fraction of 0.70, 2.6523 for the density of aluminum, the graphite fiber density is predicted to be 1.738, very close to the manufacturer's estimate of 1.76.

We determined the nine C_{ij} by measuring eighteen sound speeds on four specimen geometries described previously.[30] For brevity, we omit further description, except for a few salient details: bond--phenylsalicylate; transducers--quartz, x-cut and ac-cut; frequencies--5 to 6 MHz; specimen size--16-mm cube, or smaller, depending on specimen geometry.

Table 1 shows the study's principal results. Column 1 lists various elastic constants. Column 2 gives a set of fiber elastic constants.[32] We chose these because E_{33} agrees closely with the E_{33} for the present fiber. Column 3 shows measured results. From the measured results and the predictions from eqs. (14) through (16) together with Hill's relationships, we predicted the fiber properties shown in column 4. We used the calculational

Table 1. Measured and calculated elastic constants for graphite fiber-reinforced composite and calculated graphite fiber elastic constants. Except for dimensionless ν_{ij}, units are GPa

	Fiber[32]	Composite Measured	Fiber Calculated	Composite Calculated	Composite Calc./Meas.
C_{11}	20.02	32.18	19.09	32.18	1.00
C_{22}	20.02	32.03	19.09	32.14	1.00
C_{33}	234.77	192.24	234.99	194.14	1.00
C_{44}	24.00	21.66	19.94	21.66	1.00
C_{55}	24.00	21.51	19.94	21.66	1.01
C_{66}	5.02	9.31	5.60	9.31	1.00
C_{12}	9.98	13.55	7.89	13.56	1.00
C_{13}	6.45	16.65	10.34	16.60	1.00
C_{23}	6.45	16.65	10.34	16.60	1.00
E_{11}	15.00	25.94	15.66	25.96	1.00
E_{22}	15.00	25.81	15.66	25.96	1.01
E_{33}	232.00	180.09	227.07	180.09	1.00
ν_{12}	0.494	0.396	0.399	0.394	0.99
ν_{13}	0.014	0.052	0.026	0.052	1.00
ν_{23}	0.014	0.053	0.026	0.052	0.98
ν_{21}	0.494	0.394	0.399	0.394	1.00
ν_{31}	0.215	0.363	0.383	0.393	1.00
ν_{32}	0.215	0.366	0.383	0.363	0.99

sequence C_{44}, C_{66}, C_{11}-C_{66}, ν_{31}, and E_{33}. Column 5 shows the calculated C_{ij}. Finally, column 6 shows the ratio of column 5 to column 3.

Results of column 3 of Table I show that the studied composite shows orthotropic elastic symmetry, which is approximately transversely isotropic with the x_3-axis as the symmetry axis. The microstructure in Fig. 5 also suggests transverse-isotropic symmetry.

Concerning the first-guess graphite-fiber elastic-constant calculations, we find reasonable agreement for C_{11}, C_{22}, C_{33}, and C_{66}. Thus, the criterion of choosing a graphite elastic-constant set based on E_{33}, the axial Young's modulus, succeeds partially.

One can obtain a better, complete graphite elastic-constant set by using the model equations inversely. This is shown in column 4; the constants appearing in this column differ significantly from the first-guess values in column 2.

For the graphite-aluminum composite, Fig. 6 shows the variation with temperature of the principal elastic stiffnesses C_{ij} (i = 1,6). For comparison, the figure also shows the temperature variation of the longitudinal and shear moduli of the aluminum matrix. The graphite-aluminum C_{ij}-versus-temperature results in Fig. 6 show strong anisotropy. The largest

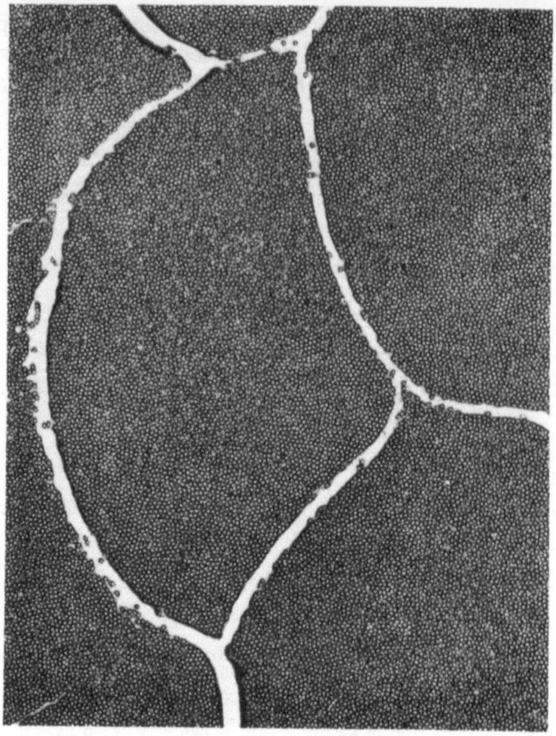

Fig. 5. Graphite-AP microstructure. Transverse section of parallel 7-mm-diameter graphite fibers distributed nearly homogeneously in aluminum matrix. The fiber volume fraction equals 70 pct. The white network represents aluminum boundary regions between fiber bundles used in manufacture.

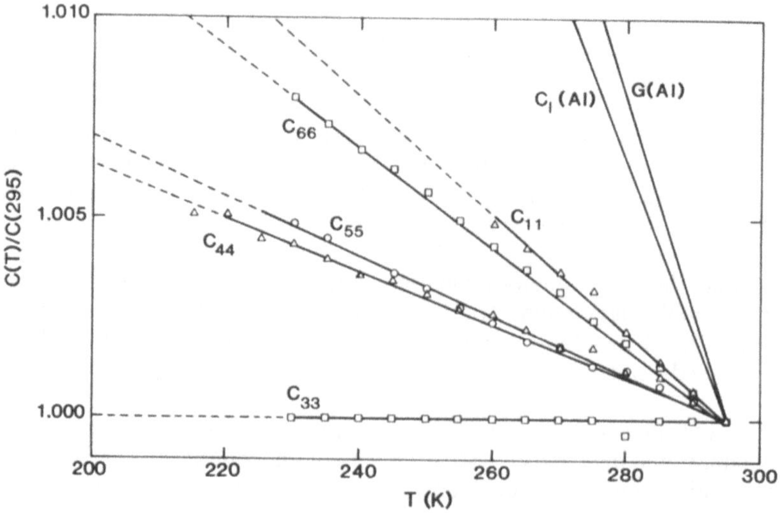

Fig. 6. For a composite consisting of 70-vol.-pct. uniaxial graphite fibers in an aluminum matrix, the variation with temperature of the principal C_{ij} elastic stiffnesses. Lines at the right show the variation of the aluminum-matrix longitudinal and shear moduli, C_ℓ and G.

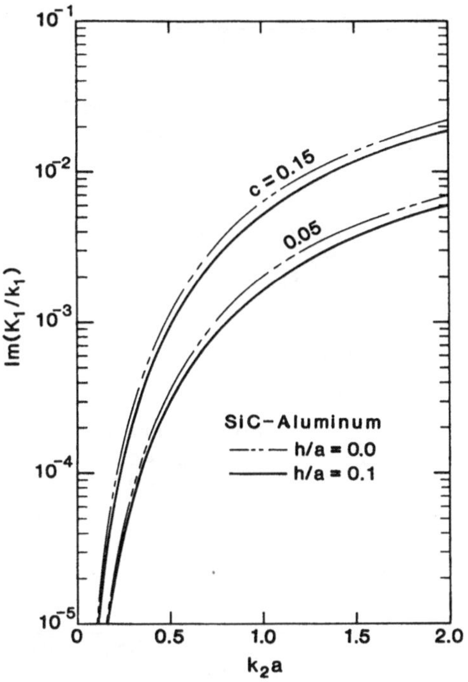

Fig. 7. Attenuation of a plane longitudinal wave in a particle-reinforced composite with and without interface layers. Volume fraction of inclusions is c.

change occurs in C_{11}, the longitudinal elastic stiffness perpendicular to the fibers. The smallest change occurs in C_{33}, the longitudinal elastic stiffness parallel to the fibers. These changes agree with the well-known high axial elastic stiffness and low axial thermal expansivity of graphite fibers. The three shear moduli--C_{44}, C_{55}, C_{66}--fall between these extremes. Among the shear moduli, C_{66} shows the largest change; this reflects the low C_{66} values for both the fiber and the matrix. Almost universally, lower elastic stiffness, C, means higher dC/dT. In Fig. 6, the near equivalence of dC_{44}/dT and dC_{55}/dT reflects the approximate transverse isotropy of this composite.

Interface Effects on Damping and Phase Velocities in an SiC-Particle Reinforced Aluminum

Equations (12) and (13) provide implicit relations for the complex wave numbers k_1^* and k_2^*, if it is assumed that the inclusion is placed in a composite medium with the effective unknown dynamic properties. Then k_1^2 and k_2^2, appearing on the right-hand sides of eqs. (12) and (13) will be replaced by k_1^{*2} and k_2^{*2}, respectively. These equations then can be solved iteratively for the unknown k_1^* and k_2^*. These results are shown in Figs. 7 through 10. Note that $\mathrm{Im}(k^*)$ measures damping and $\mathrm{Re}(k^*/k)$ measures the ratio of the phase velocities in the matrix and the composite. In these calculations the interface layer properties were assumed to vary linearly across the layer thickness from the properties of the particles to those of the matrix. The elastic properties of the particles and the matrix were taken as: $\lambda'+2\mu' = 4.742\mathrm{x}10^{11}$ N/m^2, $\mu' = 1.881\mathrm{x}10^{11}$ N/m^2, $\rho' = 3.181$ g/cm^3,

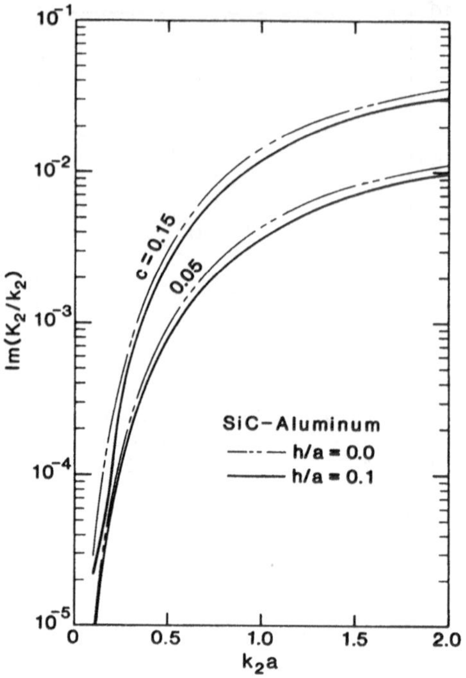

Fig. 8. Attenuation of a plane shear wave in a particle-reinforced composite with and without interface layers.

$\lambda+2\mu = 1.105\text{x}10^{11}$ N/m^2, $\mu = 0.267\text{x}10^{11}$ N/m^2, $\rho = 2.705$ g/cm^3. Calculations were performed with or without interface layers and at two volume fractions: c = 0.05, 0.15. The presence of the interface layer decreases the damping as well as the phase velocities.

Dispersion in a Sandwich Plate with Low-Velocity Interface Layers

Using the numerical technique described in earlier, we analyzed the dispersion of elastic waves in the five-layer plate shown in Fig. 3. The

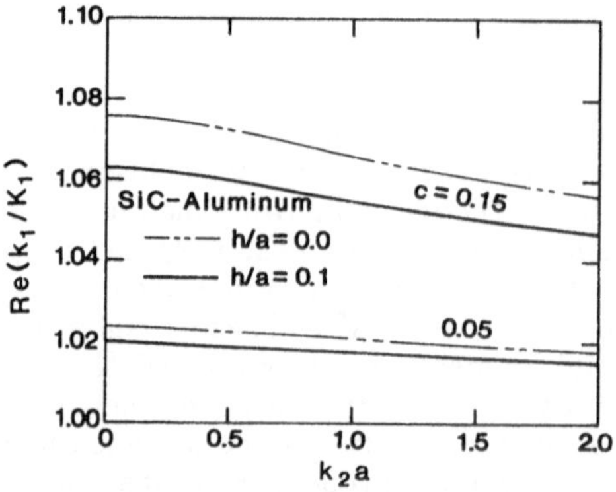

Fig. 9. Phase velocity of a plane longitudinal wave in particle-reinforced composite with and without interface layers.

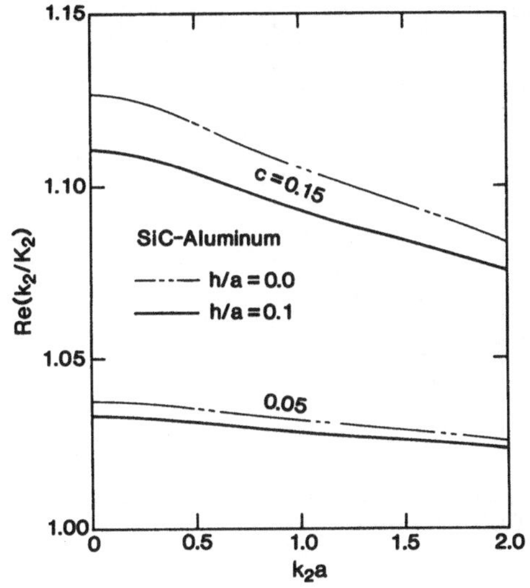

Fig. 10. Phase velocity of a plane shear wave in a particle-reinforced composite with and without interface layers.

displacement is assumed to be

$$\underline{u}(x,z,t) = \underline{f}(z)\exp(ikx-i\omega t). \quad (18)$$

If all the layers are transversely isotropic with the symmetry axes parallel or perpendicular to the direction of wave propagation (x-axis), then the equations governing the y-component of the displacement, u_y, uncouple from those governing the x and z components. The former correspond to the SH motion and the latter to plane strain motion.

Figures 11 and 12 show the disperison curves for the plane-strain motion, and Fig. 13 shows those for the SH motion. In these figures the vertical axis corresponds to $\Omega = \omega h/[\pi(\mu/\rho)^{0.5}]$ and the horizontal axis to $\zeta = kh/\pi$. Here h is the thickness of each of the two low velocity layers and $C_2 = (\mu/\rho)^{0.5}$ is the shear wave velocity in these layers. The properties of the stiff layers are taken to be $\mu' = 8.76 \times 10^9$ N/m^2, $\rho' = 1.771$ g/cm^3, $\nu' = 0.3$. Those of the soft layers are $\mu = 1.77 \times 10^9$ N/m^2, $\rho = 1.2$ g/cm^3, $\nu = 0.3$ and h/H is taken to be 0.1.

Figure 11 shows the first nine modes for small values of L and z. These dispersion curves appear very similar to those for an isotropic plate. However, the important difference is that cut-off frequencies of higher modes are lowered significantly. Figure 12 shows the first three modes over a wide range of frequency and wave number. At short wavelengths the phase velocity of the first two modes departs significantly from the Rayleigh wave velocity in the stiff layer. Dispersion curves for SH-motion (Fig. 13) also show similar features. The departure depends on the ratios of the elastic properties and of the thicknesses of the low-velocity layers and the laminae. This effect and the lowering of cut-off frequencies should lead to ultrasonic characterization of interfaces.

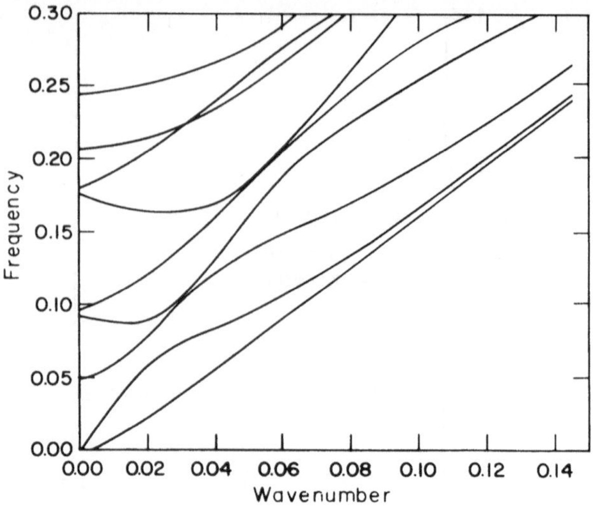

Fig. 11. Dispersion of Lamb waves in a sandwich plate at low frequencies, psv-5p.

CONCLUSIONS

We have shown that modeling and experimental observations of particle-reinforced and fiber-reinforced materials lead to property characterization of the reinforcing phases. Also, we have presented model calculations of interface effects on phase velocities and attenuation of waves in a composite medium. It is shown that, for the particular systems considered, the presence of low-velocity interface layers decreases the phase velocities as well as the attenuation.

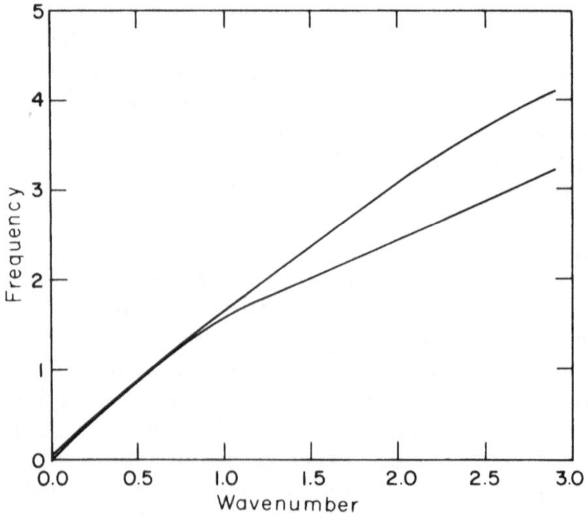

Fig. 12. Dispersion of Lamb waves in a sandwich plate at finite frequencies, psv-5p.

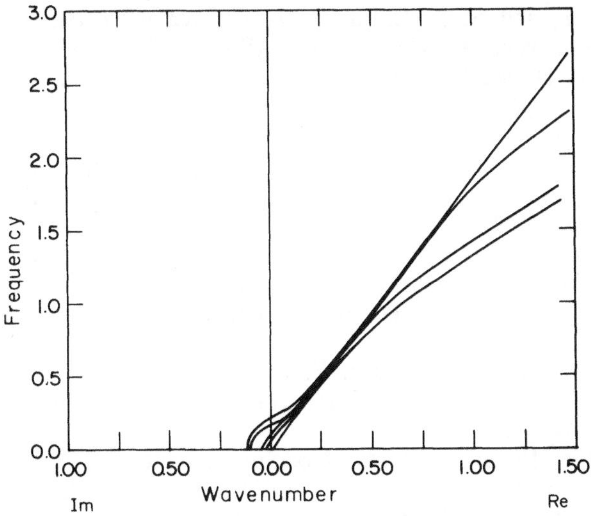

Fig. 13. Dispersion of SH waves in a sandwich plate, sh-5p.

ACKNOWLEDGMENTS

This study was supported in part by a grant from the Solid Mechanics Division (Program Manager, Dr. Y. Rajapakse) of the Office of Naval Research (ONR-00014-86-K0280), a grant from the National Science Foundation (MSM-8609813), and a grant from the Natural Science and Engineering Research Council of Canada (A-7988). Support was also received from the Office of Nondestructive Evaluation, NBS. The work was completed while the first author held a faculty fellowship from the University of Colorado and was also a Fulbright Scholar at the Technical University of Vienna.

REFERENCES

1. S. K. Bose and A. K. Mal, Elastic Waves in a Fiber-Reinforced Composite, J Mech and Phys of Solids 22:217 (1974).
2. A. K. Mal and S. K. Bose, Dynamic Moduli of a Suspension of Imperfectly Bonded Spheres, Proc of the Cambridge Phil Soc 76:587 (1974).
3. S. K. Datta, A Self-Consistent Approach to Multiple Scattering of Elastic Waves, J Appl Mech 44:657 (1977).
4. A. J. Devaney, Multiple Scattering Theory for Discrete, Elastic, Random Media, J Math Phys 21:2603 (1980).
5. J. G. Berryman, Long-Wavelength Propagation in Composite Elastic Media--I, Spherical Inclusions, JASA 68:1809 (1980).
6. J. G. Berryman, "Long-Wavelength Propagation in Composite Elastic Media--II. Ellipsoidal Inclusions, JASA 68:1820 (1980).
7. J. R. Willis, A Polarization Approach to the Scattering of Elastic Waves--II. Multiple Scattering from Inclusions, J Mech and Phys of Solids 28:307 (1980).
8. V. K. Varadan, Y. Ma, and V. V. Varadan, A Multiple Scattering Theory for Elastic Wave Propagation in Discrete Random Media, JASA 77:375 (1985).

9. H. M. Ledbetter, and S. K. Datta, Effective Wave Speeds in an SiC-particle-reinforced AP Composite, JASA 79:239 (1986).
10. D. T. Read and H. M. Ledbetter, Elastic Properties of a Boron-Aluminum Composite at Low Temperature, J of Appl Phys 48:2827 (1977).
11. V. K. Kinra, M. S. Petraitis, and S. K. Datta, Ultrasonic Wave Propagation in a Random Particulate Composite, Intl J of Solids and Struc 16:301 (1980).
12. S. K. Datta and H. M. Ledbetter, Anisotropic Elastic Constants of a Fiber-reinforced Boron-Aluminum Composite, in: "Mechanics of Nondestructive Testing," W. W. Stinchcomb, ed., Plenum, New York (1980).
13. H. M. Ledbetter, and S. K. Datta, Young's Modulus and the Internal Friction of an SiC-Particle-Reinforced Aluminum Composite, Matls Sc and Engng 67:25 (1984).
14. H. M. Ledbetter, Dynamic Elastic Modulus and Internal Friction in Fibrous Composites, in: "Nonmetallic Materials and Composites at Low Temperatures," Plenum, New York (1979).
15. H. M. Ledbetter, N. V. Frederick, and M. W. Austin, Elastic-Constant Variability in Stainless Steel, J of Appl Phys 51:305 (1980).
16. S. K. Datta, Diffraction of Plane Elastic Waves by Ellipsoidal Inclusions, JASA 61:1432.
17. J. A. Stratton, "Electromagnetic Theory," McGraw-Hill, New York (1941).
18. S. K. Datta and H. M. Ledbetter, Effect of Interface Properties on Wave Progagation in a Medium with Inclusions, in: "Mechanics of Material Interfaces," A.P.S. Selvadurai and G.Z. Voyiadjis, eds., Elsevier, The Netherlands, (1986).
19. J. E. Gubernatis and E. Domany, Effects of Microstructure on the Speed and Attenuation of Elastic Waves in Porous Materials, Wave Motion 6:579 (1984).
20. S. K. Datta, H. M. Ledbetter, and R. D. Kriz, Calculated Elastic Constants of Composites containing Anisotropic Fibers, Intl J of Solids and Struc 20:429 (1984).
21. R. Hill Theory of Mechanical Properties of Fibre-Strengthened Materials: - I. Elastic Behavior, J of the Mech and Phys of Solids 12:199 (1964).
22. A. H. Shah and S. K. Datta, "Harmonic Waves in a Periodically Laminated Medium," Intl J of Solids and Struc 18:397 (1982).
23. S. K. Datta, A. H. Shah, R. L. Bratton, and Y. N. Al-Nassar, Guided Wave Propagation in Laminated Composite Plates with Low-Velocity Interface Layers, to be published.
24. T. Okamoto, A. Kagawa, K. Kiyoshi, and H. Matsumoto, Effects of Graphite Shape on Thermal Conductivity, Electrical Resistivity, Damping Capacity and Young's Modulus of Cast Iron below 500 degrees C" J of the Japan Foundrymen's Soc 55:32 (1983).
25. D. Lohe, O. Vohringer, and E. Macherauch, Der Einfluss der Graphitform auf den Elastizitatsmodul von ferritischen Gusseisen-Werkstoffen, Zeitschrift fur Metallkunde 74:265 (1983).
26. G. R. Speich, A. J. Schwoeble, and B. M. Kapadia, Elastic Moduli of Gray and Nodular Cast Iron, J of Appl Mech 47:821 (1980).
27. L. Anand, Elastic Moduli of Gray and Ductile Cast Irons, Scripta Met 16:173 (1982).
28. E. Kroner, and H. Koch, Effective Properties of Disordered Materials, SM Archives 1:183 (1976).

29. H. Wawra, B. K. D. Gairola, and E. Kroner, Comparison between Experimental Values and Theoretical Bounds for the Elastic Constants E, G, K and μ of Aggregrates of Noncubic Crystallites, Zeitschrift fur Metallkunde 73:69 (1982).
30. H. M. Ledbetter and D. T. Read, Orthorhombic elastic constants of an NbTi%Cu Composite Superconductor, J of Appl Phys 48:1874 (1977).
31. R. D. Kriz and W. W. Stinchcomb, Elastic Moduli of Transversely Isotropic Graphite Fibers and their Composites, Exp Mech 19:41 (1979).
32. R. E. Smith, Ultrasonic Elastic Constants of Carbon Fibers and Their Composites, J of Appl Phys 43:2555 (1972).

ACOUSTO-ULTRASONIC WAVE PROPAGATION IN COMPOSITE LAMINATES

S. M. Moon and K. L. Jerina

Department of Mechanical Engineering
Washington University
St. Louis, Missouri 63130

H. T. Hahn

Dept. of Engineering Science and Mechanics
The Pennsylvania State University
University Parks, Pennsylvania 16802

ABSTRACT

In the acousto-ultrasonic technique, a wave is injected onto the surface at one location and the displacement normal to the same surface is measured at another location. Thus, understanding the propagation characteristics of the wave is essential for successful application of the technique. The main objective of the present investigation is to identify through analysis-experiment correlation how the AU wave propagates through composite laminates.

Lamb wave speeds were calculated in the low frequency region for a unidirectional graphite/epoxy laminate. For wave propagation in off-axis directions transverse displacement was assumed zero. An experimental verification was carried out by measuring the wave speeds on the upper and lower surfaces of the specimen. The changes of the wave velocity and attenuation with frequency were monitored. Fourier spectra of the received signal were obtained using a wave analyzer to study the dispersion.

It has been found that the dominant AU waves produced experimentally were Lamb waves. The wave velocity changed with the fiber direction in the predicted manner. The maximum attenuation was obtained in the 22.5° fiber direction while the minimum attenuation was observed parallel to the fibers.

INTRODUCTION

The analysis of the response of composites to elastic stress waves is involved because of the inherent anisotropy and heterogeneity. The displacements cannot be identified as being longitudinal or transverse except when the direction of propagation corresponds to a symmetry direction for the material. The heterogeneity results in higher dispersion when the wavelength is comparable to the characteristic dimension of the internal structure of the composite.

Many composite laminates can be considered as macroscopically orthotropic if the wavelength is longer than the thickness. When a wave propagates in one of the material symmetry direction, the particle displacement would be either parallel or transverse to the direction of wave propagation, just as in isotropic materials. For wave propagation in an off-axis direction, the particle displacement is neither parallel nor transverse to the wave propagation direction because of normal-shear coupling. When the wavelength approaches the ply thickness, the wave velocities depend on the frequency, i.e. dispersion results.[1]

In the acousto-ultrasonic technique a wave is injected onto the surface at one location with the help of a longitudinal transducer. The beam from the transducer is conical (beam spreading). Hence the injected wave strikes the free boundary of the specimen/structure normally and at an angle. The longitudinally polarized stress wave incident at an angle gives rise to longitudinal and shear waves on reflection from the boundary. The interaction between longitudinal and shear waves in a relatively thin isotropic material produces Lamb/plate waves.[2-6] The introduction of a longitudinal wave onto the surface of the specimen also produces leaky surface waves.[7-8] Lamb waves are dispersive in nature. The dispersion in such a situation is caused by the boundaries of the plate and is known as geometric dispersion. An experimental verification of the dispersive Lamb waves has been presented by Worlton[9] for aluminium and zirconium.

Propagation of Lamb waves in paper has been investigated previously.[6] Paper was considered an orthotropic material and the dispersion relations for symmetric and antisymmetric Lamb waves were obtained. The low frequency approximation of these relations led to the nondispersive symmetric mode and a dispersive antisymmetric mode.

For longitudinal and shear wave propagation[11-15] there exist both dispersion and attenuation in composite laminates. Further, dispersion and attenuation increase with frequency. The main contribution to this phenomenon is the viscoelastic nature of the composite material. In dispersive and attenuating medium the wave will continuously change its shape and frequency content during propagation.[12] Pulse distortion in longitudinal wave propagation decreases its amplitude and frequency of the dominant component.[16-17] Lower frequencies will result in longer wave length and lower detectability of a flaw.

The main objective of the present investigation is to understand through analysis-experiment correlation the wave propagation in composite laminates that has used acousto-ultrasonic technique.

EXPERIMENTAL PROCEDURE

The composite laminates used were unidirectional T300/5208 ($\theta = 0°$) and T300/5208 ($\theta = 90°$). These panels were fabricated at NASA Langley Research Center. The dimensions and properties of the specimen are listed in Table 1.[18]

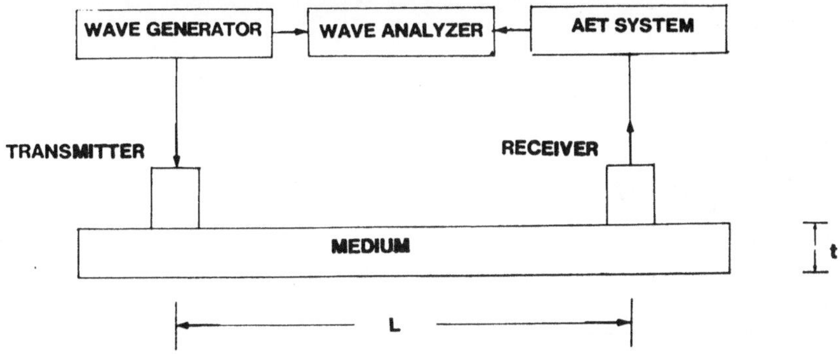

Fig. 1. Specimen Geometry/Experimental Set Up for Velocity Measurement.

TABLE 1. Material Properties and Specimen Dimension

Material	E_1 GPa	E_2 GPa	E_3 GPa	ν_{12}	ν_{23}	Density gm/cm^3
T300/5208 unidirectional	138.58	11.03	11.23	0.313	0.543	1.53

The experimental setup for the measurement of wave velocity is shown schematically in Fig. 1. The broad band AET FC-500 transducers were used as emitter and receiver. The transducers were bonded to the specimen with high vacuum silicone grease. A contact pressure of 0.072 MPa was maintained between the transducer and specimen throughout the experiment. Acoustic Emission Associates AES-1L acoustic emission simulator was used to drive the transmitting transducer. Signals having the fast rise and fast decay with peak amplitude of 8.91 volts were transmitted. The fast rise and decay rates were selected so as to generate a short duration waveform. The signals were received with a broadband transducer. An AET 5000 system was used to amplify the incoming signal. The emitted and received signals were displayed on a wave analyzer Data 6000, Fig. 2. The delay time between the emitted signal and the received signal was measured accurately. The digital wave analyzer has a minimum sampling time of 0.2 μsec. The velocity was determined, knowing the distance between the transducers and the delay time. Fourier spectra of the signals were taken.

For wave propagation in the off-axis direction the scheme as shown in Fig. 3 was used. The material used was a T300/5208 panel of dimensions 285 mm x 325 mm x 3 mm. The laminate was supported on a cellular rubber pad over a table. The emitting transducer was mounted with a silicone grease and a contact pressure of 0.072 MPa, Fig. 3. The location of the emitting transducer was selected to avoid the reflection of the wave. The perturbation originating at the emitter propagates in all directions. The receiving transducer was positioned at a distance of 56 mm and 86 mm from the emitting transducer. A contact pressure of 0.16 MPa was maintained between the receiving transducer and the laminate. Fiber orientation effects were studied by placing a transducer at various angles with respect to the fiber direction, Fig. 3. The high attenuation at fiber orientations other than 0° required the use of the high

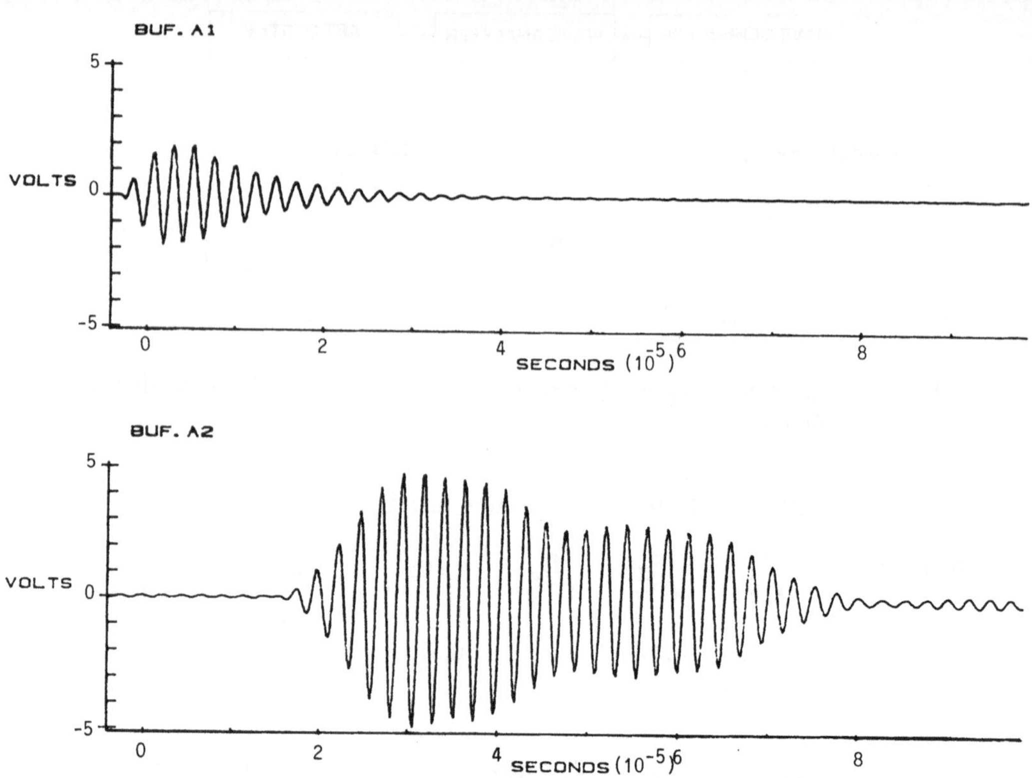

Fig. 2. Typical Transmitted and Received Signal for Velocity Measurement.

sensitivity AET MAC 425 L transducer. This transducer has flat frequency response up to 900 kHz. However, the resonant peak of the transducer is 425 kHz.

RESULTS AND DISCUSSION

Plate Wave Propagation

Two methods are commonly used to generate a Lamb wave in a plate. One of them is the wedge method and the other is the excitation of the plate by normal perturbations created by a transducer coupled acoustically to the surface of the test plate.[4,5] The second method is a broadband method where all possible modes for given frequency are excited. In this investigation the second method is used, Fig. 1. The Lamb waves traveling on the surface have particle motion parallel and perpendicular to the surface. The symmetric mode has particle motion which is symmetric with respect to the midplane, i.e. the longitudinal displacements are equal, while transverse displacements are equal in magnitude but opposite in direction. The antisymmetric mode has particle motion which is antisymmetric with respect to the midplane.[5]

In the acousto-ultrasonic technique continuous wave train/pulse are introduced into the material by a transducer in the frequency range of 100 kHz to 2 MHz. The primary interest is the low frequency response of a composite

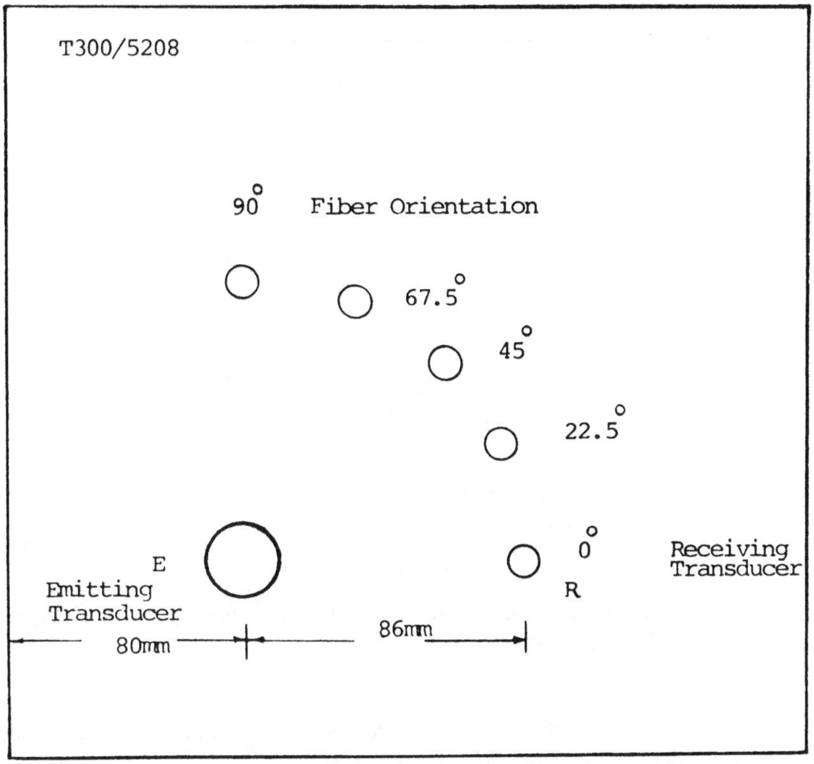

Fig. 3. The Schematic for off-axis Velocity and Attenuation Measurement.

material. Many composite laminates can be considered macroscopically orthotropic. In the low frequency range it can be assumed that damping or viscoelastic character of the composite does not effect the speed of wave propagation. The theory given by Habeger et. al.[6] and Moon[10] for a unidirectional laminate in the low frequency regime was used to evaluate the velocities of the plate wave.

The dispersion relation for the symmetric Lamb/plate wave velocity in the low frequency range is given by [10]

$$V_s = [(C_{11} - C_{13}^2/C_{33})/p]^{0.5}[1 - (C_{13}wh)^2/24yC_{33}^2] \quad (1)$$

where

C_{ij} = stiffness tensor

V_s = Symmetric mode Lamb/Plate wave velocity.

ρ = Density of material

h = thickness of the laminate

$y = (C_{11} - C_{13}^2/C_{33})/\rho$

w = Circular frequency

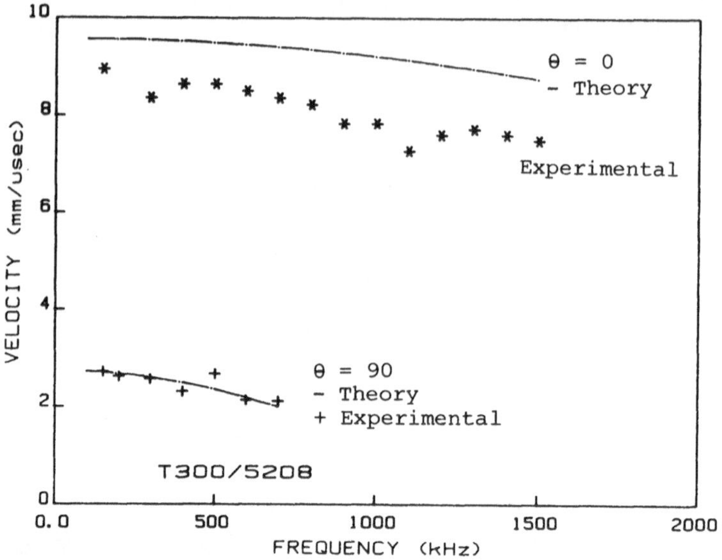

Fig. 4. Dispersion of Symmetric Plate Wave Velocity with Frequency (T300/5208 $\theta = 0°$, $\theta = 90°$).

The stiffness tensor for an orthotropic composite laminate was obtained from engineering constants using the following relations.[19]

$$C_{11} = [1-v_{23}v_{32}]/(E_2E_3D)$$

$$C_{13} = [v_{13}+v_{12}v_{23}]/(E_1E_2D)$$

$$C_{33} = [1-v_{12}v_{21}]/(E_1E_2D)$$

$$D = [1-v_{12}v_{21}-v_{23}v_{32}-v_{31}v_{13}-2v_{21}v_{32}]/(E_1E_2E_3)$$

where

E_1 = Engineering constant

v_{ij} = Poisson ratio

Subscripts 1 and 3 are associated with the wave propagation direction and the thickness direction respectively.

Similarly, velocity of antisymmetric mode in the low frequency regime is given by [6]

$$V_a = \left([(C_{11}-C_{13}{}^2/C_{33})/3\rho]^{0.5}hw/2\right)^{0.5} \quad (2)$$

The change of Lamb wave velocity with frequency for T300/5208 laminate with ($\theta = 0°$) has been calculated using eq. (1) and is shown in Fig. 4. The theoretical prediction indicates that the symmetric wave velocity decreases with frequency, whereas the antisymmetric wave velocity increases with frequency, Fig. 5. The dispersion of the symmetric mode velocity with frequency is not appreciable. There is only a 9% reduction in the velocity for the frequency

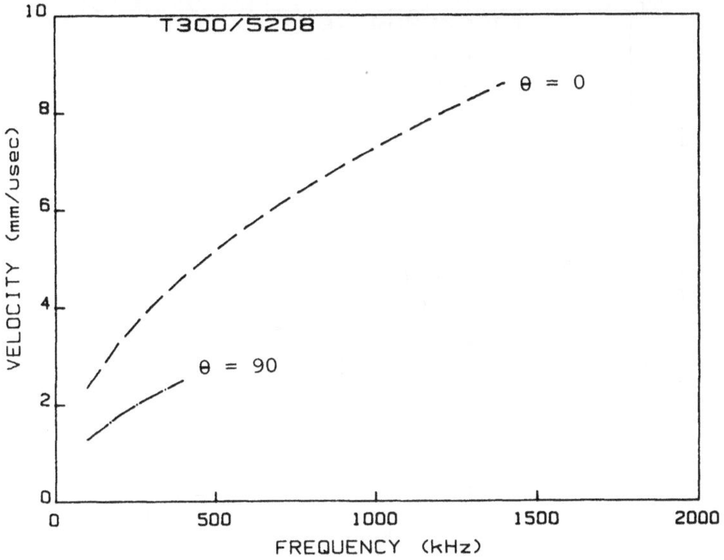

Fig. 5. Dispersion of Antisymmetric Plate Wave Velocity with Frequency.

range from 0 to 1.5 MHz. Hence, the symmetric mode is almost nondispersive in the frequency range studied. However, the antisymmetric mode velocity cannot exceed the symmetric mode velocity.

The measured velocities in T300/5208 for a frequency range of 0.15 to 1.5 MHz are shown in Fig. 4. Experimental values are lower than those predicted. The discrepancy is around 10%. However, both the theory and the experiment show the same trend. The decrease in the measured symmetric wave velocity

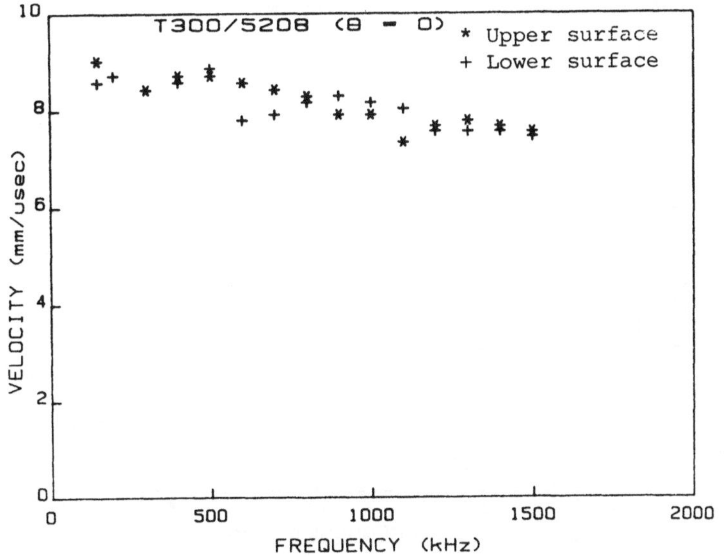

Fig. 6. Symmetric Mode Plate Wave Velocity on Upper and Lower Surface of Specimen as a Function of Frequency (T300/5208).

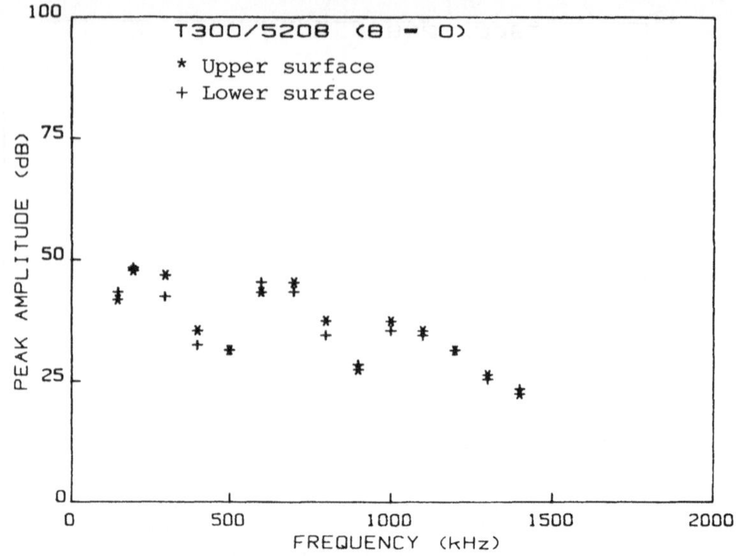

Fig. 7. Peak Amplitude of Received Wave on Upper Surface and Lower Surface of Specimen for T300/5208, $\theta = 0^{\circ}$.

as the frequency increases from 0.15 to 1.5 MHz is 16.41%. Over this frequency range the wavelength decreases from 95 mm to 5 mm. Yet, even the smallest wavelength is larger than the laminate thickness (3 mm).

A plate wave has the same out-of-plane particle motion on the upper and lower surfaces. To see if this is indeed true, another receiving transducer was bonded to the lower surface of the specimen. The same contact pressure

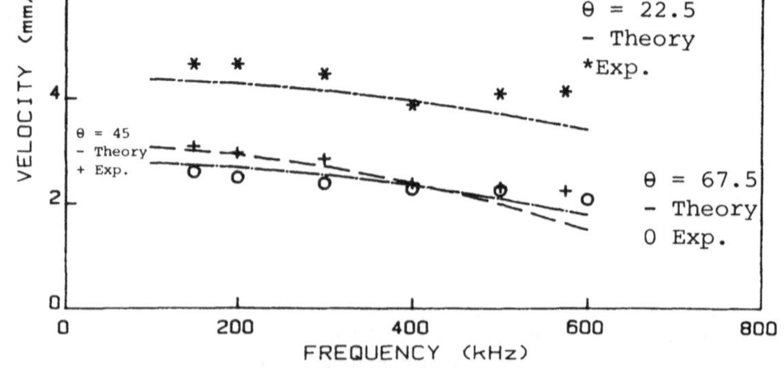

Fig. 8. Dispersion of Symmetric Mode Velocity with Frequency T300/5208 ($\theta = 22.5^{\circ}$, $\theta = 45^{\circ}$ and $\theta = 67.5^{\circ}$).

of 0.072 MPa was maintained. The measured velocities are almost equal on the lower and upper surfaces of the specimen, Fig. 6. The peak amplitudes are the same on both the lower and upper surfaces of the specimen, Fig. 7. The decrease in peak amplitude with increasing frequency is due to the attenuation. Therefore, it is concluded that a plate wave is generated under the experimental setup shown in Fig. 1.

In order to investigate the effect of fiber orientation, velocities were measured normal to the fiber direction, Fig. 4. The measured velocities are close to the computed symmetric mode velocities and twice the antisymmetric mode velocities. The decrease in the symmetric mode velocity as the frequency increases from 0 to 1 MHz is 18.51%. Hence, there is more dispersion in the transverse direction than in the longitudinal direction, Fig. 4.

The effect of fiber orientation on the wave velocity at various frequency is shown in Fig. 8. The predictions are from eq. (1) with C_{ij} representing the appropriate off-axis stiffnesses. Although eq. (1) does not take into account the possible normal-shear coupling, it agrees fairly well with experimental data. This result is especially significant in that the velocities of off-axis waves generated using the present experimental arrangement can be approximated by eq. (1).

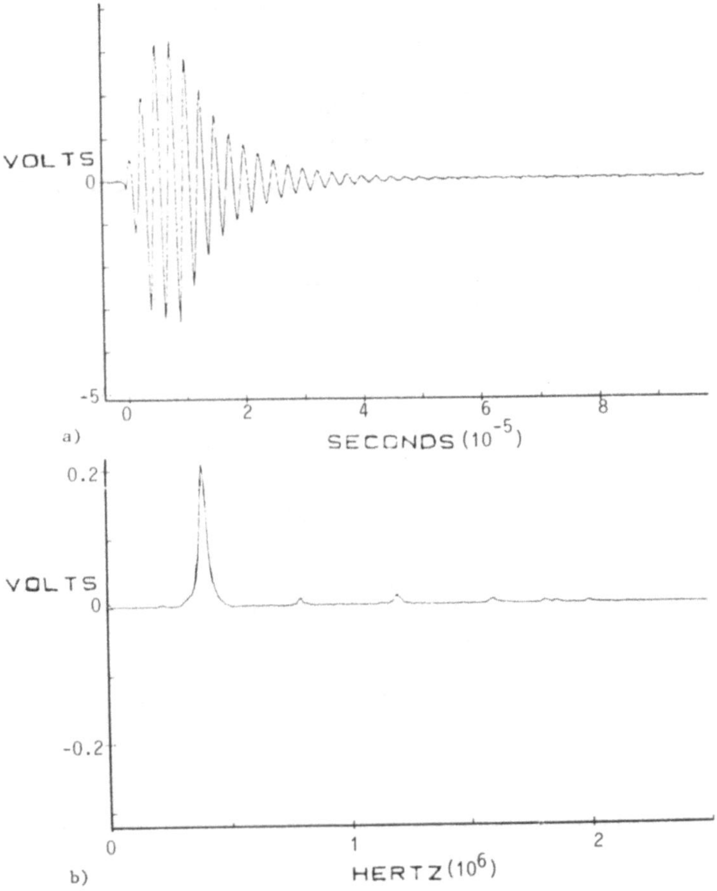

Fig. 9. The Input Signal and Its Spectrum.

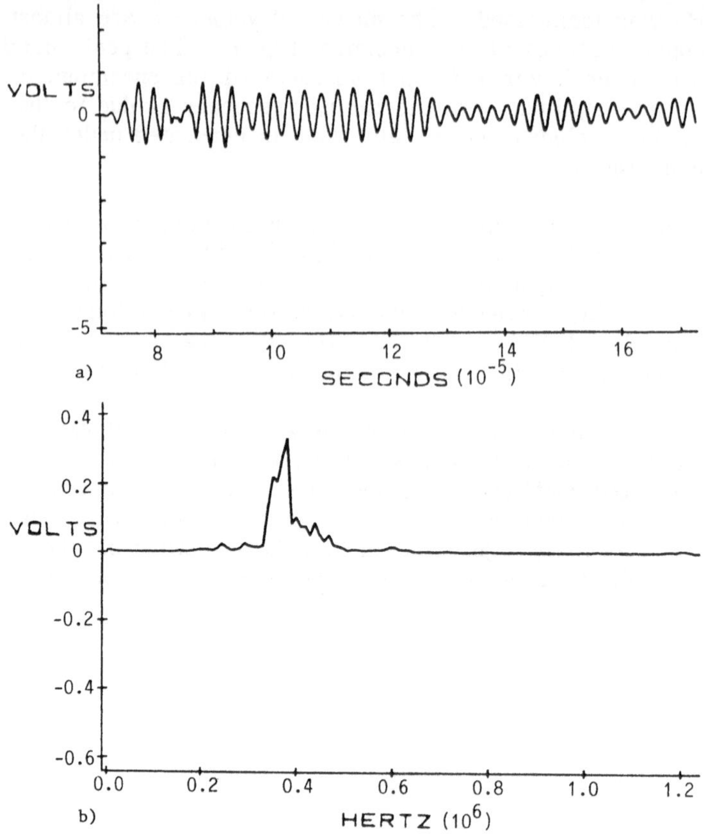

Fig. 10. The Received Signal and Its Fourier Spectrum.

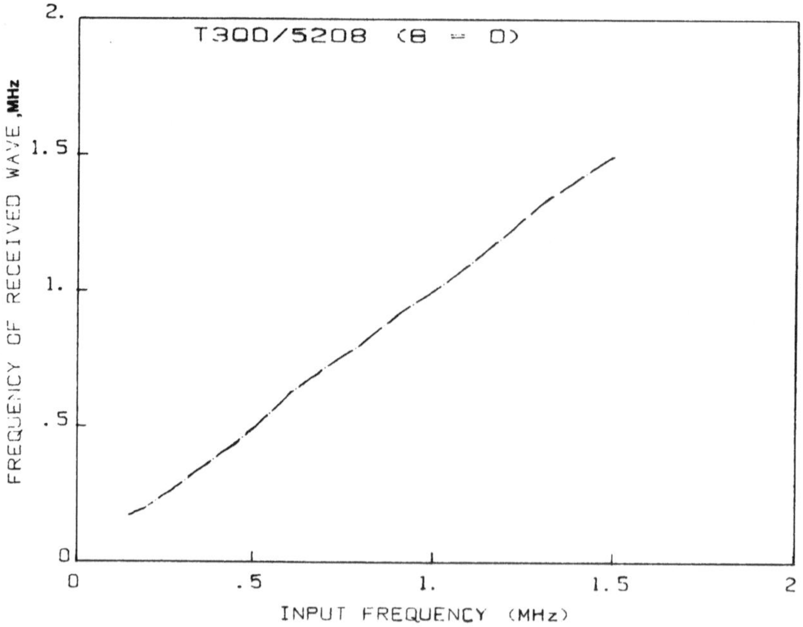

Fig. 11. Variation of the Frequency of Received Wave with Input Wave Frequency (T300/5208, θ = 0°).

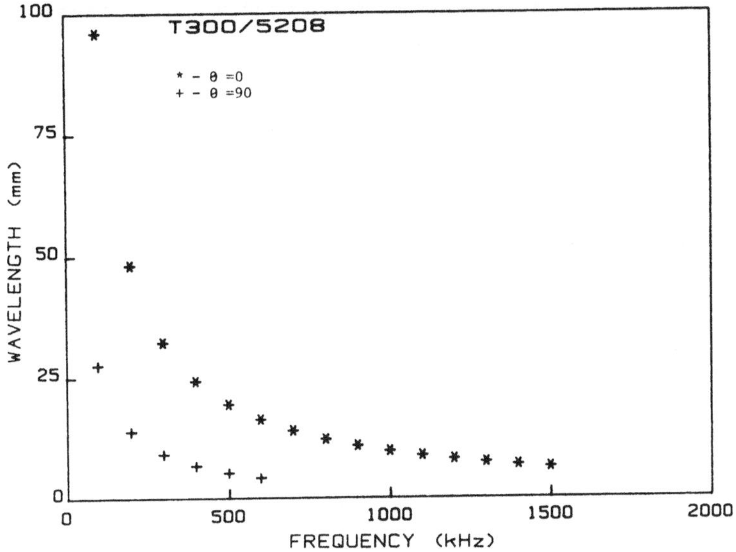

Fig. 12. Wavelength as a Function of Frequency.

Fourier spectra of the received signals were obtained with the wave analyzer. The spectrum of the received signal was similar to that of the input signal except that the former had a component corresponding to the reflected wave, Figs. 9 and 10. The central frequency of the received signal was equal to that of input signal, Fig. 11. Frequency components other than the central frequency were absent in the Fourier spectrum of the received signal. Thus, it can be concluded that the wave propagated through the material without appreciable dispersion in the frequency range studied.

Figure 12 shows wavelength from the Lamb wave decreasing with increasing frequency. At the input frequency of 500 kHz the wave lengths are 18.92 mm and 5.83 mm for the fiber and transverse directions respectively. Such information is useful in selecting a flaw detection frequency in the laminate as the flaw detection capability increases with a decrease in wavelength.

Attenuation in Laminates

The attenuation in composites is due to the absorption and scattering of the acoustic energy as the wave propagates. Figure 13 shows the variation of peak amplitude of the received signal at a distance of 86 mm from the transmitting transducer, It is seen that with an increase in fiber orientation the peak amplitude decreases initially. The peak amplitude is at a minimum at 22.5° and then increases slightly thereafter; the exact change depends on the frequency.

Attenuation in the laminate with various fiber orientations and frequencies was determined by monitoring the peak amplitudes of the received signal at 56 mm and 86 mm. To improve accuracy 92 measurements were taken at each location, frequency and fiber orientation. From the average values, attenuation coefficients were calculated using the following relation:

$$r(w) = [\ln(V_1/V_2)]/(X_2 - X_1)$$

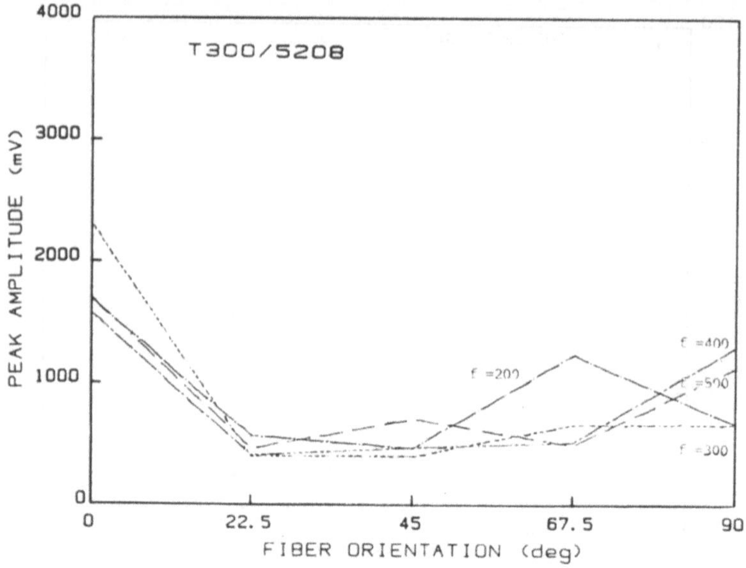

Fig. 13. Peak Amplitude of Received Wave with Fiber Orientation for Different Frequencies.

where

r = Attenuation coefficient

V_1 = peak amplitude at distance X_1

V_2 = Peak amplitude at distance X_2

It is seen that the measured attenuation coefficient depends on the frequency and fiber orientation, Fig. 14. At each fiber orientation the attenuation coefficient increases with frequency. A linear regression analysis was performed on the data of Fig. 14 at each fiber orientation. The results indicate a linear relation between the attenuation coefficient and the frequency.

At a fixed frequency attenuation is a minimum at 0° fiber orientation and a maximum at 22.5° for the fiber orientations studied, Fig. 15. The rate at which it increases is higher than in any other fiber orientation. The rate of increase in attenuation at 67.4° was next to 22.5° but it showed greater scatter in experimental data. Fiber orientations 22.5°, 45° and 90° showed less scatter in the observations. Fiber orientations of 45° and 90° showed higher attenuation at 150 kHz and a lower rate of increase in attenuation with frequency.

CONCLUSIONS

In the application of the AU technique to graphite/epoxy laminates the dominant wave produced is a symmetric Lamb wave. A plate wave with a wavelength as small as 5.81 mm for T300/5208 (θ = 0°) can be produced without much dispersion up to 1.5MHz. Since wave speed and dispersion depend on the fiber orientation, proper care must be exercised in the interpreting the data.

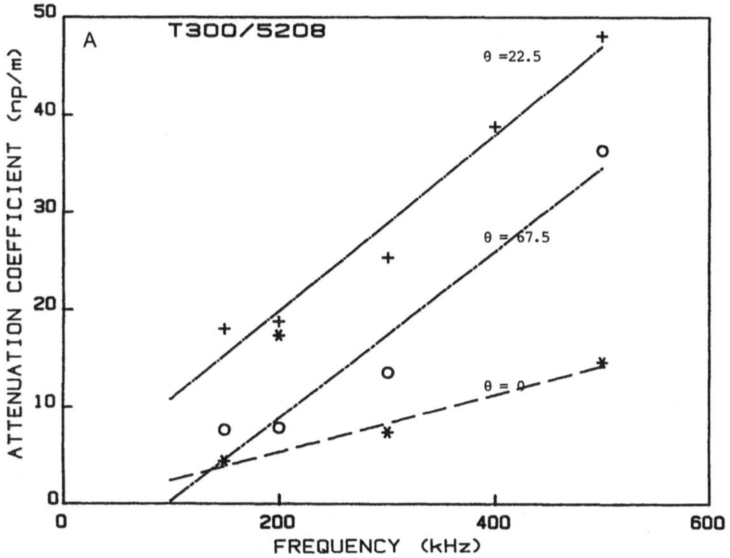

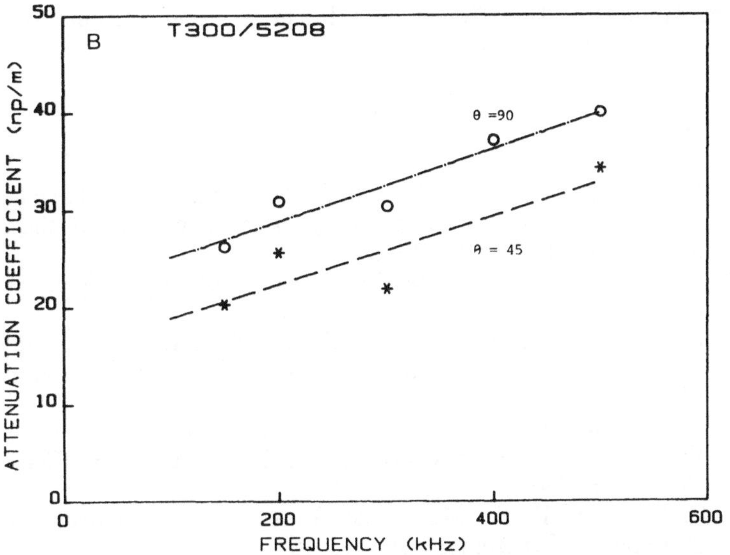

Fig. 14. Attenuation Coefficient as a Function of Frequency for Different Orientations.

The wave velocities in all directions can be predicted assuming zero displacement normal to the wave propagation direction. The attenuation coefficient was the smallest along the fibers. It reached the largest value at 22.5° and then decreased until the fiber angle increased to 67.5°. Further increase in the fiber angle to 90° yielded an increase in attenuation coefficient for the fiber orientations studied.

ACKNOWLEDGEMENTS

This paper is based on work supported by the Office of Naval Research, through Grant N0014-87-K-0143, with Y. Rajapakse as Program Director.

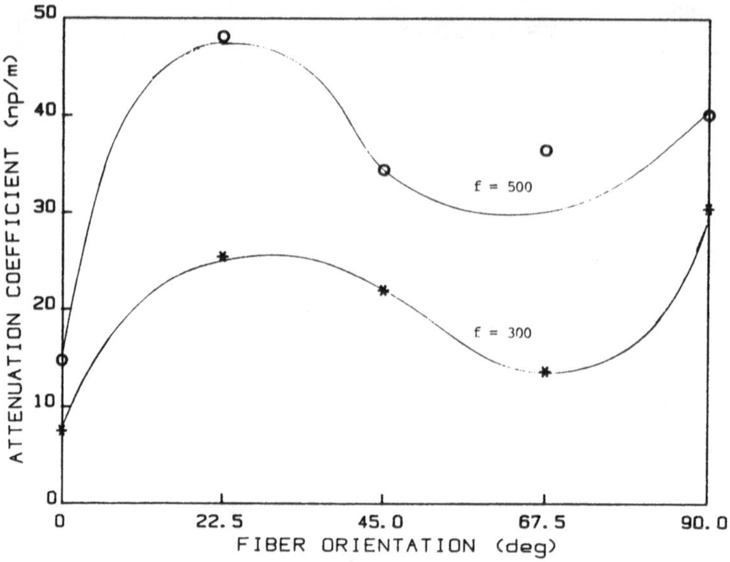

Fig. 15. Attenuation Coefficient as a Function Fiber Orientation for Different Frequencies.

REFERENCES

1. C. T. Sun, J. D. Achenbach, and G. Herrman, Time-harmonic Waves in a Stratified Medium Propagating in the Direction of the Layering, J of Appl Mech. 35:408 (1968).
2. H. Lamb, On Waves on Elastic Plates, Proc. Royal Soc. A. 93:114 (1917).
3. H. F. Pollard, "Sound Waves in Solids, Pion Limited, London (1977).
4. I. A. Viktrove, "Rayleigh and Lamb Wave," Plenum Press, New York (1967).
5. J. H. Hemann and G. Y. Baaklini, "The Effect of Stress on Ultrasonic Pulses in Fiber Reinforced Composites, NASA CR-3724," NASA, Cleveland (1983).
6. C. C. Habeger, R. W. Mann, and G. A. Baum, "Ultrasonic Plate Waves in Paper," Ultrasonics, 3:57 (1979)
7. L. J. Bond and N. Saffari, Crack Characterization in Turbine Disks, in: "Review of Progress in Quantitative Nondestructive Evaluation, Vol. 3A," D. O. Thompson and D. E. Chimenti, eds., Plenum Press, New York (1984).
8. A. Fahr, S. Johar, and M. K. Murthy, Surface Acoustic Wave Studies of Surface Cracks, in: "Review of Progress in Quantitative Nondestructive Evaluation, Vol. 3A," D. O. Thompson and D. E. Chimenti, eds., Plenum Press, New York (1984).
9. D. C. Worlton, "Experimental Confirmation of Lamb Waves at Mega Cycle Frequencies," J. Appl. Phy. 32:967 (1961).
10. F. C. Moon, Wave Propagation and Impact in Composite Materials, in: "Mechanics of Composite Materials, Vol. 7," C. C. Chamis, ed., New York (1974).
11. C. Sve, "Time-Harmonic Wave Traveling Obliquely in Periodically Laminated Medium," J. Appl. Mech. 38:477 (1971).

12. W. Kohn, "Propagation of Low-Frequency Elastic Disturbances in Composite Materials," J, Appl. Mech. 41:97 (1974).
13. J. C. Peck and G. A. Gurtman, "Dispersive Pulse Propagation Parallel to the Interfaces of a Laminated Composite," J. Appl. Mech. 36:479 (1969).
14. J. S. Whitter and J. C. Peck, "Experiments on Dispersive Pulse Propagation in a Laminated Composites and Comparison with Theory," J. Appl. Mech. 36:485 (1969).
15. J. D. Achenbach, Waves and Vibration in Composites, in: "Mechanics of Composite Materials, Vol. 2," G. P. Sendeckyji, ed., Academic Press, New York (1974).
16. S. Serabian, "Influence of Attenuation Upon the Frequency Content of Stress Wave Packet," JASA 42:1052 (1967).
17. S. Serabian, "Implication of the Attenuation-Produced Pulse Distortion Upon the Ultrasonic Method of Nondestructive Testing," Matl Eval. 26:173 (1968).
18. M. Knight, "Three-Dimensional Elastic Moduli of Graphite/Epoxy Composites," J. Comp Matl 16:153 (1982).
19. R. M. Jones, "Mechanics of Composite Materials," Scripta Book Company, Washington, DC (1975).

RAY PROPAGATION PATH ANALYSIS OF ACOUSTO-ULTRASONIC SIGNALS IN COMPOSITES

Harold E. Kautz

NASA Lewis Research Center
Cleveland, Ohio 44135

INTRODUCTION

It has been established that the stress-wave factor (SWF) calculated from acousto-ultrasonic (AU) signals is sensitive to mechanical properties in composite structures.[1-13] In particular, it was shown[1] that SWF was sensitive to interlaminar shear strength (ISS) in graphite/epoxy.

More recent studies have shown that analysis of AU wave propagation paths in relatively thick specimens taken from filament wound composite (FWC) structures is useful for predicting ISS.[14,15] This analysis led to the prediction that the relationship between SWF and ISS would be strongest in certain partitions of the AU signal in the time domain. The propagation path analysis correctly predicted, for the material and geometry under consideration, that signals arriving about 15 to 20 microseconds into the waveform would carry the least ambiguous information on laminar interfaces.

The analysis, previously reported by the author,[14,15] also included frequency domain partitioning. The ray path model employed correctly identified time segments of interest, but made no predictions about frequency. It was evident from this work, however, that there is a dependence on frequency. Indeed, partitioning of the frequency domain seemed at least as useful as that of the time domain in revealing good correlation between SWF and ISS. It seems appropriate, therefore, to examine factors that may shape the frequency spectrum of composite AU waveforms.

The most striking finding in the present work is the identification of particular propagation paths in a graphite/epoxy panel by comparison of the AU signal with that collected in an otherwise similar panel composed of the epoxy resin alone. Beyond this, the rest of this paper attempts to identify the source of other features that are commonly encountered with AU signals.

ANALYSIS

Attenuation Effects

It is evident that many factors are involved in the shaping of an AU frequency spectrum. The two most obvious are the response of the transducers and the attenuating characteristics of the composite. Materials tend to attenuate high frequency ultrasonic waves more than low frequency waves.[16] The attenuating effect is especially strong in fiber/polymer composites. For this reason one expects that the output spectrum obtained at the receiving transducer will center at a lower frequency than the input spectrum introduced by the sending transducer.

Multiple Path Effects

It has been pointed out, by Krautkramer,[16] that layered interfaces along an ultrasonic ray path give rise to multiple arrival times for propagating waves. These produce interference effects that appear as peaks in the frequency spectrum. Another source of the same kind of interference effects is the arrival of signals from along different propagation paths.[17]

In each of these interference effects all the signal arrivals at the receiver are from a common input pulse. Extremes, (peaks or troughs), in the spectrum will be centered at frequencies, f, where f is related to the characteristic time, T, between the arrivals as:

$$T=n/f \tag{1}$$

and also as:

$$T=(n+1/2)/f \tag{2}$$

where n is an integer (see Appendix A). If equation (1) produces a maximum then equation (2) will be minimum. The maximum-minimum conditions can be reversed depending on the phase relation between the pulses.

The interference effects are oscillations in the frequency domain. Just as the frequency spectrum is obtained by taking the magnitude of the Fourier transform of the original time domain signal, the Fourier transform of the spectrum can be used to obtain a function sometimes called the cepstrum. This cepstrum will exhibit resonance at the characteristic times T. This may be useful in identifying characteristic times.

EXPERIMENTAL

Materials

AU signals were collected on a 0.25 cm thick, 30 cm by 30 cm, 26 ply, graphite/epoxy resin panel. The resin matrix was 1500 molecular weight polymerization of monomeric reactants (PMR-15) with the ester of pyromellitic dianhydride (PMDE ester). The graphite lay-up was alternate 0/90 degree orientations of Celion 6000 unfinished, 7 micrometer diameter, fiber. AU signals were also collected on a 0.25 cm thick, 30 cm by 30 cm, PMR pure resin panel of the same chemistry as described composite matrix.

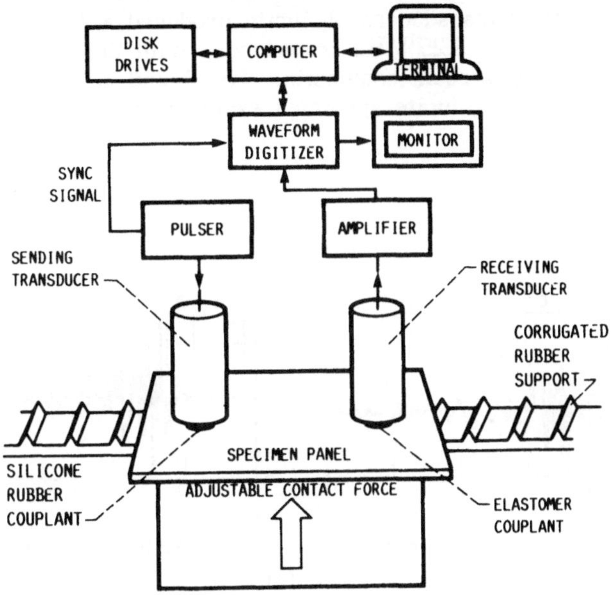

Fig. 1. Block diagram of the Acousto-ultrasonic and data processing system.

Signals were also collected on a 0.88 cm thick, approximately 12 cm by 8 cm, specimen of lucite. The lucite was used to simulate some of the ultrasonic response characteristic of graphite/epoxy without the presence of internal reflections at lamina boundaries.

Acousto-ultrasonic System and Procedures

In these experiments AU signals were collected, digitized, and stored on disc. A block diagram of the data processing system is shown in Fig. 1. Two 2.25 MHz center frequency broad band transducers were used with 0.17 cm thick silicone rubber dry couplant pads and approximately 9 N/cm^2 pressure. The transducers were the immersion type with wearplate faces. The conditions are the same as for work reported previously.[15] An extensive spectral analysis was performed with the aid of the fast Fourier transform (FFT) algorithm. This analysis included calculation of frequency spectra for segments of the time domain signal. The FFT was performed on the Hanning windowed segment.[18] The purpose of the Hanning window is to suppress artificial frequency components that are observed when a simple rectangular window is used to gate the segment of interest.

The data reported here from the composite and the resin panel were acquired by placing the AU transducers at a point at about the center of one side of the 30 cm by 30 cm panels. This was found to be typical of data acquired at other positions on the panels.

Besides the AU data on the composite panels, waveform data was also collected with various other transducer configurations. These geometries were used to compare data from through-transmission signals at normal incidence to through transmission with off-set transducers and to the AU configuration. With the lucite specimen, immersion type 1 MHz broad band sender and receiver were used. The 1 MHz transducers on lucite produced a frequency range that simulated that of a pair of 2.25 MHz transducers with the attenuating effect of a typical graphite/epoxy resin composite. Gel liquid couplant was used with the 1 MHz transducer, lucite specimen experiments.

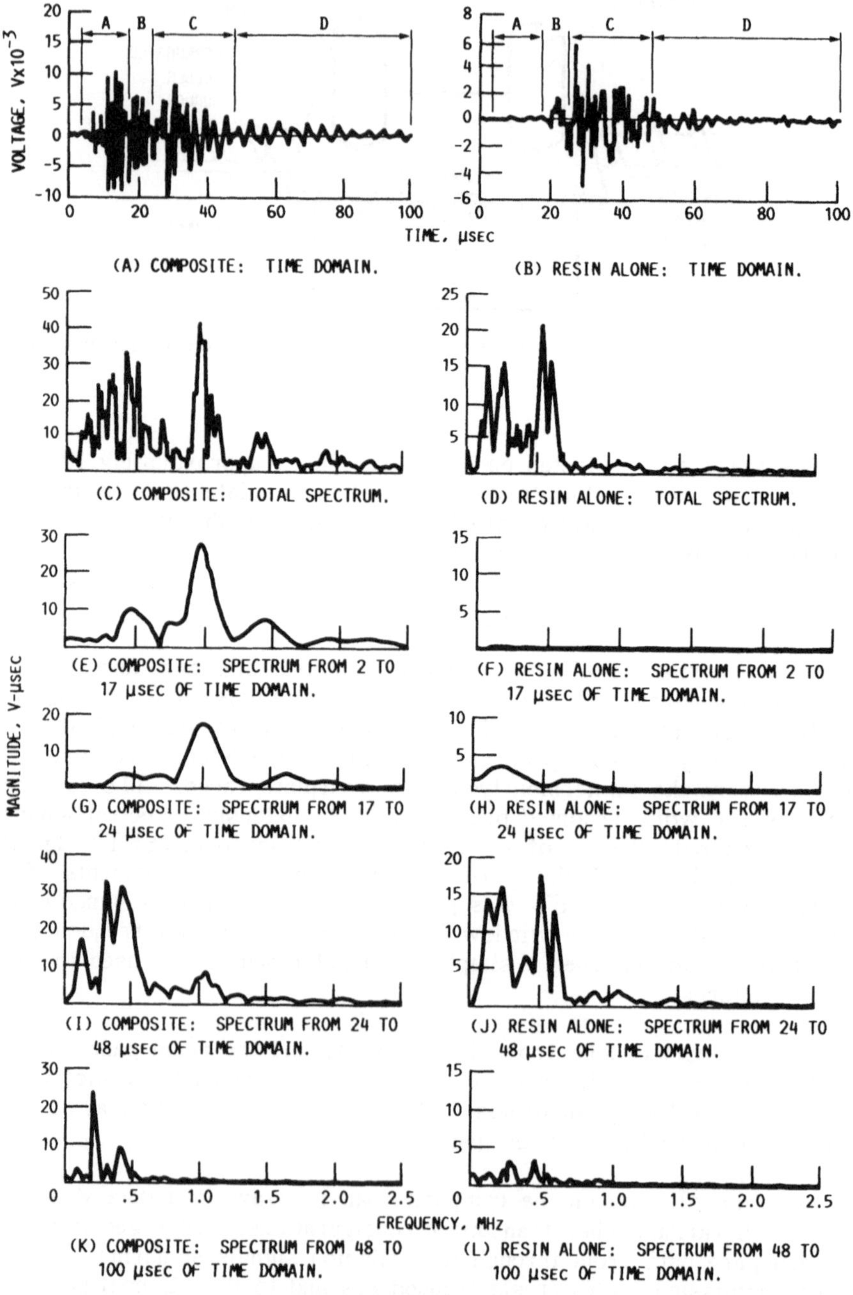

Fig. 2. Comparison of AU signal from graphite/epoxy resin composite panel with AU signal from panel of the resin alone.

RESULTS

AU Comparison of Graphite/Epoxy Composite with Pure Epoxy Resin

Figure 2 presents typical AU data for the graphite/ epoxy panel and also for the panel of the epoxy alone. Figures 2a and 2b are time domain signals. Figure 2a indicates four partitions of interest in the composite signal marked A, B, C, and D. The signal for the resin in Fig. 2b is marked with the same partitions. Figures 2c, and d are the frequency spectra of Figs. 2a and b respectively. Figures 2 e through 1 are the Hanning windowed partitions of the four time domain signals from both specimens. Partition A is 2 to 17 microseconds, B is 17 to 24 microseconds, C is 24 to 48 microseconds, and D is 48 to 100 microseconds.

Studies with Lucite and 1 MHz Transducers

Figure 3 illustrates the geometry employed for one series of through transmission experiments performed on a lucite panel with two 1 MHz transducers. In this case the transducers are lined up face to face so that the ultrasonic ray paths that pass through the lucite arrive at the receiving transducer at normal incidence.

Figure 4 is a series of through transmission signals collected in this manner. The column of figures on the left contains the time domain signals, the second column contains the frequency spectrum of signals, and the last column shows the cepstrum. In Fig. 4a the digitizer time base delay has been set to capture only the first pulse from the sender. Directly below in Fig. 4c the delay is changed to capture the first pulse plus the first echo. Below that in Fig. 4f two echos are included so that a total of three signals are captured. Finally at the bottom of the first column in Fig. 4i all the detectable pulses are brought on the screen and captured.

In the next set of experiments the geometry of Fig. 3 is altered as indicated in Fig. 5. The receiving transducer has been moved from the face to face configuration such that a direct ultrasonic ray path connecting the centers of the wear plates will make approximately a 71 degree angle with the normal to the contact plane.

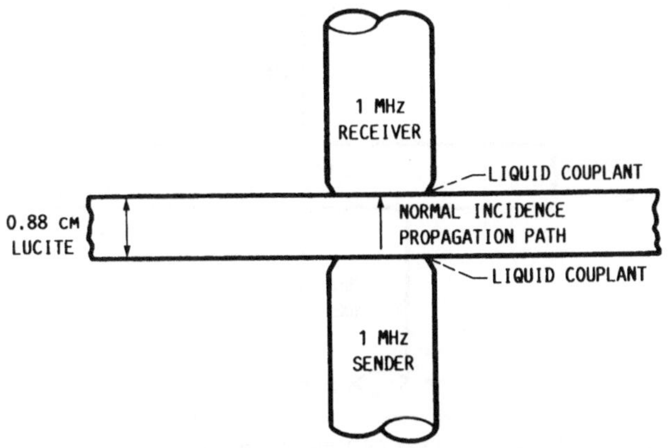

Fig. 3. Transducer arrangement used to transmit ultrasonic waves directly through a lucite panel at normal incidence.

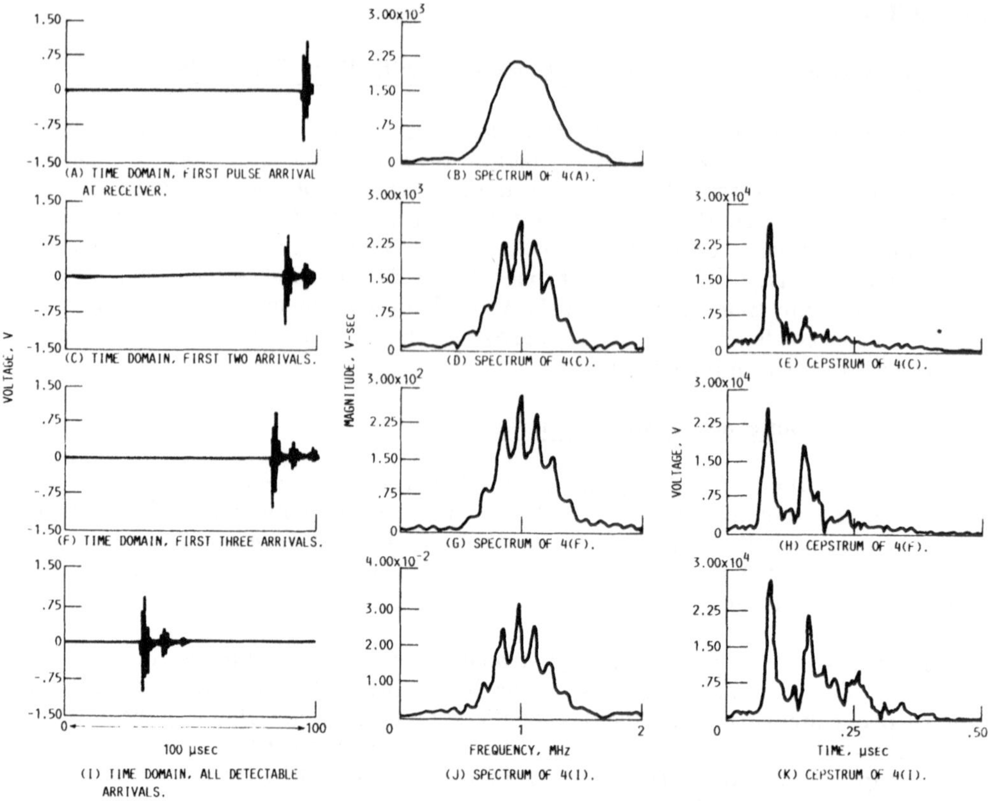

Fig. 4. Through transmission signals with 0.88 cm thick lucite, liquid couplant, and two 1 MHz transducers arranged as shown in Fig. 3.

Figure 6 shows waveforms collected with this arrangement. In Figs. 6a and 6b the digitizer time base has been set to capture only the first pulse from the sender. In Figs. 6c through 6e all the detectable pulses are brought on screen and captured. The format of the presentation is the same

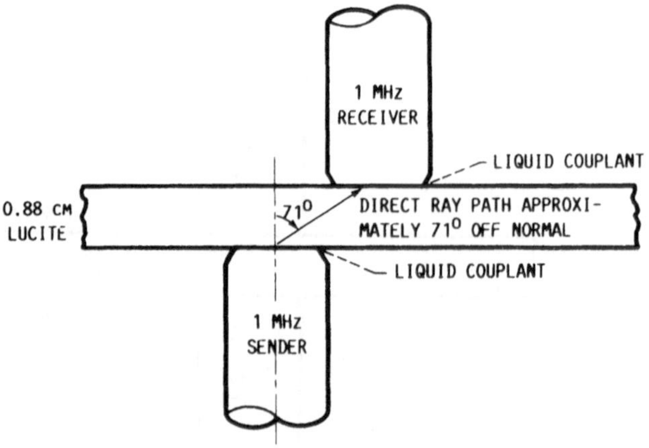

Fig. 5. Offset transducer arrangement used to transmit ultrasonic waves through lucite.

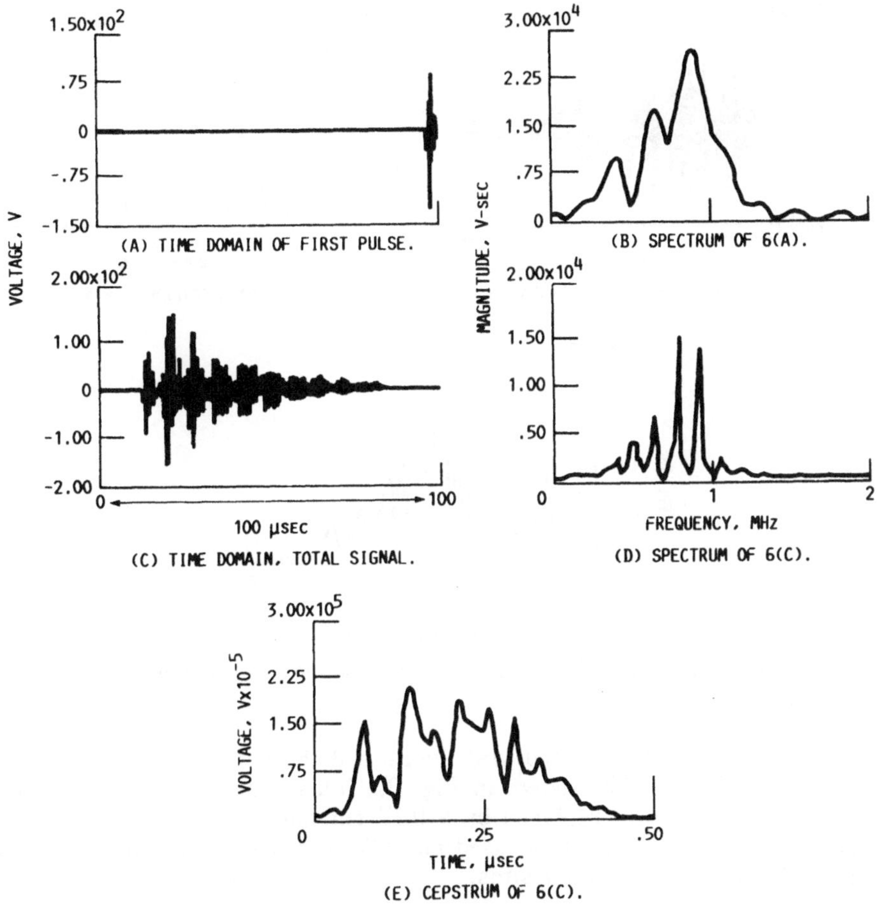

Fig. 6. Through transmission signals with the transducer arrangement shown in Fig. 5.

as in Fig. 4; the first column shows the time domain signal, the second column shows the frequency spectrum, and the last column shows the cepstrum.

The transducer arrangement in Fig. 7 was used for the acousto-ultrasonic results in Fig. 8. As with previous figures, 8a shows the time domain, 8b shows the spectrum, and 8c shows the cepstrum.

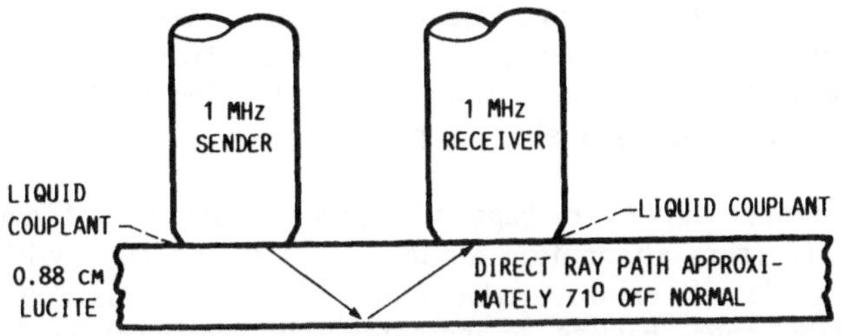

Fig. 7. Transducer arrangement used to transmit ultrasonic waves through lucite in the acousto-ultrasonic configuration.

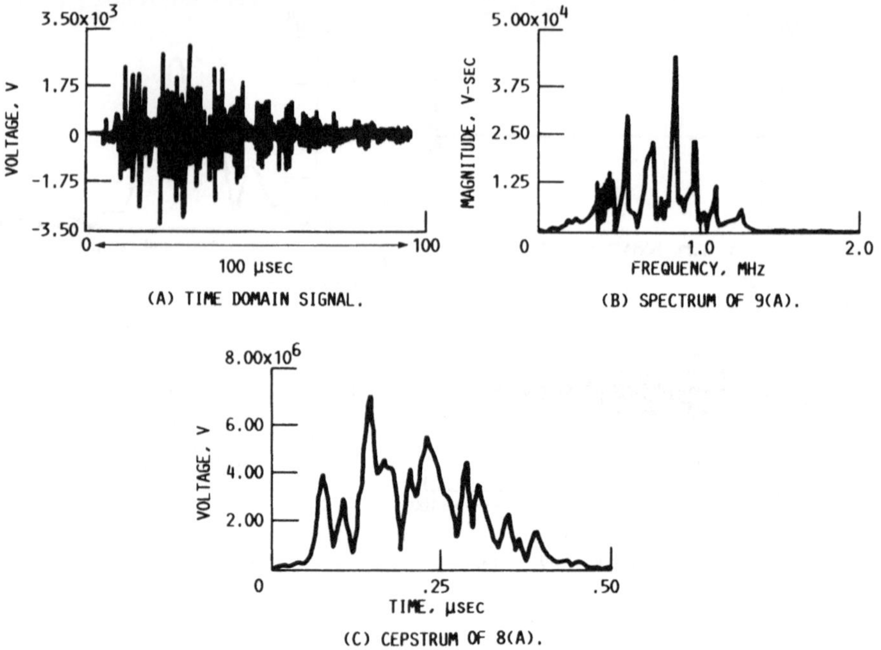

Fig. 8. Acousto-ultrasonic signal with transducer arrangement of Fig. 7.

Figure 9 data were collected with lucite and the normal incidence arrangement similar to that shown in Fig. 3. This time, however, one of the liquid couplant contacts is replaced by a 0.17 cm thick silicone rubber pad. This pad was typical of the kind used on both transducers in earlier AU work with graphite/epoxy resin composites.

DISCUSSION

Analysis of AU Spectra on Graphite/Epoxy and Pure Epoxy Resin

The time domain representation of an AU signal is the superposition of many pulse arrivals at the receiving transducer. The wide range of travel times among these pulses leads to the long time span associated with the signal. It seems likely that different time segments may have different spectra. Visual inspection of the AU signal often suggests a dominance by relatively high frequency components near the beginning with a progression to lower frequency dominance towards the end. This might be explained by considering the effect of attenuation. Higher frequency waves are attenuated more than lower frequency waves. Any wave that travels a long path length with a given rate of attenuation will be more affected by this than an otherwise similar wave traveling a shorter path.

Some AU signals, however, suggest sharper changes in frequency dominance than seems likely from attenuation alone. This is the case with the waveform presented in Fig. 2a. It exhibits what could be as many as four independent frequency distributions. Figure 2c indicates low frequency peaks at 0.3 and 0.5 MHz, and two high frequency peaks at about 1.0 and 1.5 MHz. The left hand member of each of the pairs of Figs. 2e-2k show the spectra of Hanning

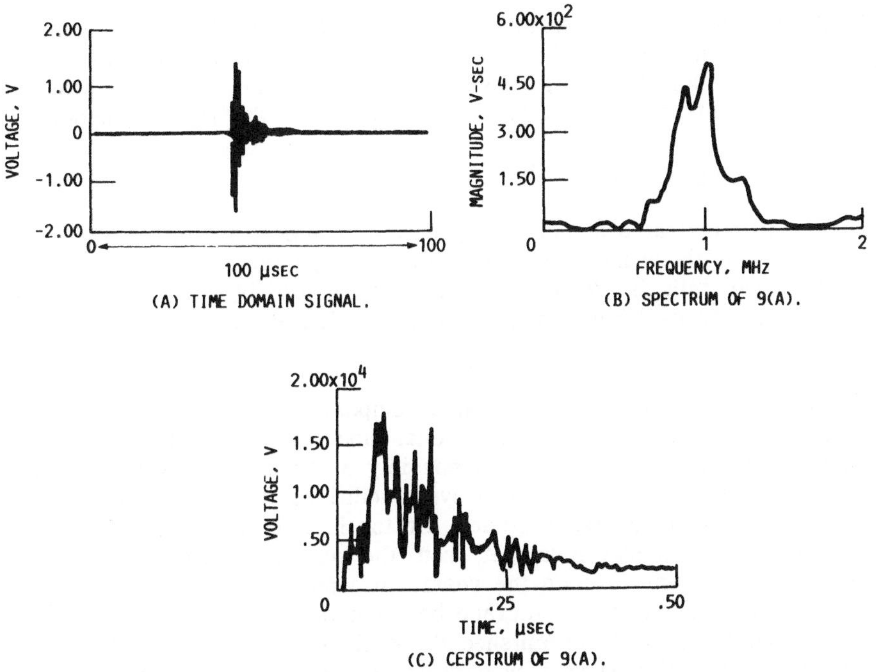

Fig. 9. Through-transmission with two 1 MHz transducers and one silicone rubber layer.

windowed partition of the time domain of the composite. Figures 2e and 2g indicate that the 2 to 17 microseconds and the 17 to 24 microseconds partitions have almost the same spectrum. They are very strong in the 1.0 and the 1.5 MHz peaks. Figure 2k shows that the 48 to 100 microsecond partition is strong in the low frequency peaks. The intermediate partition of 24 to 48 microseconds is shown in Fig. 2i to be a transition region. This result suggests at least two modes, or two mode types of propagation in this material. One mode passes through the specimen in a short time and retains frequencies up to the 1.5 MHz region. The other takes relatively long and loses frequencies above about 0.5 MHz. The composite total AU signal, (Figs. 2a and 2c), does not by itself give clear information concerning the nature of these two types of propagation.

The right hand member of each pair of graphs in Fig. 2 is an identical analysis as the left member except that it is with the 0.25 cm thick PMR epoxy resin panel. It is a specimen with no graphite fiber structure. What is most clear in the set of graphs is the absence of the high frequency, early arrival waves. This is evident by comparing Fig. 2b to Fig. 2a and then comparing Fig. 2d to Fig. 2c. One is led to infer that the high frequency, early arrival mode in the composite is due to propagation in the graphite fibers.

For comparison with the composite, the resin AU signal has been analyzed in terms of the same Hanning windowed, time partitions. Note that the bulk of the signal comes in the 24 to 48 microsecond partition of Fig. 2j. As with the composite in Fig. 2i it appears to exhibit a bimodal low frequency distribution. Unlike Fig. 2i there is little high frequency component. The similarity between the spectra of the resin and the low frequency components in the composite infers that the low frequency response in the graphite-epoxy is due to propagation through its epoxy resin component where the attenuation is high and the ultrasonic velocity is low relative to the fiber component.

Propagation Model for Composites

Figure 10 illustrates a plausible propagation model for the Fig. 2 results. Figure 10a represents propagation through the pure resin. This is the simple case with the detected signal being shaped by the specimen only through attenuation of PMR and multiple reflection at the surfaces of the panel.

Figure 10b represents the laminated composite panel. The resin path is still present but now there is an additional mode due to the fiber layers. This mode may travel in the fibers, possibly as transverse waves. Another possibility is that the fiber layers provide additional reflecting planes making the layers act as wave guides for the high frequency mode. Either way, the ultrasonic energy must get from the sending transducer to the fiber layers by first passing through the resin. It was noted during these experiments that touching the panel surface between the transducers with an energy absorbing medium strongly diminished the resin mode of propagation, but had little effect on the high frequency component. Thus it seems that once the ultrasonic energy gets into the fiber layers, the bulk of the high frequency energy travels along the fibers with some re-radiation occurring along the way.

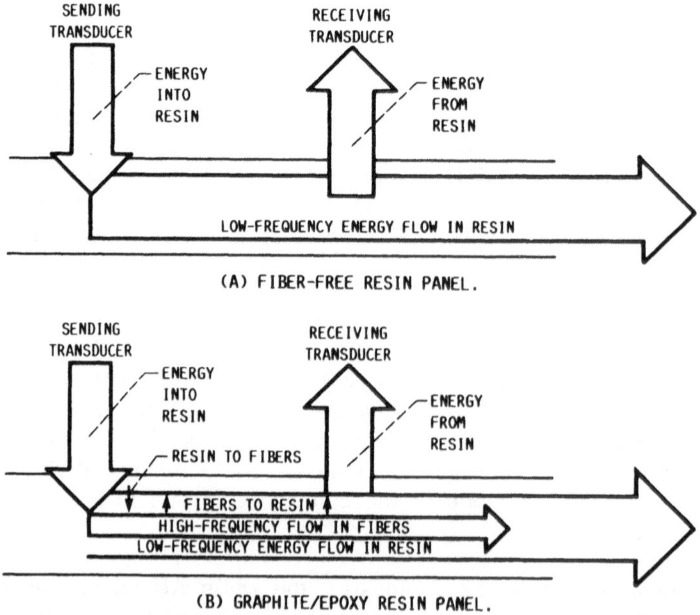

Fig. 10. Illustration of model for ultrasonic wave modes to explain acousto-ultrasonic results observed in graphite/epoxy laminates. Strength of signal diminishes (left to right) with distance from source.

Comparison of Figs. 2a and 2b shows that the resin mode, partition D, persists longer in the composite than in the resin alone. This mode is still present out to 100 microseconds for the composite, but is essentially gone by 70 microseconds for the neat resin. This seems to imply that the high frequency mode, while propagating internally, tends to radiate energy to the resin along the way thus enhancing the latter intensity. Note in Fig. 2e that the spectrum for the 2 to 17 microsecond partition has a low frequency part that is in the region of the 48 to 100 microsecond spectrum of Fig. 2k. The fibers do carry low frequency waves and re-radiate them to the resin. Figure 10 attempts to illustrate these ideas.

Not all graphite/epoxy structures may exhibit as sharp a time domain distinction of propagation modes as this one. It seems likely, however, that in principle the same propagation modes exist in all graphite/epoxy structures. Specimen geometry, especially thickness, can effect their relative importance. For example it has been shown that very thin specimens will be dominated by surface waves.[7] Awareness of the present results may be useful in analyzing other geometries.

Other Effects on the AU Spectra of Graphite/Epoxy Resin

The analysis of AU spectra in Fig. 2 was carried out with attention to the overall structure in terms of major peaks in the frequency spectrum. What was ignored was a fine structure of spike like peaks which is especially obvious in the spectra of Figs. 2c and 2d. The experiments discussed next are an investigation of factors that might produce the fine structure.

If we examine the AU spectra of the composite and the PMR epoxy resin in Fig. 2 the attenuation effects of the specimens is quite evident. The broad transducer resonance at 2.25 MHz is not at all visible in the AU spectra. The specimen AU spectra are large in various frequency regimes. The regimes depend upon which material is examined, but in each case they are well below 2.25 MHz. The overall shift to lower frequencies is expected on the basis of attenuation. What is not explained is the presence of many spikes in the spectra. Neither the transducer response nor classical attenuation theory can explain this. However, it can be understood through examination of the effects of transducer positioning and of dry couplants on AU waveforms.

Effects of Transducer Positioning

Figures 4, 6, 8, and 9 show waveforms collected with two immersion type, 1 MHZ, broad band transducers.

For Fig. 4 the transducers were in a face to face through transmission configuration. In Fig. 4a the time delay has been set to digitize only the first pulse that comes through. The spectrum in Fig. 4b approximates a slightly skewed Gaussian distribution peaking near 1 MHz. The column of Figs. 4c, 4f, and 4i show cases where progressively more echos are added to the captured signal. In Fig. 4c the time delay has been adjusted to digitize the initial arrival plus the first echo. Figure 4f contains the initial arrival plus two echos and Fig. 4i contains all the detectable arrivals. The center column of Fig. 4 shows the effect of multiple arrivals on the spectrum. The spectra display interference peaks that are reminiscent of the spikes in the AU spectra discussed earlier.

The interpretation of the peaks is simplest with the case of just two arrivals as with Fig. 4d. The height of the peaks and depth of the troughs depends upon how much the first echo is attenuated relative to the initial arrival. The locus of points for the minima is a curve of the difference in magnitudes of the two pulses as a function of frequency while the maxima fall on the curve of the sum of the amplitudes. This is illustrated by equations (A7) and (A8) of Appendix A.

The frequency interval between peaks is the reciprocal of the round trip time between the signal arrivals in the first column. These characteristic times appear as the major peaks in the cepstrums of the last column. Note that since Fig. 4c has one characteristic time between arrivals, Fig. 4e has one peak. Figure 4f with two possible characteristic times between arrivals leads to Fig. 4h with two peaks. Figure 4i with all the detectable arrivals leads to Fig. 4k with a train of diminishing peaks which presumably obey equations 1 or 2.

The above analysis of pulses and echos produces only the characteristic times that are integer multiples of the round trip duration through the thickness of the specimen. Characteristic times in AU signals will not be so simply related. In the simplest cases the times involved are for multiple reflections at non-normal incidence. The times must be solved from more the complex geometry of ray paths.[14,15,19] Besides this there is the effect of mode conversion at non-normal reflection[16]. Mode conversion means that not all rays travel with the same velocity.

Figures 5 and 6 illustrate a condition intermediate between simple normal incidence through transmission and the complex situation found in acousto-ultrasonics. Here the transducers are still on opposite faces but they are off-set such that the ray path from the center of one wear plate to the center of the other makes an angle of approximately 71 degrees from normal. In Figs. 6a and 6b, as in Figs. 4a and 4b, only the first pulse, as best as it can be isolated, is digitized. The frequency spectrum in Fig. 6b is a relatively smooth curve with interference peaks separated by large frequency increments. These interference peaks did not appear in Fig. 4b. In the present case we are probably seeing the effect of a distribution of characteristic times. The active area of the transducer wearplate is finite. For this reason there is a range of first pulse transit times for propagation between them.

For Figs. 6c through 6e the total detectable signal was digitized. The signal looks much like an AU waveform, but still there are individual pulse arrivals that can be distinguished in Fig. 6c. The frequency spectrum and cepstrum reveal that there are now many important frequency peaks and characteristic times for this signal. (It is interesting to do the experiment and watch the multiple reflections from the face to face geometry spread out in the time domain and change to different ray paths as the angle of these ray paths increases with the off-set sliding of the receiving transducer. In addition to this change there is the appearance, and gradual increase in importance, of new propagation paths associated with mode conversion at reflection.)

Figure 8 shows data obtained with the same transducers and specimen in an AU configuration, that is, with both transducers on the same side of the lucite. Now the arrival of individual pulses is much more obscure. The spectrum and cepstrum are more complex in structure. This signal, which has the spike like interference spectrum of a graphite/epoxy specimen, has been produced solely by positioning the transducers relative to each other similar to that shown in Fig. 1.

The experiments described with Figs. 4, 6, and 8 have illustrated a step by step construction of the spectrum fine structure of an AU signal. The center column of Fig. 4 shows that the presence of multiple pulse arrival times leads to multiple peaks in the spectrum. But the peaks in Fig. 4 are broader than is typical for AU spectra. However when the propagation path of the rays is made non-normal as in Fig. 6 the spectral peaks are found to become narrower. This is much more like the AU spectrum of Fig. 8b as well as the composite and resin spectra in Figs. 2c and 2d. This seems to show that this narrow, spike like, fine structure of the AU spectrum can be at least partly attributed to off-normal angles of reflection. Off-normal reflection causes generation of mode conversion and characteristic times that have no simple arithmetic relation to each other. In addition to this, discrete characteristic times become distributions of times.

Effect of Dry Couplants on AU Data

The above experiments show that the frequency spectrum obtained from AU signals can be simulated by the geometry of the specimen and transducer positions. It does not, however, prove that this is the only source of the peaks in the frequency spectrum. Figure 9 exhibits the results of an experiment designed to show the effect of silicone rubber pads. This thickness of this material was used as a dry couplant in several studies of graphite/epoxy composite structures.[14,15] Comparing Fig. 9b to Fig. 4b one can conclude that there is an interference pattern introduced by the silicone rubber layer. It is likely that this is the result of multiple reflection within the pad. The cepstrum Fig. 9c, shows there are many characteristic times present, with the most prominent being 6.7 microseconds.

CONCLUSIONS

Acousto-ultrasonic energy introduced into a laminated graphite/resin has been shown to propagate through the structure in two modes. The first mode is along the graphite fibers. This mode is the more rapid of the two. The second mode is through the resin matrix. Besides being slower, the matrix mode is also the most strongly attenuated in the higher frequency regime.

Awareness of these differences in propagation speed and frequency distribution can be of use in separating the modes for study. It is likely that the analysis is applicable to other kinds of composites as well.

Studies of AU in conjunction with material and mechanical properties should take advantage of this type of wave propagation mode analysis. Some modes will be more sensitive to particular properties than will others. Focusing on the most appropriate mode can lead to higher reliability in using acousto-ultrasonics as a predictor for the property of interest.

The fine structure of narrow peaks in the frequency spectrum, typical of AU spectra, has been shown to be a manifestation of the characteristic times associated with the arrival of individual pulses at the receiving transducer.

The times are not in general related as simple arithmetic ratios. They are not discrete but rather distributions of times. These features would seem to discourage attempts to analysis them. This fine structure is not likely to carry useful information on the material or mechanical properties of the specimen being studied.

One can contemplate that there are applications, however, where information on specimen geometry would be useful. This information can be expressed in terms of the characteristic delay times. It was shown that the cepstrum is most useful in detecting and measuring these characteristic times even in rather complex spectra. If analysis of the fine structure is considered, however, the first step is to eliminate or at least account for any fine structure produced from such factors as dry couplant pads.

APPENDIX A

The Source of Interference Peaks in the Magnitude of the Frequency Spectrum

The magnitude of the frequency spectrum, A(f), is given by:[17]

$$A(f) = (1/2\pi)\int_{-\infty}^{\infty} x(t)\exp(-i2\pi ft)\, dt \tag{A1}$$

Here f is frequency, t is time, and x(t) is the time domain signal collected at the receiving transducer.

Let the time domain signal be a pulse arrival x(t) and a pulse arrival x(t) that are components of an AU signal.

$$A(F) = (1/2\pi)\int_{-\infty}^{\infty} [x_1 + x_2]\exp(-i2\pi ft) \tag{A2}$$

Let the pulses be related by:

$$x_2(t) = Cx_1(t-T) \tag{A3}$$

The two pulses have a common source, the input pulse from the sending transducer. T is the characteristic time between them. Since they left the sending transducer at the same time, T is the difference in their transit time from sender to receiver. The factor C represents the difference in attenuation between the pulses due to the longer time of travel for one of them. C will therefore be frequency dependent. C must also contain any non-time dependent phase differences incurred by the pulses during the transit of the specimen. The only such phase differences, however, are at reflections. Reflections multiply the signal by plus or minus a positive constant. Therefore the cumulative effect of all such reflections will be plus or minus a positive constant. Then C can be expressed as:

$$C = (+/-)K(f) \tag{A4}$$

Substituting equations (A4) and (A3) into equation (A2) the amplitude of the spectrum at any frequency f is:

$$A(F) = (1/2\pi)\int_{-\infty}^{\infty} [x_1 \exp(-i2\pi ft(+/-)x_1 \exp(-i2\pi f(t+T))]dt \tag{A5}$$

or:

$$A(f) = (1/2\pi)\int_{-\infty}^{\infty} x_1(t)[1\ (+/-)K(f)\exp(-i2\pi fT)]\exp(-i2\pi ft)dt \tag{A6}$$

The term in the bracket of (A6) modulates the frequency spectrum of the x (t) pulse. The condition for equation (1) of the ANALYSIS section, T=n/f, where n is an integer reduces the bracket to:

$$[1\ (+/-)\ K(f)] \tag{A7}$$

In the case of equation 2 where T=(n+1/2)/f the bracket becomes:

$$[1\ (-/+)\ K(f)] \tag{A8}$$

The sign in eq. (A8) will always be opposite the sign in eq. (A7). Peaks occur in the frequency spectrum where the plus sign occurs and troughs occur for the minus sign.

REFERENCES

1. A. Vary and K. J. Bowles, Ultrasonic Evaluation of the Strength of Unidirectional Graphite Polyimide Composites, in: "Proceedings of the Eleventh Symposium on Nondestructive Testing," NTIAC, San Antonio (1977).
2. D. T. Hayford, E. G. Henneke, II, and W. W. Stinchcomb, The Correlation of Ultrasonic Attenuation and Shear Strength in Graphite-Polyimide Composites, J of Comp Matls. 11:429 (1977).
3. A. Vary and K. J. Bowles, Use of an Ultrasonic-acoustic Technique for Nondestructive Evaluation of Fiber Composite Strength, in: "Proceedings of the 33rd Annual Conference of the Society of the Plastics Industry," SPI, New York (1978).
4. A. Vary and R. F. Lark, Correlation of Fiber Composite Tensile Strength with the Ultrasonic Stress Wave Factor, J of Test & Eval. 7:185 (1979).
5. J. H. Williams, Jr. and N. R. Lampert, Ultrasonic Evaluation of Impact Damaged Graphite Fiber Composites, Matls Eval. 38:68 (1980).
6. E. G. Henneke, II, J. C. Duke, Jr., W. W. Stinchcomb, A. Govada, and A. Lemascon, "A Study of the Stress Wave Factor Technique for the Characterization of Composite Materials, CR-3670," NASA, Cleveland (1983).
7. J. H. Hemann and G. Y. Baaklini, "The Effect of Stress on Ultrasonic Pulses in Fiber Reinforced Composites, CR-3724," NASA, Cleveland (1983).
8. R. H. Wehrenberg, II, New NDE Technique Finds Subtle Defects, Matls Eng. 92:59 (1980).
9. S. Serabian, Composite Characterization Techniques: Ultrasonic, Mantech Journal 10:11 (1985).

10. L. Lorenzo and H. T. Hahn, "Damage Assessment by Acousto-ultrasonic Technique in Composites, WU/CCR-86/2," Office of Naval Research, Washington, (1986).
11. A. Govada, E. G. Henneke, II, and R. Talreja, Acousto-ultrasonic Measurements to Monitor Damage During Fatigue of Composites, in: "1984 Advances in Aerospace Sciences and Engineering," U. Yuceoglu and R. Hesser, eds., ASME, New York (1984).
12. E. G. Henneke and J. C. Duke, Jr., Analytical Ultrasonics for Evaluation of Composite Materials Response, Part I: Physical Interpretation, in: "Analytical Ultrasonics in Materials Research and Testing, NASA CP-2383," NASA, Cleveland (1986).
13. H. L. M. dos Reis and H. E. Kautz, Nondestructive Evaluation of Adhesive Bond Strength Using the Stress Wave Factor Technique, J of Acoustic Emission 5:144 (1986).
14. H. E. Kautz, "Ultrasonic Evaluation of Mechanical Properties of Thick, Multilayered, Filament Wound Composites, NASA TM-87088," NASA, Cleveland (1985).
15. H. E. Kautz, "Acousto-ultrasonic Verification of the Strength of Filament Wound Composite Material, NASA TM-88827," NASA, Cleveland (1986).
16. J. Krautkramer and H. Krautkramer, "Ultrasonic Testing of Materials," Springer-Verlag Inc., New York (1969).
17. J. S. Bendat and A. G. Piersol, "Engineering Applications of Correlation and Spectral Analysis," John Wiley and Sons, New York (1980).
18. D. Childers and A. Durling, "Digital Filtering and Signal Processing," West Publishing Co., St. Paul (1975).
19. H. Karagulle, J. H. Williams, and S. S. Lee, "Stress Waves in an Isotropic Elastic Plate Excited by a Circular Transducer, NASA CR-3877," NASA, Cleveland (1985).

NONDESTRUCTIVE EVALUATION OF COMPOSITE MATERIAL USING ULTRASOUND

V. K. Kinra and V. Dayal

Aerospace Engineering Department and
Mechanics and Materials Center Texas A&M University
College Station, TX 77843

ABSTRACT

Fiber-reinforced composites are finding an increasing use in the aerospace industry. Initially the FRP components constituted only the non-critical components of the structure. Now the composites are being used in the primary load bearing members. After undergoing a certain amount of usage, the mechanical, thermal and environmental loading produces a complex damage state which includes transverse cracks, longitudinal splits, delaminations, debonding, etc. Almost all the NDT techniques are geared towards the estimation of the extent of damage to the structure. From these results it is expected that the damage modelers will be able to estimate the residual stiffness and residual strength and life of the structure. We have used ultrasonic waves to study the changes in stiffness of the structure as the damage progresses. This work will help the damage modelers in furthering their analysis.

The ultrasonic waves passing through a composite specimen interact with the various defects, and in turn, these defects affect the basic ultrasonic parameters; wavespeed and attenuation. It is well known that wavespeed is directly related to the stiffness, and attenuation is a measure of the damping characteristics of the material. A very important stiffness component is the in-plane stiffness of the plate. Hence, we propagate the waves in the plane of the plate to measure the in-plane stiffness. The mode of propagation is called the Lamb wave or Plate wave mode. The plate is immersed in a fluid and the Lamb waves leaking into the fluid are called the leaky Lamb waves. These leaky waves have been used to determine the wavespeed and attenuation of the Lamb waves traveling in the plate.

Damage is gradually introduced in the composite plate. In the work presented here, we have limited the mode of damage to transverse cracks only. The changes in the wavespeed and attenuation are measured as a function of damage. We present here some results from the tests of cross-ply and angle-ply graphite/epoxy laminates. The reduction in the in-plane stiffness and an increase in the attenuation is observed as the number of transverse cracks increase.

INTRODUCTION

The growth of damage in composite materials is very different from that in homogeneous materials. In homogeneous materials once the damage is initiated, any further loading tends to increase the existing damage. On the other hand, in composites the damage relieves the stresses in its vicinity such that the next cracking takes place at some other location. It is only when the microcracks become densely populated that they initiate larger damage such as interior delaminations. The study of damage growth in cross-ply laminates due to cyclic loading[1] has shown that first the transverse cracks appear in the 90° -plies. These cracks are restrained in their growth and hence further loading results in the axial split in the 0° -plies due to the Poisson's effect. The intersection of the transverse cracks and the longitudinal splits becomes the nucleation site for the interior delamination between the plies.

When an ultrasonic wave is passed through a damaged composite, the interaction between the wave and the damage can affect the wave in two ways: (1) Damage will, in general, reduce stiffness and since wavespeed is directly proportional to the square-root of stiffness, damage will reduce the wavespeed, and (2) The attenuation will increase because the crack-wave interaction results in an incoherent scattering of the waves. Thus the effect of damage on the overall behavior of the composite can be studied by measuring the acoustic parameters of the ultrasound passed through the specimen.

We have developed two new techniques[2] for the measurement of acoustic parameters in thin laminates which are expected to be used in aerospace structures. With the first of these techniques, one is able to measure the acoustic parameters when the pulses reflected from two surfaces of a plate specimen can be separated in time domain. With the second technique one can make the measurements even when the pulses are inseparable. Here, the signal in time domain is transferred to the frequency domain by the use of fast Fourier transforms (FFT). The data in the frequency domain is used in an

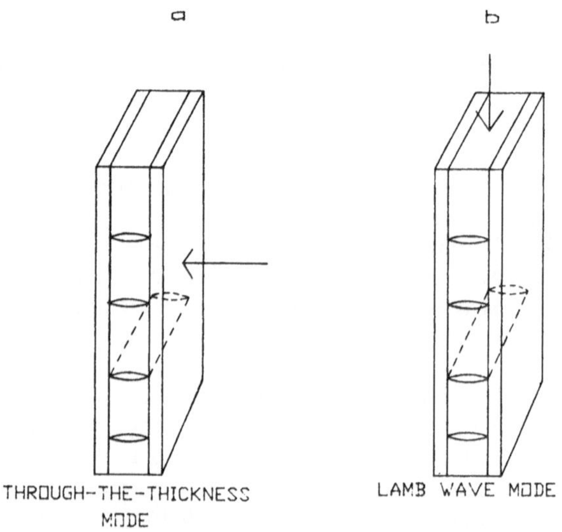

Fig. 1. Wave Propagation Direction Relative to Transverse Crack Plane.

algorithm developed by us to measure the complex-valued wavenumber $k = k_1 + k_2$, where $k_1 = \omega/c$, ω is the circular frequency of the signal used, c is the phase velocity and k_2 is the attenuation coefficient. Figure 1a shows that in this mode of wave motion the wave travels in the plane of the crack and, hence, there is a weak interaction between the waves and the crack. Results show that although the changes in attenuation are substantial, the effect on wavespeed is immeasurably small.

Figure 1b shows the Lamb wave mode where the wave travels normal to the transverse cracks. Here the interaction between the wave and the crack is stronger, and therefore, a larger effect on the acoustic parameters was observed. In this mode of wave motion, the longitudinal stiffness of the plate determines the wavespeed. Hence, the effect of the transverse cracks on the longitudinal in-plane stiffness of the composite laminates could be studied. We present here some results from the testing of the composite laminates with transverse cracks by the Lamb wave technique. The results show that both the wavespeed and the attenuation are significantly affected by the transverse cracks.

THEORY

For the Lamb wave tests we have used the fundamental symmetric mode of wave propagation. The reason behind this choice is that in this mode the wave travels with a plane wavefront. The relation between the material properties and wavespeed in the fundamental symmetric mode[3] is

$$C_L^2 = E_1/[\rho(1-\nu_{12}\nu_{21})] \qquad (1)$$

where C_L is the Lamb wavespeed, E_1 is the in-plane modulus, ν_{12} is the major Poisson's ratio, and ν_{21} is the minor Poisson's ratio.

For the composites used by us $\nu_{12} = 0.28$ and $\nu_{21} = 0.018$. Thus $\nu_{12}\nu_{21} \ll 1$ and to a first approximation eq. (1) is:

$$C_L^2 = E_1/\rho \qquad (2)$$

The relation between the angle of incidence of the wave on the plate, θ_i, and the wavespeed (C_L) of the Lamb wave is governed by the Snell's Law;

$$\sin(\theta_i)/v_w = \sin(\pi/2)/C_L \qquad (3)$$

where V_w is the wavespeed in water.

EXPERIMENTAL SETUP

The block diagram of the experimental setup is shown in Fig. 2. The specimen, the transmitter and the receiver are immersed in the water bath. The specimen is mounted on a turn table and can be rotated about a vertical axis in steps of 0.1°. This rotation is required to set the specimen for the through-the-thickness and Lamb wave measurements. The transducers are mounted on precision traveling mechanisms. In the through-the-thickness measurements the specimen is normal to the incident wave and the two transducers are in the same line.

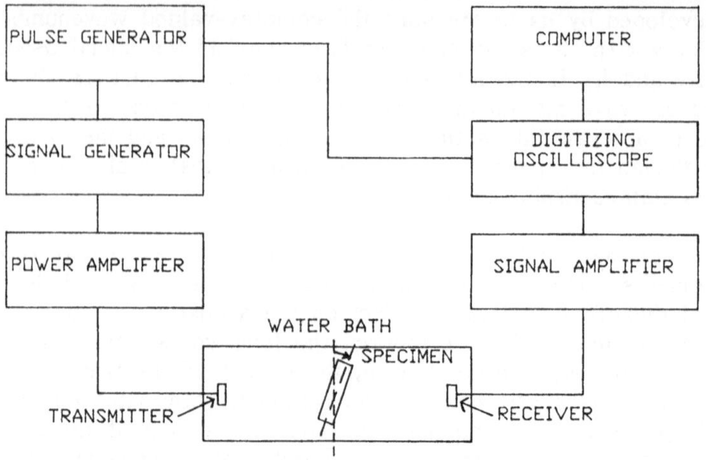

Fig. 2. Block Diagram of the Experimental Setup.

The pulse generator triggers the signal generator which produces a single cycle of sinusoidal wave for through-the-thickness measurements and a tone burst for the Lamb wave measurements. This signal is amplified by the power amplifier and the signal is fed into the wide band transmitting transducer. The wave launched into the water travels through the specimen and sensed by the receiving transducer. The signal from the receiver is amplified by the signal amplifier and fed into the digitizing oscilloscope (Data 6000 by Data Precision). The analog signal is digitized and stored in the oscilloscope. The built-in signal processor of this oscilloscope provides the computer with the amplitude and location of a characteristic point of the toneburst signal e.g. a maximum of a sine wave. The signal amplitude at different angles of incidence is recorded, and critical Lamb angle is identified by the peak in the received signal. A very simple test is sufficient to check for the correct Lamb angle. It can be shown by an elementary calculation that if the receiver is moved in a straight line perpendicular to the line joining the two transducers then there should be no change in the arrival time of the

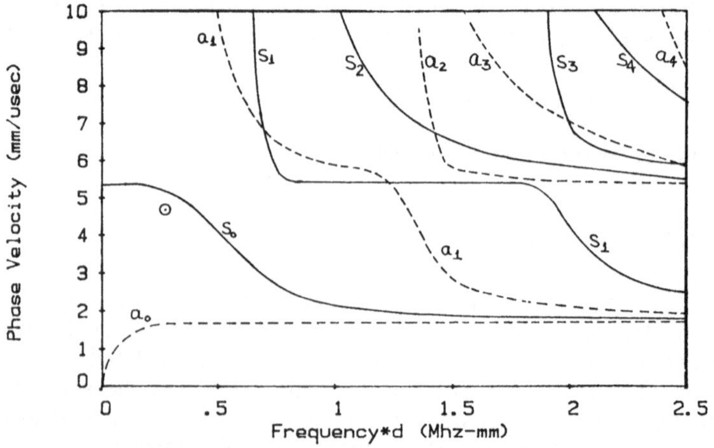

Fig. 3. Dispersion Curve for $[0/90_3]_s$ Gr/Ep Laminate.

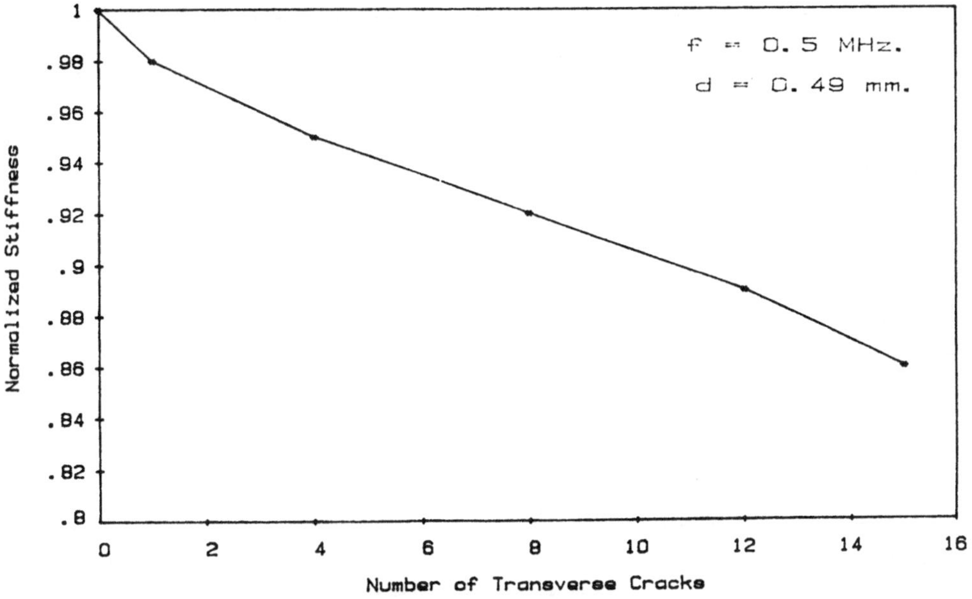

Fig. 4. Stiffness Variation with Damage in $[0/90_3]_s$ Laminate.

signal at the receiver provided the specimen is oriented at the correct Lamb angle. When the transducer is moved for this check, the wave spends more time in the specimen and less time in water and hence the attenuation can be measured.

All the specimens for which the results are presented here are made of AS4/3502 graphite/epoxy laminates. The specimens are 11" x 1" coupons.

RESULTS AND DISCUSSIONS

Now we present some results from the testing of $[0/90_3]_s$ and $[0/90_4]_s$ laminates by the Lamb wave method when transverse cracks are introduced. The cracks are generated in these specimens by displacement controlled monotonic loading at a displacement rate of .02"/min.

We have carried out a theoretical analysis of Lamb wave propagation in a symmetrical balanced composite laminate[4]. The dispersion curves reproduced here are from that work. Fig. 3 shows the dispersion curve for the $[0/90_3]_s$ specimen. Phase velocity of the Lamb waves is plotted against the product of signal frequency and d, where 2d is the plate thickness. The symmetric and antisymmetric modes are shown in solid and discontinuous lines, respectively. The frequency at which the tests were performed (0.5 MHz), is indicated by a circle; s_0 represents the fundamental symmetric mode. The normalized reduction in stiffness as the number of cracks increases in the transducer field are shown in Fig. 4. The stiffness is normalized with respect to the stiffness of the virgin specimen (i.e. no damage). There is a steady decrease in the stiffness and the overall reduction is about 12%. The increase in the attenuation for this specimen is shown in Fig. 5. A four fold increase in attenuation can be observed in this test. Figure 6 shows the dispersion curve for $[0/90_4]_s$ laminate and the tests were conducted at

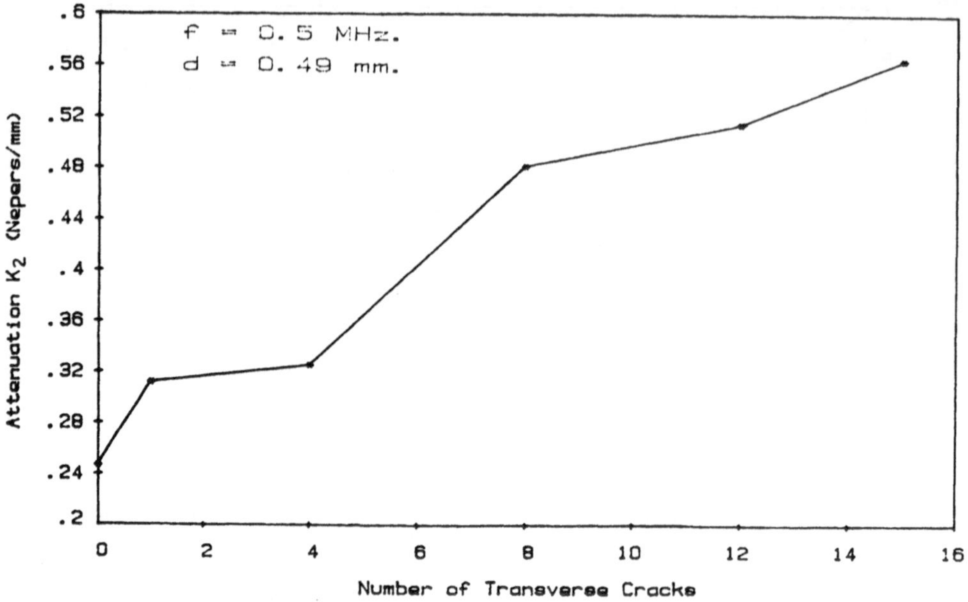

Fig. 5. Attenuation Variation with Damage in $[0/90_3]_s$ Laminate.

the frequency indicated by the circled point. The line diagram of the state of damage is shown in Fig. 7. The location of the transmitter (TR) and the receiver (R) are shown in the figure. The reduction in the stiffness of this laminate when the transverse cracks are introduced, is shown in Fig. 7. Observe that going from the damage state 3 to 4 though there was a substantial increase in the number of cracks in the specimen, the number of cracks in the local region interrogated by the transducer did not increase and hence the changes in the stiffness of the specimen were not observed. This is very reassuring for it demonstrates that our measurement reflects local changes in the stiffness. For this specimen the reduction in stiffness of about 30% was observed. The increase in the attenuation is shown in Fig. 8 which shows almost a six fold increase in attenuation.

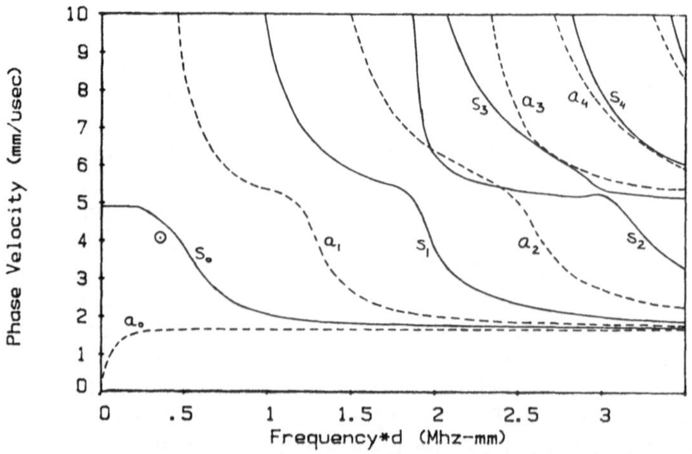

Fig. 6. Dispersion Curve for $[0/90_4]_s$ Gr/Ep Laminate.

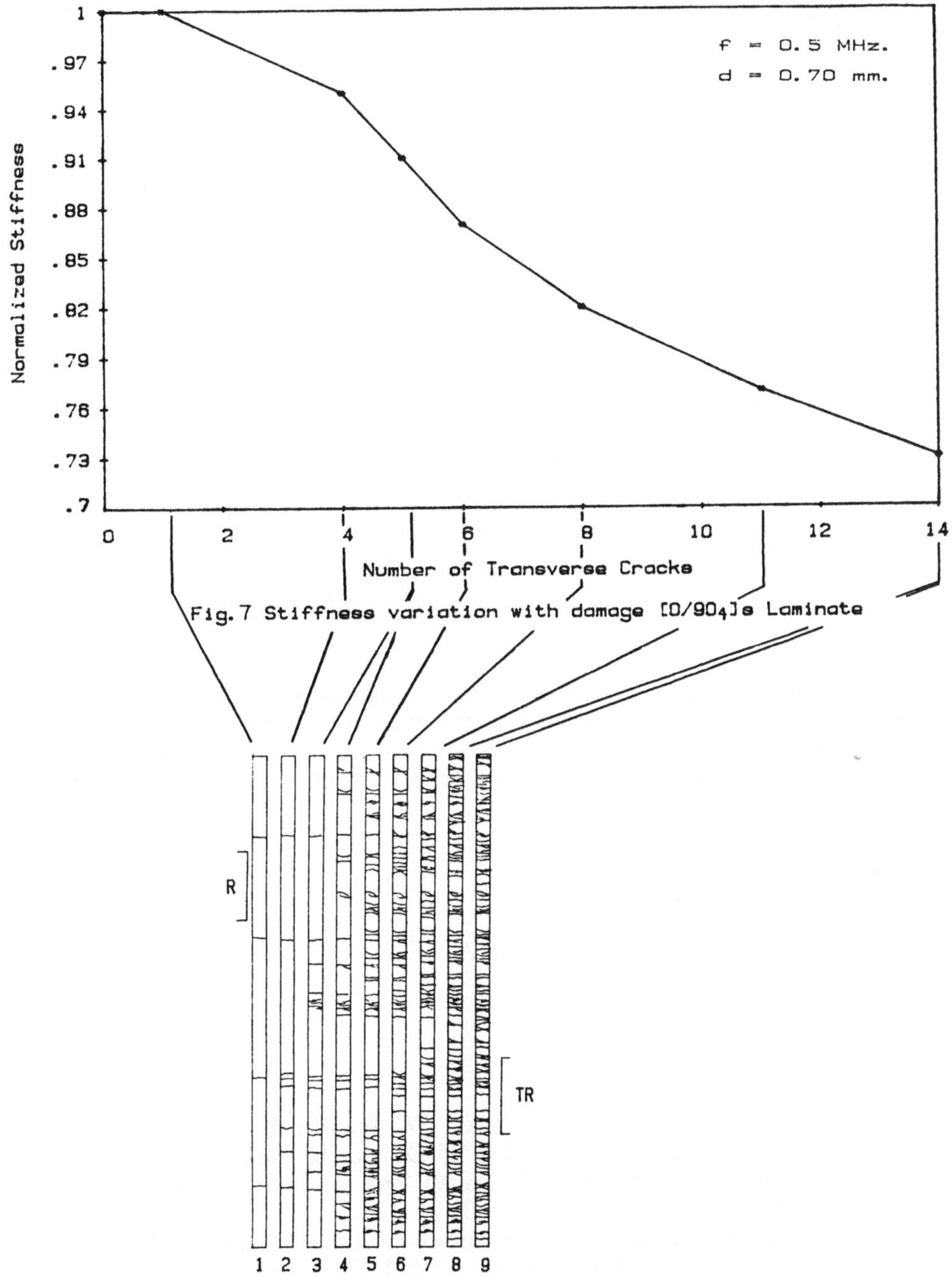

Fig. 7. Stiffness Variation with Damage for $[0/90_4]_s$ Laminate.

CONCLUSIONS

The use of two techniques for ultrasonic nondestructive evaluation of damage (transverse cracking) in laminated composites has been demonstrated. In the first, a longitudinal wave is propagated in the thickness direction. Here the crack-wave interaction is weak. As expected, the wavespeed does not change measurably while the attenuation increases with transverse cracking.

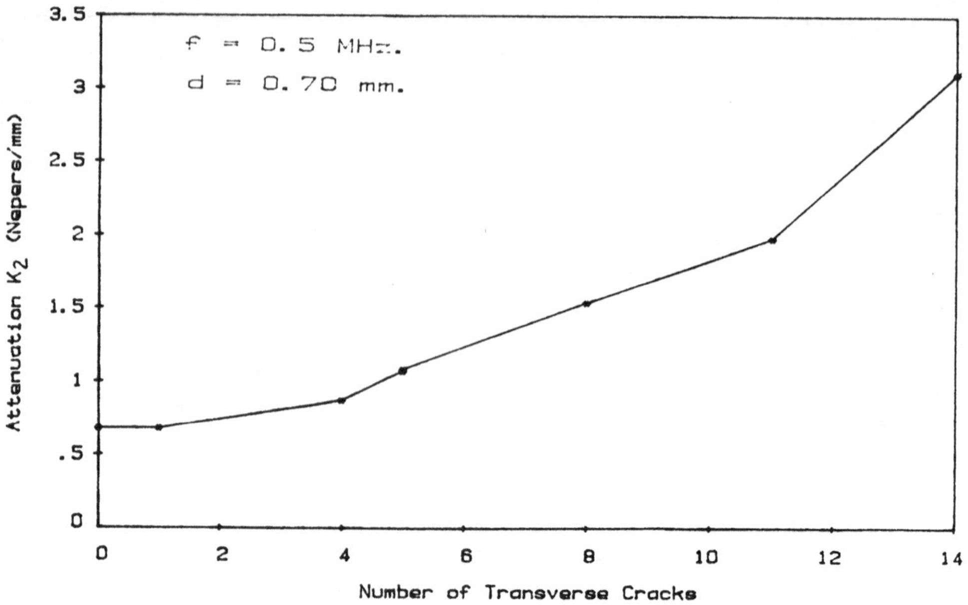

Fig. 8. Attenuation Variation with Damage $[0/90_4]_s$ Laminate.

In the second technique, Lamb waves are propagated along the length of the specimen. Here, the crack-wave interaction is the strongest; both the wavespeed and the attenuation change appreciably with damage. The Lamb wave method, therefore, is a much more effective method for the detection of transverse cracks.

ACKNOWLEDGEMENTS

This research is supported by the Air Force Office of Scientific Research Contract No. F49620-83-C-0067.

REFERENCES

1. K. L. Reifsnider, E. G. Henneke, W. W. Stinchcomb, and J. C. Duke, Damage Mechanics and NDE of Composite Laminates, in: "Mechanics of Composite Materials," Z. Hashin and C. T. Herakovich, eds. Pergamon Press, New York (1983).
2. V. Dayal and V. K. Kinra, Ultrasonic NDE of Composite Transverse Cracking, in: "Proceedings of SEM Fall Conference," Society for Experimental Mechanics (1986).
3. C. C. Habegar, R. W. Mann, and G. A. Baum, Ultrasonic Plate Waves in Paper, Ultrasonics 3:57 (1979).
4. V. Dayal and V. K. Kinra, "Lamb Waves in Anisotropic Plates Immersed in a Fluid--An Exact Numerical Solution," (unpublished results).

EXPERIMENTAL AND THEORETICAL ANALYSIS OF BACKSCATTERING MECHANISMS IN FIBER-REINFORCED COMPOSITES

R. A. Roberts

Material and Components Technology Division
Argonne National Laboratory
Argonne, IL 60439

J. Qu and J. D. Achenbach

The Technological Institute
Northwestern University
Evanston, IL 60208

INTRODUCTION

For the purpose of ultrasonic backscatter measurements, a material can be considered homogeneous if the signals received from scattering by microstructure are below the detection sensitivity of the ultrasonic instrumentation. This condition exists when 1) microstructural inhomogeneities in density and/or elastic properties are sufficiently small (weak scattering), or 2) all the characteristic dimensions of the internal structure are sufficiently small (long wavelength scattering). The ultrasonic inspection of homogeneous materials for abnormalities is generally straightforward; if scattering is detected, it may be assumed that a flaw is present. The inspection of structurally inhomogeneous materials such as fiber reinforced composites, however, is complicated by the need to discriminate between scattering from non-defective microstructure, and from defects such as cracks, porosity, inclusion, etc. Thus the development of ultrasonic NDE techniques for fiber reinforced composites requires first a careful examination of the scattering characteristics of the defect-free composite microstructure, followed by an examination of the changes in the defect-free scattering characteristics caused by the introduction of defects.

This paper examines experimental measurements of polar backscatter following the techniques first employed by Bar-Cohen and Crane,[1] and later by others.[2-6] Experimental measurements of the angular dependence of ultrasonic backscatter are presented for fiber reinforced composite specimens containing various levels of porosity. This data demonstrates how backscattered signals from porosity are inseparably intertwined with signals scattered by the internal fiber-related structures of the composite. Differences in the angular dependence of backscatter are compared for specimens having differing levels of porosity, from which empirical rules describing the effect of porosity on the angular backscatter characteristics are postulated. The experimental results are compared with the predictions of an analytical model for backscatter from a submerged fiber-reinforced layer containing a distribution of flaws. The analysis of scattering by the fibrous microstructure emphasizes structural inhomogeneities with characteristic dimensions comparable to the ultrasonic wavelength. The ultrasonic wavelength in the experiment is, in general, much longer than the reinforcing fiber diameter, and, as a consequence, the model considers the composite to be a transversely isotropic continuum with inhomogeneities in elastic properties and density of the same length scale as the incident wavelength. The analysis models scattering from porosity by coherently summing the scattering from a distribution of flaws contained throughout the submerged inhomogeneous, transversely isotropic layer.

The transmission characteristics at the water/composite interface are of fundamental importance to the theoretical analysis of the backscatter experiment. The transmission characteristics of the composite specimens were measured experimentally, and a comparison is made with theoretically evaluated transmission coefficients for a fluid-loaded transversely isotropic halfspace. The close agreement observed in this comparison demonstrates the validity of the transverse isotropy assumption used in modeling the backscatter from the fiberous microstructure.

COMPOSITE SPECIMENS

The composite specimens examined in this study consist of 16 plies of epoxy-impregnated carbon fiber fabric pressed under high temperature. The constituent plies of these specimens were oriented so that the reinforcing fibers in all plies run in the same direction. Each ply is ~125 μm thick, resulting in a total specimen thickness of ~2 mm. The reinforcing fibers are 8 μm in diameter and they constitute 60-65% of the total volume of the specimen.

Optical micrographs of the composite microstructure reveal that the reinforcing fibers are not uniformly distributed. Regions with abnormally low concentrations of fibers are found throughout the specimen volume. An example is shown in Fig. 1 of two micrographs obtained from two different regions of the same specimen. The micrographs show cross-sections of fiber distributions, with the fiber orientation perpendicular to the plane of the micrograph. Fig. 1a displays a fairly uniform concentration of fibers, whereas Fig. 1b displays regions of abnormally low fiber concentrations. The total field of view in these micrographs is 250 μm, which is comparable to the ultrasonic wavelength in the experiment. It is evident that the local static elastic properties of the material in Fig. 1a on a ~200 μm length scale would be considerably different than those of Fig. 1b. It is this type of inhomogeneity in elastic properties which the analysis discussed later exploits in modeling the backscatter experiment.

Various levels of porosity were introduced in the specimens under study by appropriately modifying the pressing environment at the time of manufacture. Results are presented here for specimens containing 0.2, 1.1, 2.0, and 6.5 volume percent porosity, measured by acid digestion. Morphological studies of the porosity in these specimens revealed elongated pores with

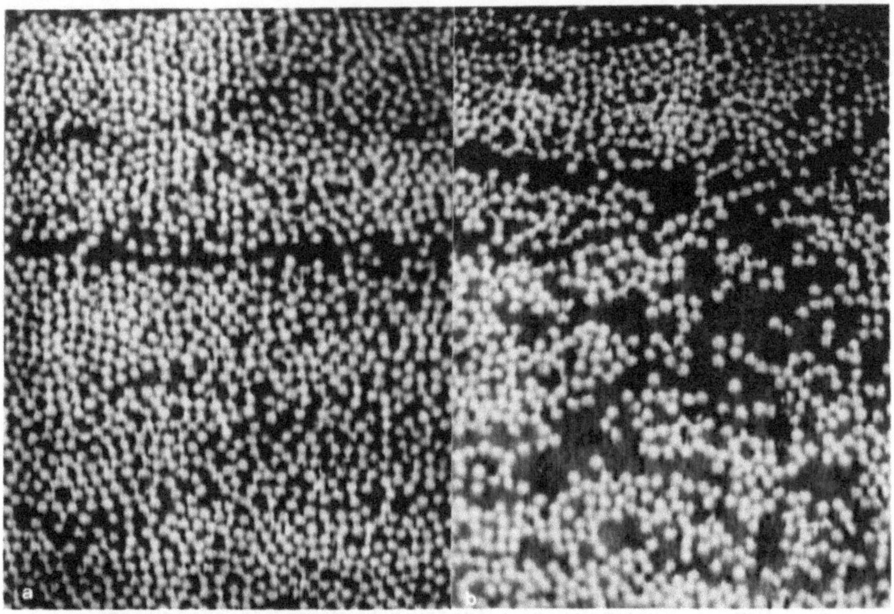

Fig. 1. Micrographs showing fiber distribution in composite specimens a) uniform distribution, b) nonuniform distribution with region of low fiber concentration.

length-to-width aspect ratios which increase significantly with the volume percent of porosity.[7] Small pores appear spherical, whereas the formation of larger pores appears to follow the fiber orientation. Pore diameters range from 40 to 100 μm, whereas pore lengths can range from 40 μm to > 1 cm for extremely porous specimens. In the extremely porous specimens (> 6 vol. %) the pores often coalesced, resulting in voids which resembled planar delaminations, rather than pores. The porosity in these extreme cases is generally distributed between the plies.

ULTRASONIC MEASUREMENT

The spatial configuration for the angular ultrasonic measurements is shown in Fig. 2. The composite specimen is placed in a submerged fixture which rotates the specimen through an azimuthal angle ϕ about an axis perpendicular to the composite/water boundary. An ultrasonic transducer insonifies the specimen at a polar angle θ measured from the aximuthal rotation axis. For the measurement of ultrasonic backscatter, a single transducer is used in a pulse-echo mode. For pulsed through-transmission measurements, a second transducer is placed on the far side of the composite plate in a position for optimal reception of the through-transmitted pulses. Computerized measurements are performed in which either the specimen or the transducer(s) are stepped through discrete angular or translational positions. At each position, the received signals are digitized, averaged, and stored on a mainframe computer for further analysis.

A typical signal received in a pulse-echo backscatter experiment is shown in Fig. 3. This signal was received in response to the incidence of a pulse

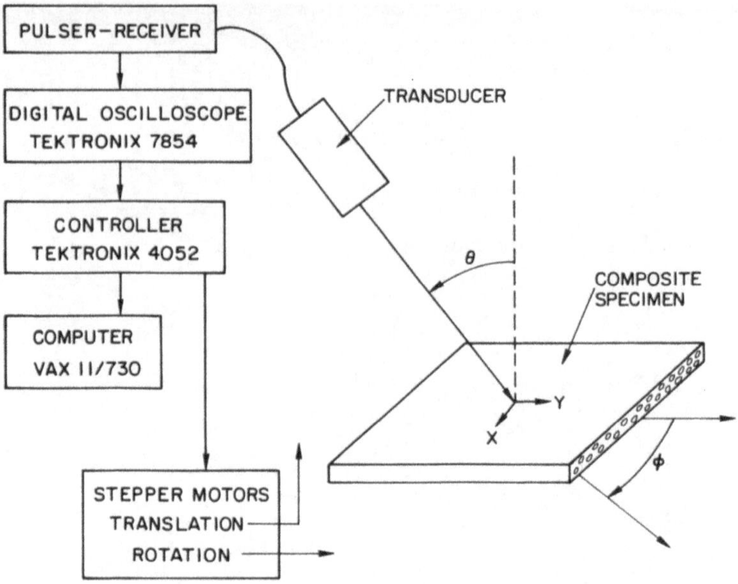

Fig. 2. Schematic of ultrasonic backscatter experiment.

of about 1.5 cycle duration and 10 MHz center frequency, on a nonporous (0.2 vol. % porosity) specimen with θ = 35° and ϕ = 0° (parallel to the fibers). The phase information of such backscattered signals undergoes severe and seemingly random fluctuations as small changes are made in either angular or translational positioning, which is indicative of scattering from a random distribution of scattering centers. As a consequence, the frequency spectrum of the backscattered signals likewise fluctuates in a severe and random manner as small changes are made in positioning. Thus the direct comparison of experimental frequency spectra and time harmonic theoretical calculations

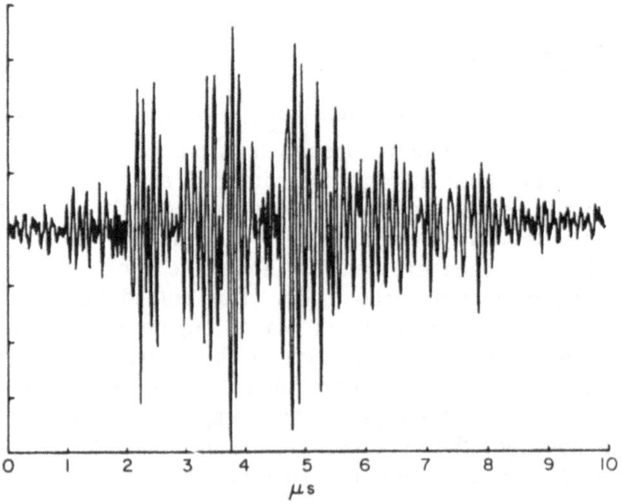

Fig. 3. Ultrasonic backscatter from composite with 0.2 vol. % porosity, 10-MHz transducer at θ = 35°.

is not practical. For that reason the parameter derived from the backscattered signal v(t), such as shown in Fig. 3, which is used for comparison of theory and experiment is the parameter E defined by

$$E \equiv \int_{-\infty}^{\infty} v^2(t)\, dt = \int_{-\infty}^{\infty} v^*(\omega)^2\, d\omega \qquad (1)$$

where $v^*(\omega)$ is the Fourier transform of the backscattered signal v(t). The parameter E is proportional to the total energy carried by the waveform. Comparison of theory with experiment is performed by calculating

$$E_{th} = \int_{-\infty}^{\infty} \|g^*(\omega)f^*(\omega)\|^2 d\omega \qquad (2)$$

where $f^*(\omega)$ is the time-harmonic theoretical calculation, and $g^*(\omega)$ is the spectral response of the ultrasonic pulse-echo system when a suitable reference scatterer is insonified. Since the evaluation of E_{th} involves integration over the bandwidth of the transducer, the choice of reference scatterer is not as critical as when attempting system deconvolution. Reflection from a plane surface is used as the reference in this work.

For a homogeneous spatial distribution of scattering centers and a sufficiently broadband incident pulse, the parameter E should be relatively insensitive to random fluctuations in the positioning of the scattering centers. However, a certain degree of inhomogeneity in the spatial distribution of the scattering centers is generally found in the composite specimens, hence it is desirable to obtain a spatial average of E. The angular measurements of backscatter presented in section V were obtained by spatially averaging measurements of E obtained on a rectangular grid centered about the axis of aximuthal rotation keeping polar and azimuthal angles fixed. The measurements of section V were obtained by averaging E over a 6 x 6 grid covering a 2.4 cm x 2.4 cm area of the composite surface.

Through-transmission measurements presented in section V were obtained by recording the peak amplitude of through-transmitted pulses at discrete angular positions of ϕ and θ. In general, the wave speeds of the threee independent wave modes of the transversely isotropic composite are different. Hence, by using sufficiently short ultrasonic pulses, it is possible to selectively time-gate either the quasi-longtidinal (QL) or the horizontally polarized quasi-transvese (QTH), or vertically polarized quasi-transverse (QTV) wave modes. Sufficient pulse separation was obtained for the 2 mm thick specimens of this study using broadband 10 MHz transducers. Through-transmission data was collected for only the non-porous (0.2 vol. % porosity) specimen, since it is only the transverse isotropic elastic behavior of the fiber reinforced matrix which is of interest in this through-transmission study. Apparent attenuation of the through-transmitted pulses was recorded, but the pulse shape was relatively unaffected by transmission, hence no attempt was made to account for dispersion effects in the peak amplitude measurement.

THEORY

Defect-free fiber reinforced composite laminates contain structural inhomogeneities on several different length scales. Obvious characteristic dimensions which are relatively simple to quantify include the ply thickness, reinforcing fiber diameter, and mean fiber spacing. Structural inhomogeneity due to random nonuniformity in fiber spacing, however, is far more difficult to quantify, and, in general, generates a spectrum of characteristic lengths. A problem faced in modeling ultrasonic scattering in composites is to determine which characteristic lengths and associated structures are dominant in the scattering process.

One approach to the problem would be to determine the scattering functional describing scattering by a single fiber in a matrix, and then incorporate this scattering functional in a self-consistent multiple scattering relation. Expressions for the expectation of the incoherent scattered field (e.g. backscatter) could then, in principle, be asympotically evaluated for long wavelength. The accuracy of this approach would depend on the accuracy of the statistical description of the spatial fiber distribution, and the validity of the assumptions used in evaluating the multiple scattering relation.

An alternative approach is to impose the long wavelength limit prior to the formulation of the problem by assuming that the material properties can be described locally through an "effective modulus" theory. The effective modulus theory replaces the two-constituent composite with an "effective" homogeneous material with elastic properties that approximate those of the composite in an averaged sense. The effective moduli of a fiber reinforced composite are functions of the fiber/matrix volume fraction, hence local perturbations in the fiber/matrix volume fraction as shown in Fig. 1 are readily translated into local perturbations in elastic properties. The long-wavelength scattering model based on effective modulus theory assumes that the received backscatter signal is dominated by inhomogeneities in fiber distribution on a length scale considerably larger than the fiber diameter. In particular, the long-wavelength scattering formulation based on effective modulus theory predicts zero backscatter when the fibers are uniformly spaced.

It is reasonable to assume that the local perturbations in elastic properties due to nonuniform fiber distribution are small. Hence scattering by these local perturbations in elastic properties may be accurately modeled by the Born approximation.

The theoretical results presented in this paper were obtained using a model which combines an effective modulus theory and Born approximation for scattering by a structural abnormality such as shown in Fig. 1b, with a model for scattering by a distribution of elongated flaws. A description of the procedure for the model development is presented in this section. The mathematical details of the model development can be found in references.[8,9]

The effective elastic moduli for a fiber reinforced composite are those of a transversely isotropic solid, for which there are five independent elastic constants. The symmetry axis of the equivalent transversely isotropic continuum coincides with the fiber direction, which is taken to run parallel to the x_1 spatial coordinate. An inhomogeneity in elastic properties, such as represented in Fig. 1b, is assumed to occupy an elongated tube-like volume V. The elastic properties and density within V are denoted by C'_{klmn} and ρ', whereas the elastic properties and density exterior to V are denoted by C_{klmn} and ρ. The signal received due to scattering by this inhomogeneity can be written

$$\delta\Gamma = -(i\omega/4P)\int\left(\varepsilon_{kj}\Delta C_{kjmn}\varepsilon_{mn}{}^{in} - \omega^2\Delta\rho u_k u_k{}^{in}\right)dv \tag{3a}$$

where

$$\Delta\ C_{kjmn} = C_{kjmn} - C'_{kjmn} \tag{3b}$$

$$\Delta\ \rho = \rho - \rho' \tag{3c}$$

$$\varepsilon_{kj} = (u_{k,j} + u_{j,k})/2 \tag{3d}$$

The displacement field in the effective continuum in the absence of the inhomogeneity is represented by $u_k{}^{in}$ The wave motion is assumed time harmonic with circular frequency ω. The parameter P is related to the incident energy of the ultrasonic field. Implementation of the Born approximation consists of replacing the wavefield u_k, ε_{kj} in eq. (1) by the incident field $u_k{}^{in}$, $\varepsilon_{kj}{}^{in}$. The incident field in this case corresponds to the ultrasonic field transmitted through the water/composite interface, and is approximated by

$$u_j{}^{in} = f(x_1,x_2)\sum_\alpha D_\alpha(x_3)T_\alpha d_j{}^\alpha \exp(ik_j{}^\alpha) \tag{4}$$

$$\alpha = QL,\ QTH,\ QTV$$

The coefficient T_α is a plane wave transmission coefficient for transmission from water into wave type α, $d_j{}^\alpha$ is the unit amplitude displacement vector describing the direction of transmitted displacements, $k_j{}^\alpha$ is the wave vector of type α, and D_α is an exponential decay factor describing attentuation in the composite material. The function $f(x_1, x_2)$ describes the amplitude profile of the incident beam, and is assumed in the form

$$f(x_1,x_2) = \left\{\ 1 + b^2(x_1{}^2 + x_2{}^2)\right\}^{-0.5} \tag{5}$$

where the beam width is controlled by the parameter b. The Born evaluation of eq. (3) was carried out for an inhomogeneity with cross-sectional area S in the x_2 - x_3 plane centered at coordinates $x_2 = y_o$, $x_3 = z_o$. The inhomogeneity is assumed infinite in the x_1 dimension. Using eq. (4) in eq. (3) with the Born approximation yields for the received signal due to scattering by the inhomogeneity

$$\delta\Gamma_2 = -(i\omega^3 S\pi/8Pb^2)\sum_\alpha D_\alpha(2z_o)B_\alpha \exp\left(-2k_w(y_o{}^2 + b^{-2})^{0.5}\sin\theta\cos\phi\right)\ x$$

$$\left(2k_w\sin\theta\cos\phi + b/(1 + b^2y_o{}^2)^{0.5}\right)/\left(1 + b^2y_o{}^2\right) \tag{6a}$$

$$B_\alpha = -T_\alpha{}^2\left(\Delta\rho + \Delta C_{klmn}d_m{}^\alpha d_k{}^\alpha d_n{}^\alpha k_l{}^\alpha/\omega^2\right)\exp 2i(k_2{}^\alpha y_o + k_3{}^\alpha z_o) \tag{6b}$$

where θ, ϕ are the incident polar and azimuthal angles.

For the purposes of analysis, the elongated pores found in the composite specimens are approximated by elongated cracks, with negligible dimension in the x_3 direction (perpendicular to the plies). It is assumed that the width of the crack d is much shorter than the wavelength, whereas the length of the crack is much longer than the wavelength. Therefore, scattering by the crack can be modeled by a quasi-static (long wavelength) approximation in the x_2 dimension, and by a Kirchhoff (short wavelength) approximation in the x_1 dimension. In implementing these approximations for scattering by a crack, it is convenient to convert the volume integral of eq. (3) to a surface integral over the crack faces. In the limit where the opposing crack faces mathematically coincide, eq. (3) takes the form

$$\delta\Gamma = (i\omega/4P)\int_{A^+} \sigma_{k\ell}^{in}\Delta u_k^{s} n_\ell dA \tag{7}$$

where A^+ is the insonified crack face, n_ℓ is the inward pointing normal, and Δu_k^s the crack opening displacement, i.e, the difference in displacement of the opposing crack faces. Calculations were carried out for a single crack with length a in the x_1 direction and width d in the x_2 direction, for $d \ll \lambda \ll a$, where λ is the wavelength. The crack is centered at coordinates $x = r_o$, $x_1 = x_o$, $x_2 = y_o$, $x_3 = z_o$. Proceeding with the analysis as described by Qu et. al.,[8] the scattered signal for a single crack is determined as

$$\delta\Gamma = -(i\omega/4P)\sum_{\alpha} [f(x_o,y_o)D_\alpha(z_o)T_\alpha C_{55}\tau^{\alpha}_{j3}]^2 V_j \exp(2ikr_o) \tag{8a}$$

where

$$V_1 = [(\pi^2 ad^2 V_o)/C_{55}]\{1 - (1/4)\varepsilon^2 \ln\varepsilon + \varepsilon^2[*]\} \tag{8b}$$
$$[*] = [(1/12) - N_o/\pi - iE/4 + (\pi/8)(i + \ln 2)]$$

$$V_2 = [(\pi^2 ad^2 V_o)/(C_{22} - C_{23})(1 - \tau^2)]\{1 - A_2\varepsilon^2\ln\varepsilon + \varepsilon^2[B_2 + A_2/4]\} \tag{8c}$$

$$V_3 = [(\pi^2 ad^2 V_o)/(C_{22} - C_{23})(1 - \tau^2)]\{1 - A_3\varepsilon^2\ln\varepsilon + \varepsilon^2[B_3 + A_3/4]\} \tag{8d}$$

also

$$V_o = 2J_1(X)/X \tag{8e}$$

$$X = 2\varepsilon(a/d)p_1^{in}(C_{55}/\rho_s)^{0.5}/c_w \tag{8f}$$

$$\tau = (C_{22} - C_{23})^{0.5}/2C_{33} \tag{8g}$$

$$A_2 = (3 + 2/\tau^4)/8(1 - \tau^2) \quad , \quad A_3 = (3 - 4\tau^2 + 3\tau^4)/8(1 - \tau^2) \tag{8h}$$

$$B_2 = 4/(1 - \tau^2)\{3N_2 - N_o + (\pi/8)[(1 + \ln\tau)/\tau^4 - 1 - \ln 2 + A_2(E - \pi i/2)]\} \tag{8i}$$

$$B_3 = [(5 - 12\tau^2 + 11\tau^4)/(8(1-\tau^2))]\ln 2 - [(2 - 4\tau^2 + 3\tau^4)/(8(1-\tau^2))]\ln\tau$$
$$-(2/\pi)[N_o + (1 + \tau^2)N_2] - (i\pi/2)A_3((2E/\pi) - 1) \tag{8j}$$

$$N_{2n} = \int_0^1 x^{2n} (1 - x^2)^{0.5} \ln x \, dx \; ; \; n = 0,1 \tag{8k}$$

$$E = \text{Euler's constant} \tag{8l}$$

$$\varepsilon = (\rho_s \omega^2 d^2)/c_{55})^{0.5} \tag{8m}$$

Scattering from a distribution of cracks throughout a layer of thickness h oriented parallel to the plies is modeled by coherently summing the contributions from the individual cracks. It is assumed that the volume density of the cracks is sufficiently small so that multiple scattering need not be considered. Following the procedure discussed by Qu et. al.,[8] the summation of the scattered signals from the individual cracks yields the total scattered signal

$$(\delta\Gamma)_{total} = -(i\omega N/4p)\sum_{\alpha} (2\pi/Q_\alpha k_w \sin\theta/b^3\Delta_\alpha)K_1(2k_w\sin\theta/b)(\exp(\Delta_\alpha h)-1] \quad (9a)$$

where K_1 is the modified Bessel function of the second kind, and

$$Q_\alpha = (C_{55}T_\alpha \gamma_{j3}{}^\alpha)^2 V_j \quad (9b)$$

$$\Delta_\alpha = 2ik_3{}^\alpha - 2k_m{}^\alpha \gamma^\alpha S_m{}^\alpha / S_3{}^\alpha \quad (9c)$$

The total scattered field emerging from within the composite material is given by the sum of eqs. (6) and (9).

The approximate form of the incident beam in the composite given by eq. (4) assumes a displacement at the water/composite interface of the form

$$u_3{}^{in}(x_3=0) = f(x_1,x_2)\cos\theta\exp(ik_1x_1 + ik_2x_2) \quad (10)$$

An actual displacement of the form given by eq. (8) would radiate a scattered wavefield in all directions in the water half-space, and, in particular, in the backscattered direction. There is currently some debate as to whether or not this backscattered energy from the water/solid interface represents a physical phenomenon, or is simply the result of the error in approximating the incident beam. The current model includes this signal backscattered from the water/solid interface. The expression for this signal has been derived[8] as

$$\delta\Gamma_1 = -(i\omega/4p)(2i\pi k_w\lambda_w R)2k_w b^{-3}\sin\theta\ K_1(2k_w b^{-1}\sin\theta)\cos\theta \quad (11)$$

where k_w and λ_w are the wavenumber and the compressibility of the water.

The total signal received in the backscatter experiment is therefore modeled as the sum of eqs. (6), (9), and (11). Experimental data do appear to indicate a backscattered signal from the water/solid interface, but it is conceivable that surface and/or subsurface irregularities play a dominant role rather than the beam effect considered here.

COMPARISON OF THEORY AND EXPERIMENT

First we consider the transmission coefficients T_α referred to in eq. (4). These coefficients are evaluated analytically in a straightforward manner by requiring that 1) normal displacements and tractions be continuous across the water/composite boundary, 2) shear tractions vanish on the boundary surface, and 3) reflected and transmitted energy propagates away from the boundary while evanescent field displacements decay away from the boundary. The validity of the transverse isotropy assumption is examined by comparing the calculated transmission coefficients to ultrasonic through-transmission measurements described in "ULTRASONIC MEASUREMENT". It is noted that the through-transmission experiment measures the relative change in amplitude of

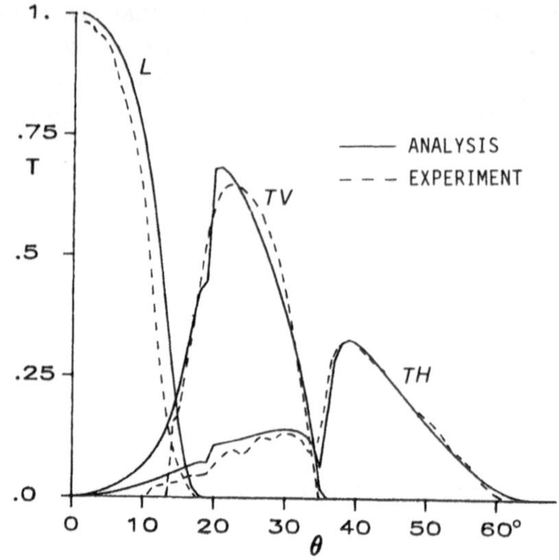

Fig. 4. Transmission coefficients for $\theta_w = 35°$, $\phi = 60°$.

an ultrasonic pulse which is transmitted through two water/composite boundaries (i.e., into and out of the specimen). Hence the comparison of theory with experiment requires the calculation of two coefficients T_α^{in} and T_α^{out}, for each wave type. It can be shown by use of a reciprocity relation that T_α^{out} is related to T_α^{in} through a multiplicative constant which is independent of incident angle. The quantity compared with experiment is then

$$M_\alpha = T_\alpha^{in} D_\alpha(h) T_\alpha^{out} \tag{12}$$

where $D_\alpha(h)$ is the attenuation factor associated with a specimen of thickness h. A comparison of $M_\alpha(\theta,\phi)$ with experiment is presented in Fig. 4. The experiment measured the peak amplitude of through-transmitted 10 MHz broadband pulses as a function of polar angle θ for fixed azimuthal angle ϕ. The attenuation factor $D_{QL}(h)$ was established through the alignment of transmitted amplitudes in theory and experiment at $\theta = 0°$. Estimates at D_{QTH} and D_{QTV} were then obtained. Very good agreement is observed between theory and experiment for all three wave types.

Consideration is now given to the polar backscatter experiment. The experimental data presented here was obtained using a 5-MHz broadband transducer in a pulse-echo configuration with a fixed polar angle of $\theta = 35°$. Spatially averaged values of $E(\phi)$ were measured at discrete 1° increments in azimuthal angle. The aximuthal scans were carried out on each of four unidirectionally reinforced composite specimens containing 0.2, 1.1, 2.0, and 6.5 vol. % porosity, respectively. Results of these scans are presented in Fig. 5. In Fig. 5, $\phi = 0°$ corresponds to incidence parallel to the fibers. Since the angular variation in backscatter amplitudes covers a wide dynamic range, the data is presented as a decibel plot in which the value plotted is

$$P(\phi) = 10 \log_{10}[E(\phi)/E_o(90)] \tag{13}$$

where $E_o(90)$ is the backscatter measurement for the nonporous (0.2 vol. % porosity) specimen with $\phi = 90°$ (incidence perpendicular to the fibers). As

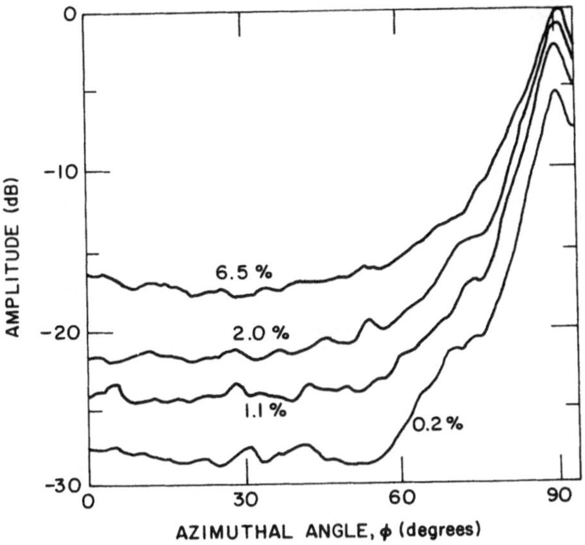

Fig. 5. Experimental results for polar backscatter as a function of azimuthal angle for $\theta_w = 35°$, 5 MHZ center frequency and various values of the pore-volume fraction.

previously reported,[1-6] a scattering peak is observed at $\phi = 90°$. In the context of the present theory, this peak is explained by observing that the internal structure of the composite is considerably more inhomogeneous in the plane perpendicular to the fibers. The introduction of porosity is seen to increase the backscatter at all azimuthal angles. It is interesting to note that the decibel increase in backscatter is rather uniform over azimuthal angle ϕ, especially when comparing the 0.2, 1.1, and 2.0 vol. % porosity specimens. This implies that the angular scattering characteristics of the added porosity resembles the angular scattering characteristics of the defect-free composite. This observation is consistent with the cylindrical morphology of the porosity in these specimens, i.e. scattering from cylindrical porosity should have angular characteristics similar to scattering by the fiber-related structures.

For the purposes of porosity estimation, the availability of scattering data such as Fig. 5 for a particular material would make conceivable the estimation of low levels of porosity (<2 or 3 vol. %) to within 1/2%.

Theoretical predictions of backscatter corresponding to the experiment of Fig. 5 are presented in Fig. 6. Calculations were performed for a fiber-related inhomogeneity having $\Delta C_{ijkl}/C_{ijkl} = 0.01$, $\Delta\rho/\rho = 0.01$, and $S = 0.314\ mm^2$. Porosity was modeled by cracks having $a = 0.6$ mm and $b = 0.4$ mm. Porosity volume was related to the crack size by assuming the crack faces were separated by 0.02 mm. Volume densities of cracks were prescribed as 1.04, 5.94, 10.63, and 34.01 cracks/mm^3, which corresponds to 0.2, 1.14, 2.04, and 6.53 vol. % porosity. The beam width parameter was taken as $b = 0.22\ mm^{-1}$. Using these parameters, the total backscattered signal $\delta\Gamma(\omega)$, given by the summation of Eqs. (6), (9), and (11), is calculated over a range of frequencies ω. The parameter E_{th} for comparison with experiment is calculated according to eq. (4), where $f(\omega) = \delta\Gamma(\omega)$, and $g(\omega)$ is assumed to

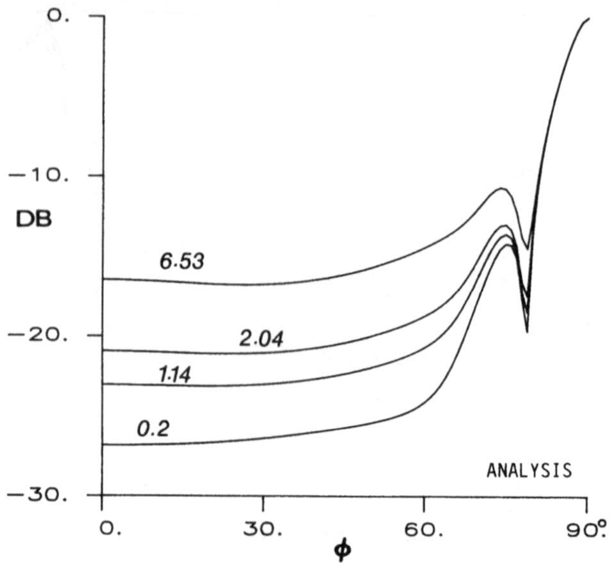

Fig. 6. Analytical results for polar backscatter as a function of azimuthal angle for $\theta_w = 35°$, 5 MHZ center frequency and various values of the pore-volume fraction.

be the Gaussian

$$\hat{g}(\omega) = \exp\left[-(\omega-\omega_o)^2/2\sigma^2\right] \tag{14a}$$

$$\omega_o = 2\pi\ 5\text{MHz} \tag{14b}$$

$$\sigma = 0.5\ \text{rads/s} \tag{14c}$$

As with Fig. 5, the value plotted in Fig. 6 is

$$P(\phi) = 10 \log_{10} \left[E_{th}(\phi)/E_{th:o}(90°)\right] \tag{15}$$

where $E_{th:o}(90)$ is the backscatter calculation at $\phi = 90°$ in the case of zero porosity.

The general agreement between theory (Fig. 6) and experiment (Fig. 5) is quite good. The peak in scattering at $\phi = 90°$ due to the theoretical fiber-related inhomogeneity closely resembles the corresponding peak seen in Fig. 5. The increases in backscatter at azimuthal angles $\phi < 45°$ due to the introduction of porosity appear to be of quite comparable magnitude to those seen in the experiment. A discrepancy between theory and experiment is noted at $\phi = 90°$, however, with regard to the introduction of porosity. The theory shows all four curves meeting at $\phi = 90°$, which implies that backscatter is being totally dominated by scattering from the fiber-related inhomogeneity. This discrepancy with experiment may conceivably be explained by the fact that the theory assumes an inhomogeneity of infinite extent in the direction of the fibers, thus overemphasizing the directionality of the angular scattering behavior. It may be more appropriate to consider a distribution of finite length inhomogeneities, as was done for the distributed cracks. This problem is currently being examined.

The theoretical curves of Fig. 6 display sharp minima at ~80°, whereas the experimental curves display only a mild break in curvature at the corresponding azimuthal orientation. Such discrepancies are common when comparing theoretical results which assume plane-wave incidence with experimental results employing incident beams with relatively broad angular spectra. This discrepancy between theory and experiment can be lessened by assuming a finite width incident angular spectrum, rather than plane-wave incidence, which will effectively average the theoretical curves over aximuthal angle. Future work will consider the implementation of this feature in the theoretical model.

ACKNOWLEDGEMENT

This work was sponsored by the Center for Advanced Nondestructive Evaluation for the Air Force Wright Patterson Aeronautical/Materials Laboratory, under subcontract SC-87-121 (J. Qu and J. D. Achenbach) and subcontract W-7405-ENG-82 (R. A. Roberts) with Ames Laboratory.

REFERENCES

1. Y. Bar-Cohen and R. L. Crane, Backscattering Imaging of Subcritical Flaws in Composites, Matls Eval. 40:970 (1982).
2. M. Azimi and A. C. Kak, On the Estimation of Porosity in Composites by Oblique Angle Illumination and Normal Reception, in: "Review of Progress in Quantitative Nondestructive Evaluation, Vol. 3b," D. O. Thompson, ed., Plenum Press, New York (1984).
3. E. D. Blodgett, L. J. Thomas, II, and J. G. Miller, Effects of Porosity on Polar Backscatter from Fiber Reinforced Composites, in: "Review of Progress in Quantitative Nondestructive Evaluation, Vol. 5b," D. O. Thompson and D. E. Chimenti, eds., Plenum Press, New York (1986).
4. D. E. Yuhas, C. L. Vorres, and R. A. Roberts, Variations in Ultrasonic Backscatter Attributable to Porosity, in: "Review of Progress in Quantitative Nondestructive Evaluation, Vol. 5b," D. O. Thompson and D. E. Chimenti, eds., Plenum Press, New York (1986).
5. W. J. Murri, D. W. Sesmon, and L. H. Pearson,Ultrasonic Backscatter Studies of Impact Damage in Graphite/Epoxy Composite Laminate Materials, in: "Proceedings 15th Symposium on Nondestructive Evaluation," NTIAC, San Antonio (1985).
6. R. A. Robert, Porosity Characterization in Fiber Reinforced Composites by Use of Ultrasonic Backscatter, in: "Review of Progress in Quantitative Nondestructive Evaluation, Vol. 6," D. O. Thompson and D. E. Chimenti, eds., Plenum Press, New York (1987).
7. D. K. Hsu and K. M. Uhl, A Morphological Study of Porosity Defects in Graphite/Epoxy Compatites, in: "Review of Progress in Quantitative Nondestructive Evaluation, Vol. 6," D. O. Thompson and D. E. Chimenti, eds., Plenum Press, New York (1987).
8. J. Qu, J. D. Achenbach, and R. A. Roberts, Theoretical and Experimental Analysis of Backscatter in Composites, to be submitted for publication.

9. J. Qu and J. D. Achenbach, Analytical Treatment of Polar Backscattering from Porous Composites, in: "Review of Progress in Quantitative Nondestructive Evaluation, Vol. 6," D. O. Thompson and D. E. Chimenti, eds., Plenum Press, New York (1986).

STATISTICAL EVALUATION OF QUALITY IN COMPOSITES USING THE STRESS WAVE FACTOR TECHNIQUE

A.Madhav

Engineering Science and Mechanics

J.A.Nachlas

Industrial Engineering and Operations Research
Virginia Polytechnic Institute and State University
Blacksburg, Virginia USA

ABSTRACT

The growth in the extent of applications of composite materials, particularly in commercial products, has been dramatic and carries an implied mandate for effective methods for material quality evaluation. The cost of composite materials dictates that nondestructive test methods be used. At the same time, the nature of composites limits the use of conventional techniques such as radiography, eddy-current or ultrasonics. Recently, a new technique known as the Stress Wave Factor (SWF) technique, has been developed and appears to hold promise as a method for the evaluation of composite material quality.

Implementation of the SWF method is examined using the zeroth moment method suggested by Talreja et. al.[6] The behavior of the response to specimens of known quality is investigated statistically. It is found that the transformed/actual readings follow a Beta distribution and that specimens of different quality are readily distinguishable using the statistical analysis of the SWF response. Reasonable future steps for translating these findings into efficient quality evaluation methods are suggested.

INTRODUCTION

Composites are materials of the future. The wide usage of these materials for various applications at present reflects the truth of the statement. The extensive use of composites for space, military and commercial applications carries a mandate for the use of existing or new methods for assuring reliability and quality control. Nondestructive Testing methods are being used for quality control in many industries. However due to the nature of composites, there are many limitations to the use of conventional nondestructive testing techniques such as ultrasonics and radiography. A new technique known as the Stress Wave Factor (SWF) technique, which has been developed recently, may prove to be an effective method for nondestructive evaluation of composite materials.

The specific SWF quantification procedure used in this study was developed at Virginia Tech and is known as the zeroth moment method. In this technique, the material is quantified and qualified in terms of numbers known as "Stress Wave Factors." These numbers can serve as a quality index of a particular material. Comparisons of the values of SWF at different points in the same specimen or with that of different specimens yields the basis of discrimination between the two specimens.

In this study a statistical approach to discriminate between specimens of the same material using the SWF technique is explored. The values obtained are modified using a linear relation. A statistical distribution is assumed and various parameters relating to this distribution calculated. A Beta distribution is assumed for this purpose and the validity of this model is investigated. Further, the feasibility of using the SWF technique combined with statistical methods for quality control of composites is investigated.

THE TECHNIQUE

Nondestructive testing is being used as a powerful technique for defect detection and evaluation of composite materials. Of the various methods, the SWF technique has significant promise in this application. A very important benefit of this technique is the fact that measurement of SWF yields a parameter which one might be able to use as a quantitative indicator of the mechanical quality of the material.

The SWF technique was developed by Vary et. al.[1] and is also known as the "conventional" method. A schematic diagram of the setup of this technique is as shown in Fig. 1.

In this method, stress (ultrasonic) waves are input to the material using an ultrasonic pulser with the help of a piezoelectric transducer known as the sending (transmitting) transducer. A receiving transducer placed on the same side of the specimen detects these stress waves which are measured by conventional acoustic emission instrumentation. The SWF is defined[3] as the product of the number of times the voltage level of a single signal exceeds a set threshold level, the pulse repetition rate and the length of the predetermined measurement time interval. The schematic describing this is shown in Fig. 2. The SWF measure is given by:

$$E = NGR \qquad (1)$$

where

N = Number of threshold crossings contained in each burst. (Quantity)
G = Predetermined time interval. (Time)
R = Pulse repetition rate. (Quantity / Time)

Note that the E is unitless.

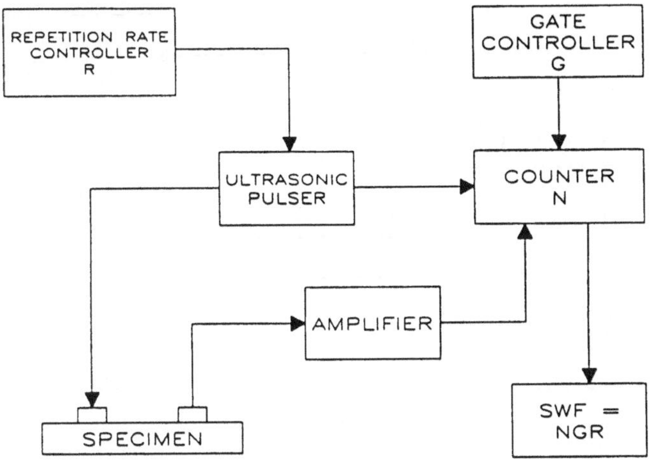

Fig. 1. Schematic of the experimental setup for measuring SWF suggested by Vary.

The value of the SWF is not unique for each material and structural configuration but is dependent on transducer characteristics, mounting pressure, type of coupling agent used between transducer and specimen, distance between transducers and the propagation characteristics of the material. Thus this method is instrumentation dependent and repeatability may be affected.

An alternate technique was suggested by Govada et. al.[2] This differs from the conventional method in the way SWF is defined. The method is very similar to the conventional method, in that ultrasonic waves are passed through the material with the help of a sending transducer and the resulting stress waves are detected by a receiving transducer. The two transducers are mounted on the same side of the specimen and the distance between them held constant. The various parameters like mounting pressure on the transducers, coupling agent etc. are kept constant. The schematic of the experimental setup is as shown in the Fig. 3. Spectral analysis is then performed on the received signal. Talreja et.al.[6] have noted three parameters which are necessary to evaluate the material. They are the location, shape and scale parameters which represent the frequency spectrum and can be defined by considering the spectrum to be a plane figure that is closed on the frequency axis. The shape parameter is of

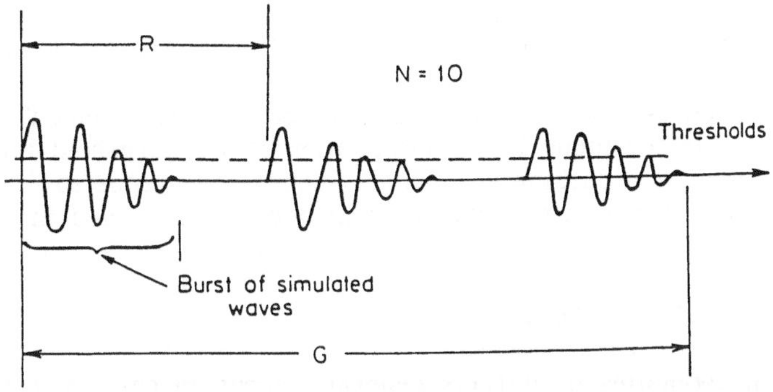

Fig. 2. Schematic showing the various parameters used to calculate SWF.

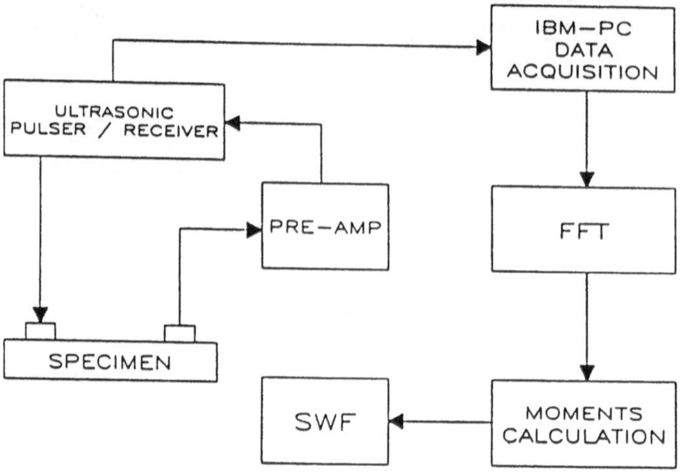

Fig. 3. Schematic of the experimental setup used in this study.

particular interest and can be evaluated from the frequency spectrum, by taking various moments of the frequency spectrum about the amplitude axis. Thus one can take the formula for area under the curve and multiply it by frequency raised to the nth power, inside the integral. Thus the nth moment can be defined by

$$M_n = \int_0^{f_m} s(f)(f^n)df \qquad n = 0,1,2, \ldots \tag{2}$$

where s(f) is power spectral density, f the frequency. By taking n=0, one can get the zeroth moment which has been used as an alternate measure of SWF. The AUF is defined as the root mean square value of the power spectral density or square root of the zeroth moment. That is AUF is given by

$$AUF = (M_o)^{0.5} \tag{3}$$

where

$$M_o = \int_0^{f_m} s(f)df \tag{4}$$

and is the area under the frequency spectrum curve (zeroth moment), Fig. 4.

This method is also termed as the "zeroth moment method". A major advantage of this technique is the elimination of the threshold level and several other instrumentation parameters. This ensures the repeatability of AUF values to a very large extent and thus this technique is used in this study.

STATISTICAL APPROACH

Material anomalies or defects generally occur in composite materials at random. They are caused by variation of feed stock materials, the production control parameters and the chemical behavior of the composite constituents.

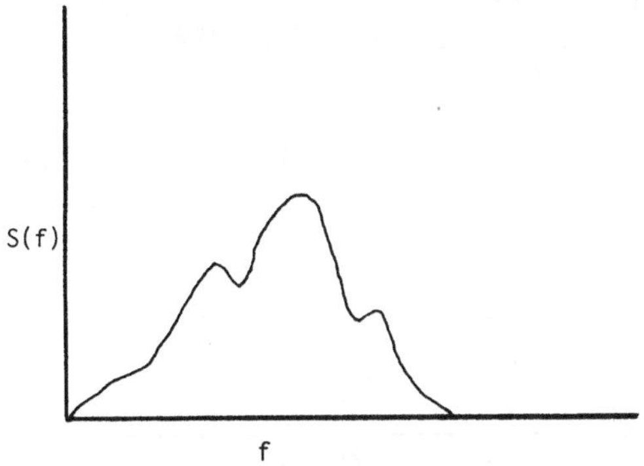

Fig. 4. Idealized power spectral density distribution

Composite structure performance depends upon the volume and distribution of defects in the material. It is suggested here that statistical methods may be used to infer material quality from the AUF measure of defect distribution in a composite material. There are many statistical methods that may be used to characterize the behavior of sample data and to support inference about the process from which the data is obtained. Of the many methods, two that are meaningful for the present application are:

1. Method of moments.

2. Method of maximum likelihood.

It is hypothesized that AUF readings conform to a Beta distribution. Data readings are taken using known materials and the mean and the variance calculated for the observed data. The above methods are used to determine parameters of a Beta distribution. The reasons for assuming the Beta distribution are as follows:

1. A Beta distribution is applicable for continuous variables.

2. It is a two parameter distribution, hence yielding great flexibility in data fitting.

3. The AUF variables lie between the values zero and one.

4. For a large subset of conceivable parameter values, the Beta distribution is moderately asymmetric and skewed.

The method of moments and method of maximum likelihood are used to estimate the parameters of the Beta distribution. Then the Kolmogorov-Smirnov goodness-of-fit test is used to evaluate the fit of the Beta distribution to the AUF data.

Beta Distribution

A random variable X is said to be distributed as the Beta distribution if the density function is given by

$$f(x) = \frac{\Gamma(\alpha + \beta)}{\Gamma(\alpha)\Gamma(\beta)} x^{(\alpha-1)}(1-x)^{(\beta-1)} \quad (5)$$

where α and β are parameters with $\alpha > -1$, $\beta > -1$. The mean and the variance are related to α and β as follows

$$\text{Mean} = \mu = \frac{\alpha}{\alpha + \beta} \quad (6)$$

$$\text{variance} = \sigma^2 = \frac{(\alpha\beta)}{(\alpha + \beta)^2(\alpha + \beta + 1)} \quad (7)$$

There are two methods for the estimation of α and β as described in the earlier section. They are described briefly here:

Method of Moments

The mean and the variance are the most commonly used descriptors of the random variable. These are related to the parameters of the distribution. Thus on the basis of the relationships between moments of a distribution, it follows that parameters of the distribution may be determined by first estimating the mean and variance of the random variable. In the Beta distribution, the parameters α and β are computed by inverting eqs. (6) and (7). Thus,

$$\alpha = \mu^2 \frac{(1 - \mu)}{\sigma^2} - \mu \quad (8)$$

$$\beta = \frac{\alpha(1 - \mu)}{\mu} \quad (9)$$

This method provides a procedure for deriving the point estimator of the parameter directly. Consider a random variable X with density function $f(x, \theta)$, in which θ is the parameter. The question is, among the possible values of θ what is the value that will maximize the likelihood of obtaining the set of observations? Such is the rationale underlying the maximum likelihood method of point estimation.

The likelihood of obtaining a particular sample value x_i can be assumed to proportional to the value of the probability density function estimated at x_i. Then assuming random sampling, the likelihood of obtaining n independent observations $x_1 ....x_n$ is $L(x_i, ...,x_n;\theta) = f(x_1;\theta)f(x_2;\theta)...f(x_n;\theta)$ which is the likelihood function. The maximum likelihood estimator is then the value of θ that maximizes the likelihood function. Due to the multiplicative nature of the likelihood function, it is often more convenient to maximize the logarithm of the likelihood function.

In our case, which has two parameters, the above rationale can be extended to find the likelihood function. The final forms of the equations from which the parameters α and β are calculated are shown below

$$\psi(\alpha + \beta) - \psi(\alpha) = \frac{-1}{n} \sum_{i=1}^{n} \ln(x_i) \quad (10)$$

$$\psi(\alpha + \beta) - \psi(\beta) = \frac{-1}{n} \sum_{i=1}^{n} \ln(1 - x_i) \quad (11)$$

where

$$\psi(z) = \frac{1}{\Gamma} \frac{d}{dz} \Gamma(z) \quad (12)$$

is an integral equation that has been analyzed. It has been shown that

$$\psi(z) = \psi(1 - z) + \frac{1}{(z - 1)} \quad (13)$$

and using this result, the ψ functions can be evaluated.

The maximum likelihood estimator is better than that of the method of moments, particularly for a large sample size n, since it has minimum variance (asymmetrically). Further, the former derives the estimator directly and the latter evaluates a parameter by first estimating the moments (mean and variance). Both methods have been used in our case.

Goodness-of-Fit Test

A prior knowledge or belief plays an important role in the application of statistical reasoning to experimental observations. Sometimes we may be unsure whether or not a particular probability distribution, or a particular form of probability distribution, which we believe to be appropriate for our observations, really is appropriate. There are certain procedures by which we may test this belief and these procedures tell us how well our data fit the probability distribution in question by comparing the observed and the expected frequencies of occurrence of a series of consecutive sets of possible values. Two such tests commonly used are the Kolmogorov-Smirnov (K-S) and Chi-Squared goodness-of-fit tests. A K-S test is performed in our case and is described briefly:

Kolmogorov-Smirnov Test. This is one of the most widely used goodness-of-fit tests. In this method, the experimental cumulative frequency and the assumed theoretical distribution are compared. If there is a discrepancy larger than normally expected for a given sample size, the theoretical model is rejected.

A brief outline of the K-S test follows: For a sample size n, the set of observed data is arranged in an ascending order. From the above ordered sampled data, a step wise cumulative frequency function is developed as follows:

$$S_n(x) = \begin{cases} 0 & x < x_1 \\ k/n & x_k \leq x \leq x_{(k+1)} \\ 1 & x \geq x_n \end{cases} \quad (14)$$

where $x_1, x_2, \ldots x_n$ are the values of the ordered sample data and n is the sample size. If $F(x)$ is the proposed theoretical distribution function, then maximum difference between $S_n(x)$ and $F(x)$ over the entire range of X is the measure of the discrepancy between the observed data and the theoretical model. If the maximum discrepancy is denoted by D_n then

$$D_n = \max \|F(x) - S_n(x)\| \tag{15}$$

For a specified significance level s, the K-S test compares D_n with the critical value $D^s{}_n$, which is defined by

$$P(D_n \leq D^s{}_n) = (1 - s) \tag{16}$$

The critical values for $D^s{}_n$ at various significance levels and for various values of n can be obtained from the statistical tables. If $D_n < D^s{}_n$, then the proposed distribution is acceptable at the specified significance level.

EXPERIMENTAL DETAILS

A schematic of the experimental setup is as shown in Fig. 3. Two transducers, one acting as a transmitter and the other as a receiver are mounted on the same side of the specimen, the distance between them being one inch center to center and is held constant by a plexiglass fixture. The transducers are matched broadband, with 5 MHz center frequency and half inch diameter (Panametrics Model V109). A Panametrics Model 5052A ultrasonic analyzer is used as the pulser/receiver. A stress wave is input to the material using a pulsing/receiving unit via the sending transducer (transmitter). An ultrasonic couplant (Ultragel, Echo Laboratories) is used to reduce an impedance mismatch, and thus insures good contact between transducer and specimen. The stress wave, after passing through the material, is detected by a receiving transducer. This signal is then amplified using a preamplifier (Panametrics Model 5660 ultrasonic preamp) and directed to the pulsing/receiving unit. The amplified signal is then digitized by means of an A/D converter for further processing. The signal is digitized using a PCDAS data acquisition board[4] and software, and recorded on an IBM PC. This board performs data capture, storage and display of the signal by converting the screen to a virtual oscilloscope. A program[5] is then used for spectral analysis. This program is capable of performing Fast Fourier Transform (FFT) and uses this result to calculate the area under the frequency spectrum which gives the value of AUF.

The tests were performed on graphite-epoxy and two carbon-carbon composites. The carbon-carbon composites are 3" X 3" and one of the specimens has an additional coating of carbon on one of the sides. This is sufficient to alter the AUF values and the two specimens were distinguishable from one another. The other specimen is an unidirectional graphite-epoxy $[0_6]$ laminate measuring 12" X 12". The specimen is divided into four quadrants and simulated flaws are introduced in three quadrants. The severity of flaws is varied in each quadrant. A one-inch crack is introduced in the first, a one-inch square delamination in the second, a half-inch crack and a half-inch square delamination in the third, and no flaw is introduced in the fourth. The purpose of this specimen is to check distribution of AUF in each quadrant and to observe any particular pattern it follows. Thus a number of AUF values are recorded in each case.

Statistical Tests

The AUF values obtained may be modified linearly by the following equation:

$$AUF_m = (AUF_a - C)N \tag{17}$$

where AUF_m is the modified AUF value, AUF_a is the actual AUF value, C and N are arbitrary constants. Note that this is done to keep the α and β values within a certain range. The modified AUF values (X) which are obtained are used to find the parameters α and β of the Beta distribution. Both the methods of parameter estimation (method of moments, method of maximum likelihood) are employed. The various percentage points for Beta distribution are obtained using an IMSL[7] subroutine MDBETI. This subroutine is used to find the cumulative distribution frequency F(X). A K-S goodness-of-fit test is then performed on the data and the validity of the assumption of the distribution is checked.

RESULTS AND DISCUSSION

The results of the statistical analysis of the observed AUF responses, are shown in Table 1. Note that the values of D_n are less than D_{crit} at a significance level of 0.2. Thus the Kolmogorov-Smirnov test indicates that the Beta distribution provides a very close fit to the observed data. Therefore, it appears that the Beta distribution is a representative model of the AUF response. In addition, the particular Beta distribution that models the response of the data from different types of specimen/regions is clearly distinct. This is shown in Fig. 5 and Fig. 6. Thus, the quality of the materials and their differences are effectively portrayed by the Beta models.

It is suggested that the results obtained point to the development of an efficient quality evaluation method for commercially produced composite

Table 1. Statistical Analysis of the Observed AUF Responses

	Mean (μ)(modified)	Variance $\sigma^2 10^{-2}$	Alpha α	Beta β	D_n	D_{crit}†
Carbon						
Uncoated	0.3139	1.0340	11.56	15.16	0.0454	0.2281
Coated	0.4325	0.8850	16.60	25.90	0.0773	0.2281
GREP-1						
Region 1	0.4211	0.5568	18.02	24.76	0.0750	0.1691
Region 2	0.4097	0.4783	20.31	29.25	0.1000	0.1691
Region 3	0.3905	0.5472	16.60	25.90	0.0750	0.1691
Region 4	0.3627	0.4761	17.25	30.30	0.1000	0.1691

† Value taken from statistical tables for significance level of 0.2 and number of AUF observations n.

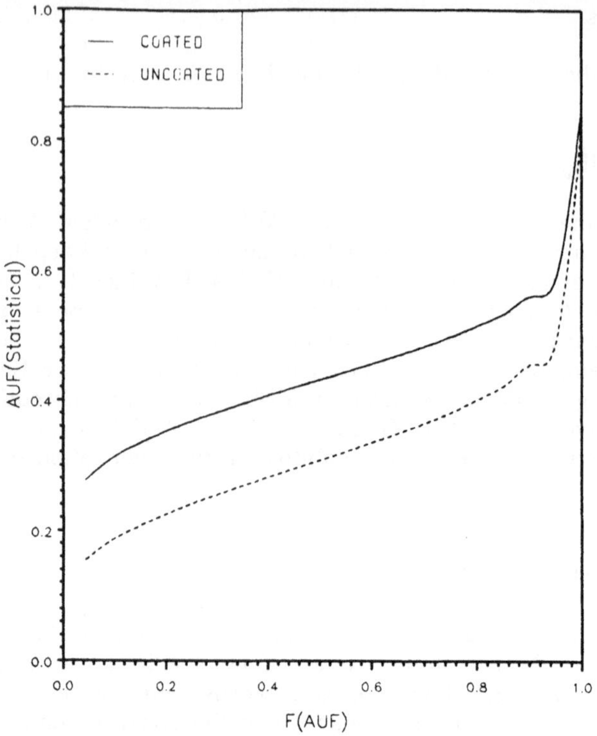

Fig. 5. Modified AUF versus Cumulative Distribution Frequency F(AUF)

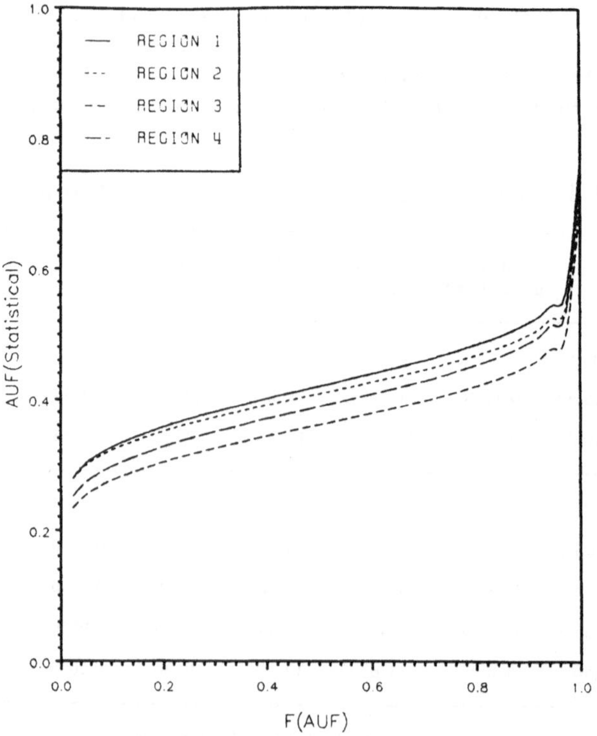

Fig. 6. Modified AUF versus Cumulative Distribution Frequency F(AUF)

materials. Minor modifications of the experimental apparatus will assure the repeatability of the AUF responses. The design of dedicated software for transforming the AUF readings into quality data will make the method more efficient. Further testing is required to identify the Beta distributions that represent desirable material quality. Finally, the definition of statistical hypothesis tests for the desired levels of quality discrimination will complete the construction of the method. These efforts should yield an efficient and dependable quality surveillance tool.

ACKNOWLEDGEMENT

The authors wish to thank Professors J. C. Duke, Jr. and E. G. Henneke, II for their suggestions in conducting the experiments. The help from M. T. Kiernan and S. W. Bartlett is gratefully acknowledged.

REFERENCES

1. A. Vary and K.J. Bowles, " Ultrasonic evaluation of the strength of unidirectional graphite/polyimide composites, X-73646," NASA, Cleveland (1977).
2. A. K. Govada, J. C. Duke, Jr., E. G. Henneke II, and W. W. Stinchcomb, " A Study of the Stress Wave Factor Technique for the Characterization of Composite Materials, CR-174870," NASA, Cleveland (1985).
3. C. R. L. Murthy, M.N.R. Rao, A. K. Rao, A. Madhav, Kishore, and B. Dattaguru, Nondestructive Evaluation of Delaminations in GFRP composites by Acousto-Ultrasonic Technique, in: "Proceedings World Conference on Composites," Beijing (1986).
4. PC-DAS User's Manual, General Research Corporation, McLean, Va., (1985).
5. M. T. Kiernan, "An Acousto-Ultrasonic System for the Evaluation of Composite Materials," Masters Thesis, College of Engineering, Virginia Polytechnic Institute and State University, Blacksburg (1986).
6. R. Talreja, A. Govada, and E. G. Henneke II, Quantitative Assessment of Damage Growth in Graphite Epoxy Laminates by Acousto-Ultrasonic Measurements, in: Review of Progress in Quantitative Nondestructive Evaluation, Vol. 3B, D. O. Thompson and C. E. Chimenti, eds., Plenum Press, New York (1984).
7. International Mathematics and Statistics Library
8. J. R. Green, and D. Margerison, "Statistical Treatment of Experimental Data," Elsevier Scientific Publishing Company, New York (1978).
9. "Handbook of Tables for Probability and Statistics," W. H. Beyer, ed., The Chemical Rubber Company, Ohio, (1968).
10. A. H. S. Ang and W. H. Tang, "Probability Concepts in Engineering Planning and Design, Vol.1," John Wiley and Sons, New York (1976).

APPLICATION OF ACOUSTO-ULTRASONICS TO QUALITY CONTROL AND DAMAGE ASSESSMENT OF COMPOSITES

Ramesh Talreja

Department of Solid Mechanics
The Technical University of Denmark
2800 Lyngby, Denmark

ABSTRACT

The paper deals with application of acousto-ultrasonics to two areas of composites engineering: quality control of manufacturing composite components and assessment of damage incurred due to service loads.

The basis of analysis presented is a set of factors, called the stress wave factors (SWFs), proposed earlier by the author. The technique developed to obtain these factors is described and its application is illustrated on two specific examples: 1) quantitative evaluation of the quality of fabrication of ring-shaped, filament-wound specimens of graphite-epoxy and 2) quantitative assessment of the extent of impact damage in a graphite-epoxy laminate.

INTRODUCTION

The acousto-ultrasonic technique proposed by Vary and Bowles[1,2] for quantitative nondestructive evaluation of composite materials strength is based on measurement of a certain property of the ultrasonic stress wave which has propagated a certain distance in the material and is received by a probe in contact with the specimen surface. The stress wave property is given by a number, called the stress wave factor (SWF), which is a measure of the rate at which the stress wave exceeds a preset (but arbitrarily chosen) threshold level. The factor has been interpreted as a measure of the efficiency of the stress wave energy transmission[3] on the basis of a conceptual analogy to acoustic emission from microstructural inhomogenities. The factor has been shown to give empirical correlation with the composite materials strength properties[1,4].

Talreja et al.[5] studied correlation between acousto-ultrasonic measurements and damage development in graphite-epoxy laminates. The measurements were quantified by using a frequency analysis of the stress wave signals and by applying a random signal analysis method based on power spectral moments[6]. The parameters provided an alternative way of analyzing the acousto-ultrasonic measurements for quantitative nondestructive evaluation of composites. In particular the measurements showed that one of the parameters (the RMS value of the stress wave) correlated with the laminate stiffness. Since laminate stiffness is related to the internal state of damage in composites[7,8], the observed correlation provided a basis for quantitative assessment of damage in composites.

The present paper reports further advance in application of the acousto-ultrasonic technique along the lines initiated in Ref. 5. Firstly, results are reported of a systematic investigation of the variation of the ultrasonic stress wave characteristics with the fiber angle in a unidirectional composite. The correlation between the RMS value of the stress wave and the in-plane elastic modulus of the composite is demonstrated. Results are then reported of two investigations aimed at applying the technique to quality control of fabricated composite components and to assessment of damage in composite laminates. Both investigations demonstrate successful application of the technique and point to the high potential of the technique for quantitative nondestructive evaluation.

THE APPARATUS

Figure 1 shows a schematic diagram of the apparatus for the acousto-ultrasonic measurements. A typical measurement is made by placing two ultrasonic transducers on one face of a specimen, sending a pulse into the specimen by one transducer and by receiving the transmitted stress wave by the other transducer. Both transducers are coupled to the specimen surface by a coupling fluid and are held in a fixture so that the transducer centers are maintained a fixed distance apart.

The transducer used for the measurements were of broad band (0.8 - 10 MHz) type with the center frequency at 5 MHz and the contact surfaces of the transducers were of 12.5 mm diameter. The received signals were amplified, digitized and stored in a transient recorder. The data was then transformed to the storage of a microcomputer for further processing. A set of menu-type programs was used to do gating, to calculate power spectral density (by using a Fast Fourier Transform algorithm) and to compare the set of SWFs (defined below). A typical measurement was reduced to plots of the gated signal and the power spectrum and a table of the SWF values.

THE SWF SET

As pointed out earlier, Vary's SWF represents the frequency at which the received stress wave exceeds a fixed threshold value. The basis for using this stress wave characteristic is the contention that the transmitted stress wave is like a simulated acoustic emission from microstructural entities. We have, however, chosen to look at the stress wave as a signal of random nature. This broader look suggests that all wave characteristics may reflect microstructural properties. A convenient way to obtain integrated average measures of a random signal (of stationary type) is to determine these from the power spectral density distribution of the signal. A particular way was suggested by the present author in a previous work on random load fatigue[6].

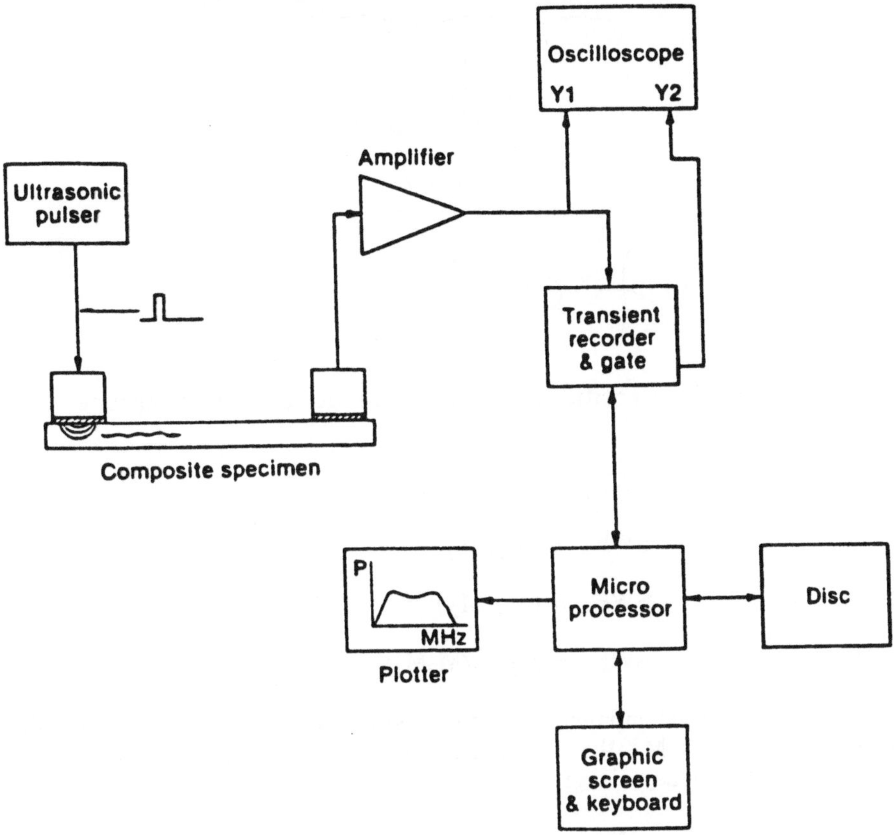

Fig. 1. A schematic diagram of the acousto-ultrasonic technique.

The concept used was that a distribution in a two-dimensional space may be characterized by a location parameter, a scale parameter and a set of shape parameters. For a power spectral density distribution the parameters were expressed in terms of the spectral moments. We use the procedure to define a set of 5 parameters and name them as stress wave factors, given below.

$$SWF_1 = M_0$$

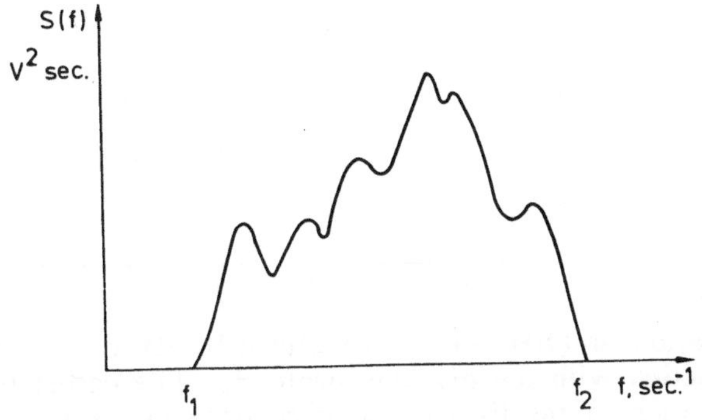

Fig. 2. A power spectral density diagram of a stress wave.

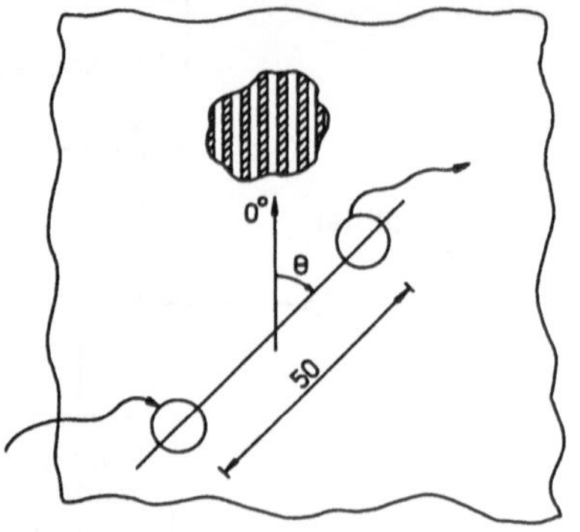

Fig. 3. Transducer positioning on a unidirectional fiber composite.

$$SWF_2 = M_1/M_0$$

$$SWF_3 = (M_2/M_0)^{0.5}$$

$$SWF_4 = (M_4/M_2)^{0.5}$$

and

$$SWF_5 = M_3M_0/M_2M_1$$

M stands for a spectral moment and the subscript on it represents the order of the moment. Thus M_r, the rth moment of the power spectral density $S(f)$, where f is the frequency, is given by

$$M_r = \int_{f_1}^{f_2} S(f)f^r df$$

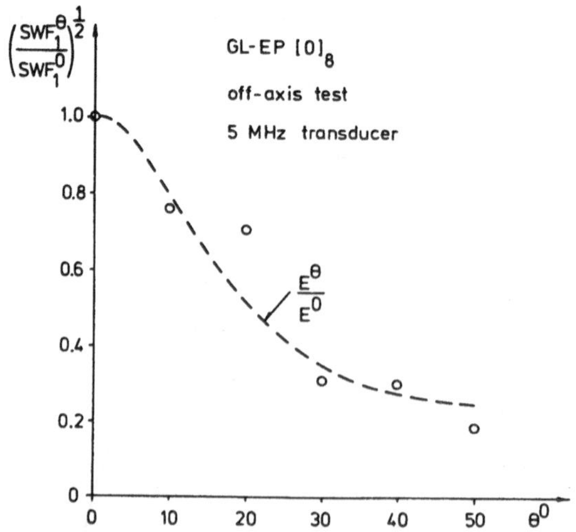

Fig. 4. Variation of $(SWF_1)^{0.5}$ normalized by its value along the fiber direction with the off-axis angle θ. The dotted line shows the variation of the E-modulus normalized by its value in the fiber direction.

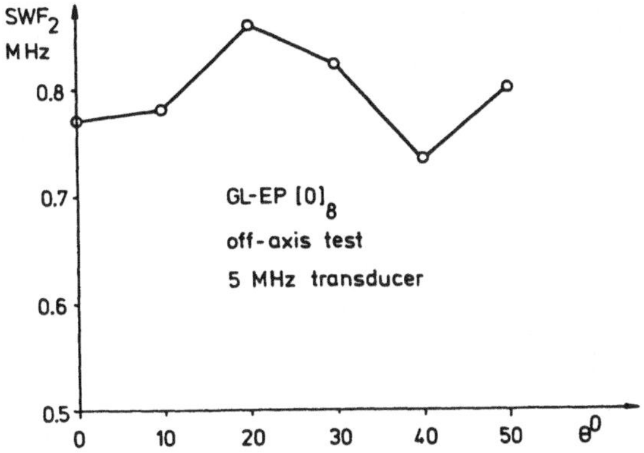

Fig. 5. Variation of SWF_2 with the off-axis angle θ.

where f_1 and f_2 are the lower and upper cut-off frequencies, respectively (see Fig. 2).

The first factor in the set, SWF_1, is the area under the power spectrum and is therefore the mean square value of the stress wave. The second factor, SWF_2, is the centroidal location of the power spectrum and is thus the central frequency of the signal. SWF_3 and SWF_4 are, for a stationary Gaussian random signal, the mean values of the frequency of mean value up-crossings and the frequency of peaks in the signal[9]. The ratio SWF_3/SWF_4 is, then, the so-called irregularity factor. SWF_5 is the skewness factor representing a measure of the bias in the spectral density distribution. The irregularity factor has the value one for sinusoidal signals and the skewness factor for uniformly distributed power spectrum is also one.

ACOUSTO-ULTRASONIC MEASUREMENTS

The measurements were made in three parts and will be described below under the headings defining the purpose.

Directional Dependence of SWFs in a Unidirectional Composite

Figure 3 illustrates the mode of the ultrasonic measurements. The straight line joining the centers of the transmitting and the receiving transducers makes an acute angle θ with respect to the fiber direction in a unidirectional composite. The distance between the transducer centers is fixed at 50 mm.

Figures 4-7 show the variation of the SWFs as functions of θ. Of significant interest is the variation of $(SWF_1)^{0.5}$ which in Fig. 4 has been normalized by its value at $\theta = 0$. Also shown in the figure is the calculated variation of the in-plane E-modulus normalized by its value at $\theta = 0$ (dotted line). The data points fall in good agreement with this variation. Thus a correlation between the RMS value of the received stress wave and the elastic stiffness of the material in the direction of the stress wave propagation appears to exist. Figure 5 shows the variation of the central frequency (SWF_2) with the off-axis angle θ. The data points appear to scatter about a constant frequency of about 0.8 MHz. The same trend was shown by the

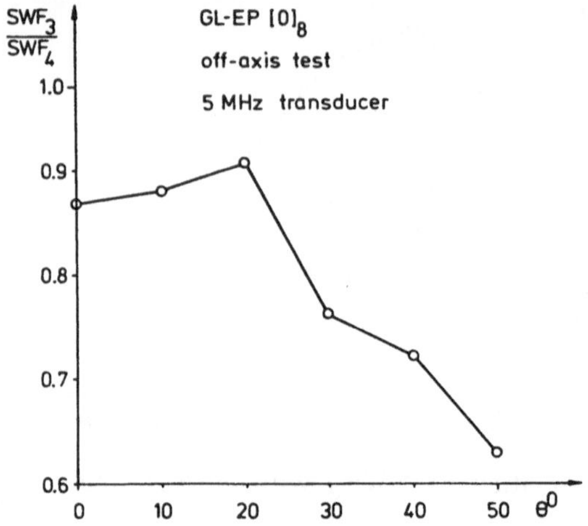

Fig. 6. Variation of SWF_3/SWF_4 with the off-axis angle θ.

frequency of the mean value crossings (SWF_3). It appears therefore that the basic frequencies of the stress wave are largely maintained in all directions of the stress wave propagation in the specimen plane. The ratio SWF_3/SWF_4, which reflects the stress wave irregularity, is plotted in Fig. 6. A general decrease in this ratio with the off-axis angle is displayed. Thus, the stress wave components traveling in the off-axis directions become increasingly irregular (i.e. non-sinusoidal) with increasing off-axis angle. The last factor, SWF_5, is shown plotted in Fig. 7. The trend here indicates that the skewness of the stress wave increases with increasing off-axis angle.

The data reported above were obtained on a glass-epoxy specimen using transducers of 5 MHz central frequency. The measurements were taken for off-axis angles up to 50° since, for larger angles, the energy of the received signal was so low that excessive amplification would have been necessary. Since this might cause distortion of the signals, measurements were restricted to the 50° angle. However, it was noted that the frequencies of the signals received were generally below 1 MHz. In order to maximize the

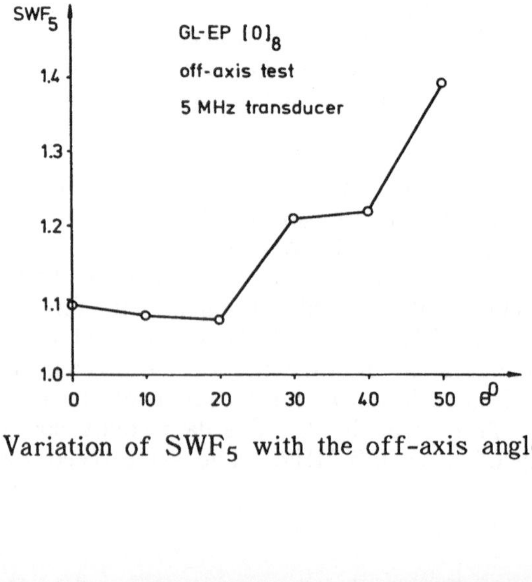

Fig. 7. Variation of SWF_5 with the off-axis angle θ.

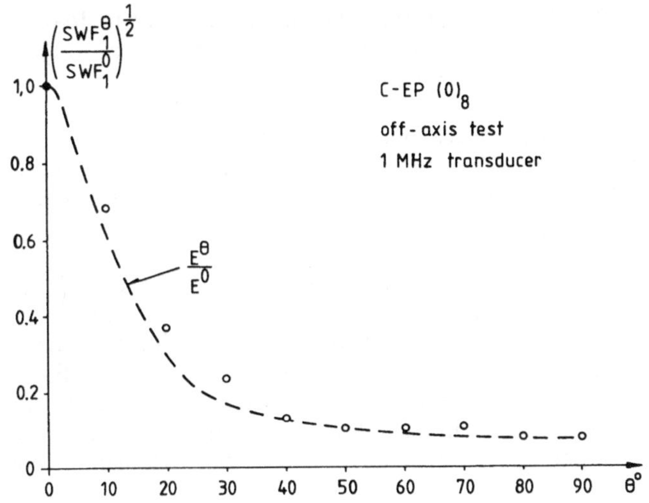

Fig. 8. Variation of $(SWF_1)^{0.5}$ and the E-modulus with the off-axis angle for a unidirectional graphite-epoxy composite.

strength of the signal, therefore, transducers of 1 MHz central frequencies were used and the measurements repeated, this time on a graphite-epoxy sample. All basic trends already seen in glass-epoxy specimen were repeated. As an illustration, the important correlation between the stress wave RMS value and the in-plane E-modulus for graphite-epoxy is shown in Fig. 8.

Quality Control of Filament Wound Ring-Shaped Specimens

Three ring-shaped specimens of graphite-epoxy were prepared by filament winding in the circumferential direction. The inner diameter of the specimens was 80 mm and the cross-section was of 20 mm x 20 mm dimensions. One specimen each was prepared in three controlled fabrication conditions: a) Normal, i.e. curing in an autoclave with recommended temperature and pres-

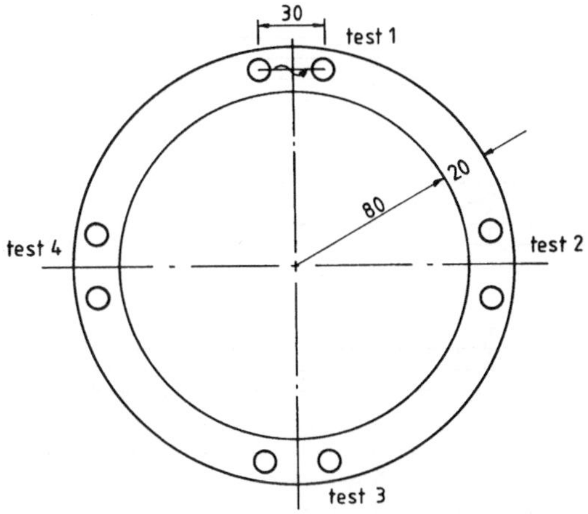

Fig. 9. Positioning of transducers on the ring-shaped specimen for acousto-ultrasonic measurements.

sure variations, b) Without vacuum, the intention being to leave some air bubbles in the cured specimen, c) With chopped pieces of teflon mixed in the epoxy prior to wet filament winding.

Four sets of acousto-ultrasonic measurements were made on each specimen at four equally spaced test positions shown in Fig. 9. The distance between the transducer centers was kept at 30 mm and the transducers were placed on a flat surface of the specimen.

Figure 10 shows the SWFs taken as average of the four measurements. The horizontal bars in the figure are labelled N, A, and T referring to the fabrication conditions a), b) and c) respectively. The horizontal lines at the end of the bars indicate the range of scatter in the SWFs. The RMS value of the stress wave in the specimen with teflon is seen to be slightly less than that in the normal specimen. The remaining SWFs show a consistent trend of being affected more by fabrication in condition c) than in condition b).

The properties of the specimen in the circumferential (fiber) direction are fiber dominated and therefore would not be expected to change much by the fabrication conditions described above. However, the properties in the radial and the thickness directions may be affected significantly by the abnormal fabrication. To investigate this a set of measurements were made in the through-transmission mode as shown in Fig. 11. The SWFs are shown in

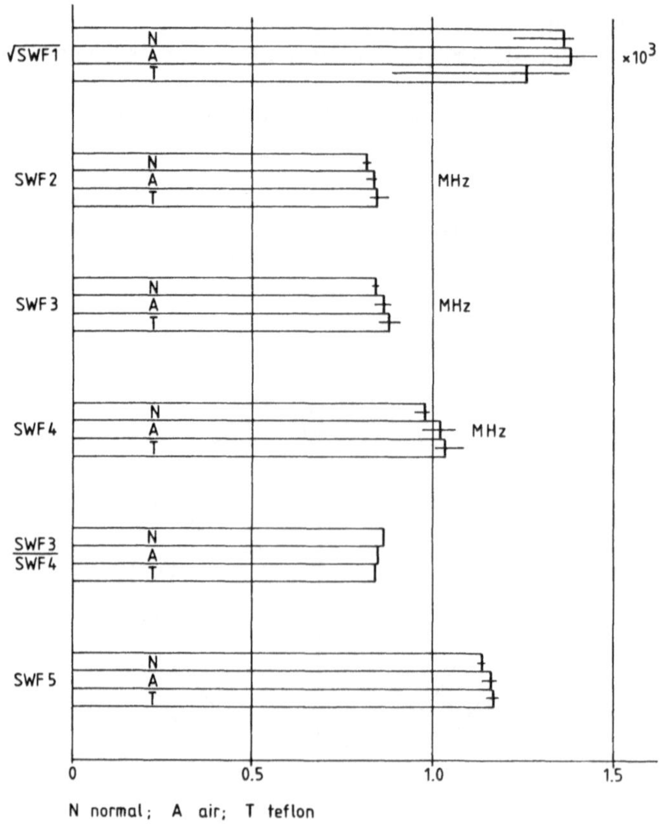

Fig. 10. SWF measurements in acousto-ultrasonic mode for three fabrication conditions of the ring-shaped specimens.

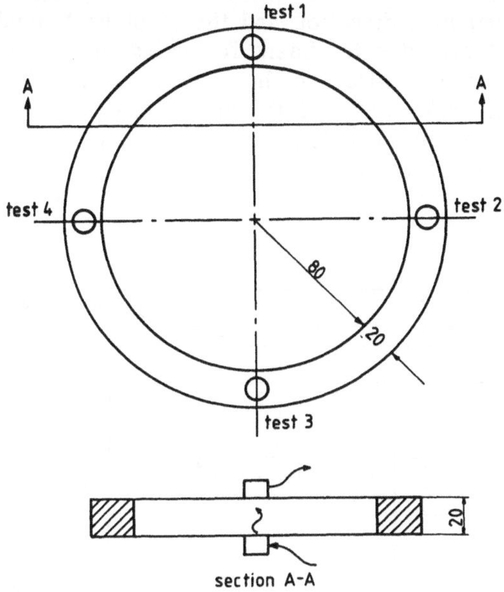

Fig. 11. Positioning of transducers on the ring-shaped specimen for through-transmission measurements.

Fig. 12. The clearest trend is shown by the RMS value of the stress wave traveling in the thickness direction. This value is less than 25% for fabrication conditions b) and by 35% for fabrication condition c) as compared to the normal fabrication.

Assessment of Impact Damage

A laminated graphite-epoxy plate with fibers in 0°, ±45° and 90° directions and with thickness of 5 mm and surface dimensions of 100 mm x 150 mm was subjected to an impact at the center by a sphere with 17 Joules of energy. The plate was examined by ultrasonic C-scan which showed a delaminated area around the point of impact.

Acousto-ultrasonic measurements were made on the specimen at 12 positions; see Fig. 13. The figure also shows the delaminated area indicated by the C-scan. The SWFs at various positions are plotted against the distance d of the measurement position from the upper edge of the plate, and are shown in Figs. 14-19. The indication that a damage zone exist is given most clearly by the RMS value of the stress wave, Fig. 14, which shows a significant drop in going from the undamaged to the damaged area. The irregularity factor SWF_3/SWF_4 (Fig. 18) and the skewness factor SWF_5 (Fig. 19) are also good indicators of the damage.

DISCUSSION OF RESULTS

The first set of measurements concerning the directional dependency of the SWFs in the plane of a unidirectional fiber composite is illustrative of the nature of stress wave propagation in an anisotropic medium. An important result from those measurements is the proportionality of the energy of the stress wave transmitted in a direction and the in-plane E-modulus in the direction. This result provides the basis for using this property to assess defects and damage as these influence the in-plane E-modulus. Indeed the correlation between the RMS value of the stress wave and the E-modulus is the strongest basis for using the acousto-ultrasonic technique for quantitative nondestructive evaluation of composites.

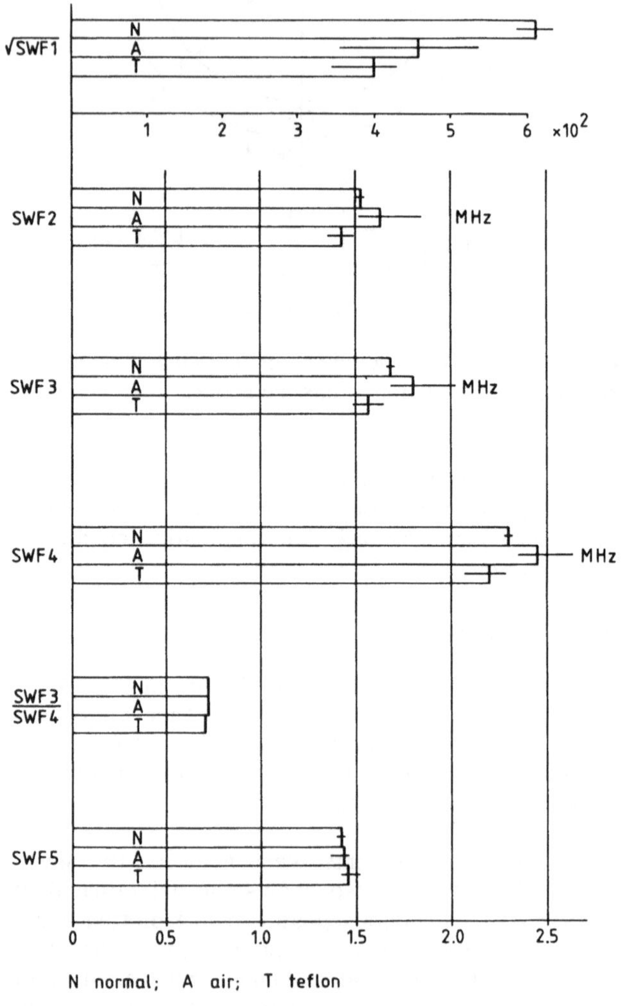

Fig. 12. SWF measurements in through-transmission mode for three fabrication conditions of the ring-shaped specimens.

The increases of the irregularity and of the skewness of the stress wave with the off-axis angle are additional indicators of the changes in the stress wave characteristics with the direction of propagation. These indicators are, however, subject to more scatter than the RMS value of the stress wave. The basic frequencies of the stress wave, given by the central frequency and the frequency of mean value up-crossings, appear to be not significantly affected by the direction of stress wave propagation.

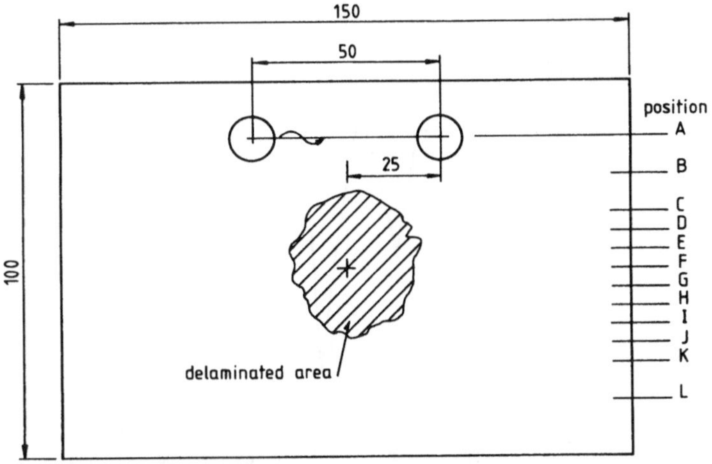

Fig. 13. Positioning of transducers on the impact-damaged specimen for acousto-ultrasonic measurements.

The second set of measurements concerning indication of the fabrication quality show that the RMS value of the stress wave is the most usable indicator. This SWF shows that the composite properties in the fiber direction are only slightly affected by the abnormal fabrication conditions. The matrix governed properties across the fiber direction are, however, significantly affected, as indicated clearly by the RMS value measurements. The reductions of 25% and 35% in the RMS value in the two abnormal fabrication conditions are in the range of loss in the E-modulus that would be expected due to the defective fabrication of the composite.

The final set of measurements concerning the impact damage assessment show that the SWFs, in particular the RMS value, the irregularity factor and the skewness factor provide usable indicators of damage. Assuming that the

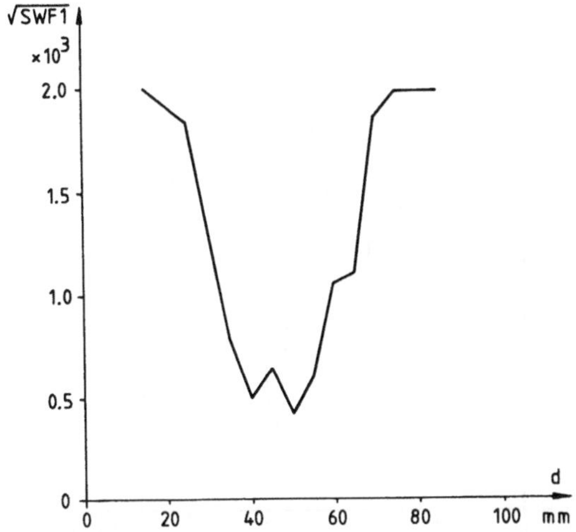

Fig. 14. Variation of $(SWF_1)^{0.5}$ with the distance from the edge of impact-damaged specimen.

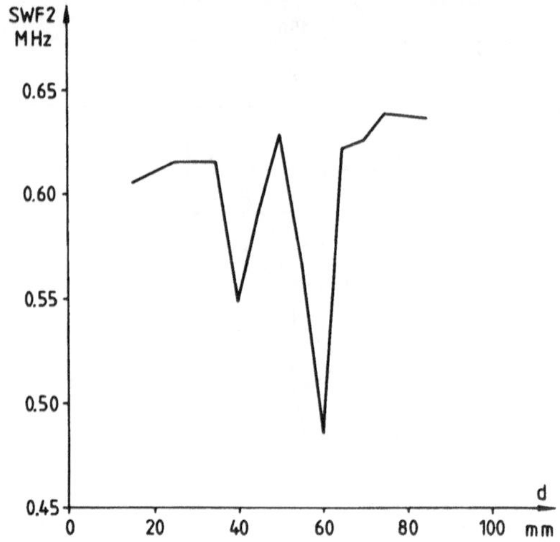

Fig. 15. Variation of SWF_2 with the distance from the edge of impact-damaged specimen.

correlation between the RMS value and the E-modulus holds also for this type of damage, a quantitative measure of the damage induced and its distribution in the impacted component is provided by the acousto-ultrasonic technique.

CONCLUSION

The investigations reported here demonstrate that the acousto-ultrasonic technique, based on the set of SWFs proposed here, is capable of conducting quantitative nondestructive evaluation of composite materials. Specifically, it has been shown that indicators of the fabrication quality and the extent and severity of impact damage can be provided by the technique.

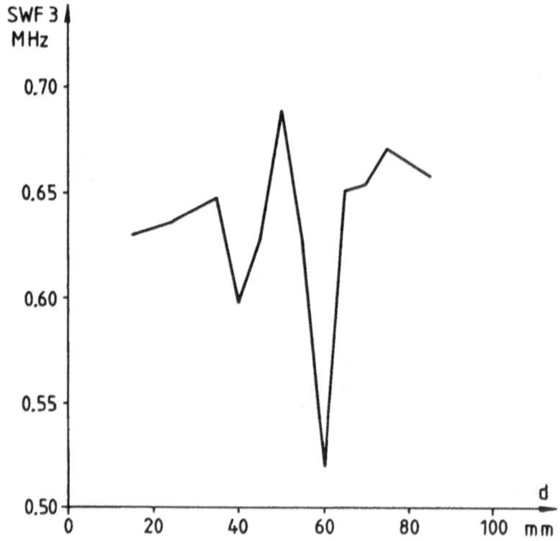

Fig. 16. Variation of SWF_3 with the distance from the edge of impact-damaged specimen.

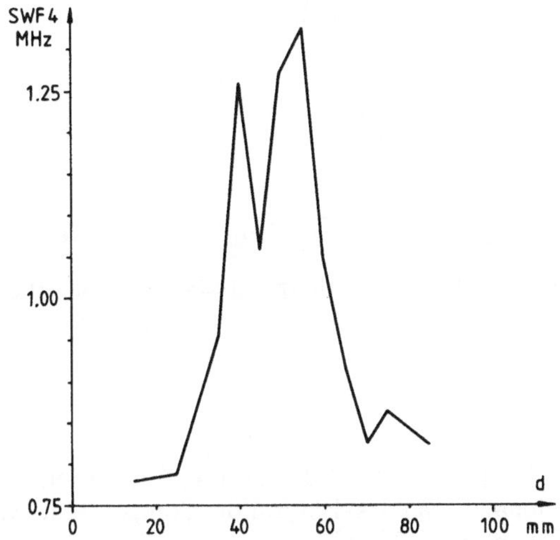

Fig. 17. Variation of SWF_4 with the distance from the edge of impact-damaged specimen.

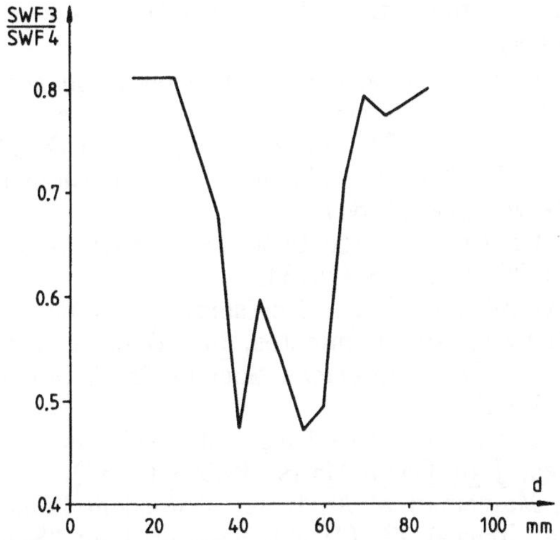

Fig. 18. Variation of SWF_3/SWF_4 with the distance from the edge of impact-damaged specimen.

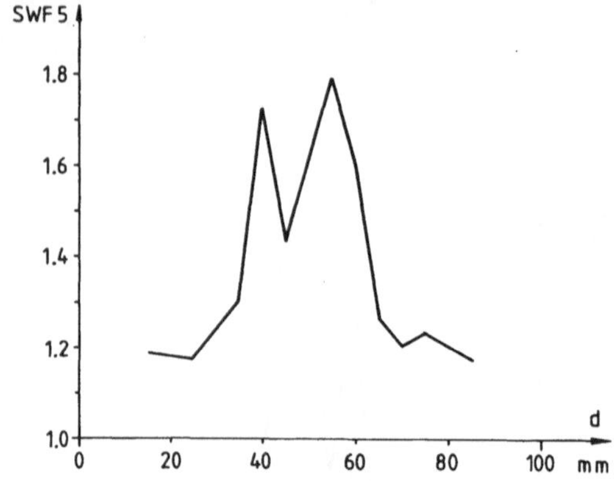

Fig. 19. Variation of SWF_5 with the distance from the edge of impact-damaged specimen.

REFERENCES

1. A. Vary and K. J. Bowles, "Ultrasonic Evaluation of the Strength of Unidirectional Graphite-Polyimide Composites, NASA TM X-73646," NASA, Cleveland, (1977).
2. A. Vary and K. J. Bowles, An Ultrasonic-Acoustic Technique for Nondestructive Evaluation of Fiber Composite Quality, Poly Eng & Sc. 19:373 (1979).
3. A. Vary, Concepts and Techniques for Ultrasonic Evaluation of Material Mechanical Properties, in: "Mechanics of Nondestructive Testing," W. W. Stinchcomb, ed., Plenum Press, New York (1980).
4. A. Vary and R. F. Lark, Correlation of Fiber Composite Tensile Strength with the Ultrasonic Stress Wave Factor, J of Test & Eval. 7:185 (1979).
5. R. Talreja, A. Govada, and E. G. Henneke, Quantitative Assessment of Damage Growth in Graphite-Epoxy Laminates by Acousto-Ultrasonic Measurements, in: "Review of Progress in Quantitative Nondestructive Evaluation," D. O. Thompson and D. Chimenti, eds., Plenum Press, New York (1984).
6. R. Talreja, On Fatigue Life Under Stationary Gaussian Random Loads, Eng Frac Mech. 5:993 (1973).
7. A. L. Highsmith and K. L. Reifsnider, Stiffness-Reduction Mechanisms in Composite Laminates, in: "Damage in Composite Materials, ASTM STP 775," American Society for Testing and Materials, Philadelphia (1982).
8. R. Talreja, Transverse Cracking and Stiffness Reduction in Composite Laminates, J of Comp Matls. 19:355 (1985).
9. S. O. Rice, Mathematical Analysis of Random Noise, Bell Systems Technical Journal 23: (1944), Reprinted in: "Selected Papers on Noise and Stochastic Processes," Dover Publications, New York (1963).

PREDICTING DAMAGE DEVELOPMENT IN COMPOSITE MATERIALS BASED ON ACOUSTO-ULTRASONIC EVALUATION

J. C. Duke, Jr. and M. T. Kiernan

Materials Response Group
Engineering Science and Mechanics Dept.
Virginia Polytechnic Institute and State University
Blacksburg, VA 24061-0219 USA

INTRODUCTION

The stress wave factor nondestructive evaluation method introduced by Vary et al.[1] embodied a number of features, some of which were attractive regarding practical application, and others of major significance regarding material characterization. The technique as originally described for application to fiber reinforced composites allowed for local assessment of mechanical integrity in the direction of service loading, and required only access to one side of the component.

Experimental evaluation of materials by means of this technique has been correlated with various aspects of mechanical behavior.[2] Now although these findings indicate that this technique is of significant practical use, it is the fundamental method of material characterization that serves as the basis for the work reported here. The use of a transient mechanical disturbance for probing a material component, acousto-ultrasonics (AU), is the aspect to be explored. The development of this basic idea has been guided by the notion that the ability of a material to accommodate damage is intimately related to the ability of the material to transfer strain energy associated with transient mechanical disturbances. The consideration of such a methodology is motivated by fiber reinforced composite materials. Unlike most conventional engineering materials, failure of components of composite materials does not proceed from a single critical imperfection that develops in a self-similar fashion. Rather, these materials develop an intricate damage condition, "characteristic damage state," that eventually culminates in failure of the component, through a sequence of events that are presently not possible to definitively describe. Consequential efforts to nondestructively evaluate such materials in order to discover and describe a critical defect are inadequate. The assessment of this complex damaged condition in a way that will enable an evaluation of the component as regards its ability to sustain service loads is the objective of the AU method development.

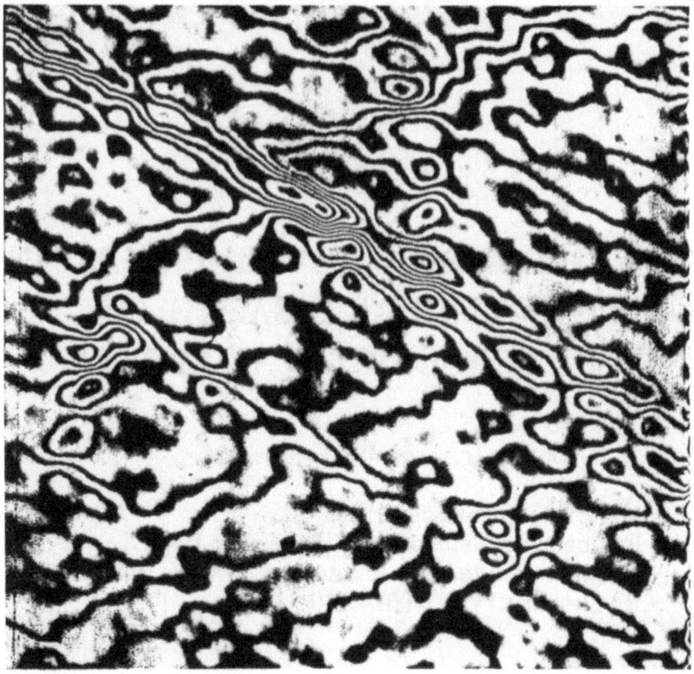

Fig. 1. Nonuniform or anomalous part of the transvers displacement field, in a graphite/epoxy $[0_2/\pm60/0_2/\pm60]_s$ with a uniform specimen width of 25 mm, resulting from tensile loading. The method of moiré interferometry was used to obtain this data the virtual reference grating pitch was adjusted to subtractthe uniform displacement field. (Photograph courtesy of the VPISU Photomechanics Laboratory, D. Post, director.)

BACKGROUND

Continuous fiber reinforced composite materials are anisotropic and inhomogeneous, although as regards certain properties, the scale of measurement may allow these characteristics to be insignificant; this is never the case for damage initiation and development. The nature of the composite material architecture involving materials with dissimilar mechanical properties results in local triaxial states of stress. In actual materials this state of stress may be "locally" nonuniform even from point to point in the same ply under nominally uniform external loading, Fig. 1. In fact, such nonuniformity is certainly the cause of failure occurring nonuniformly, resulting in nonplanar fracture morphologies.

At VPI&SU, a systematic study of the AU behavior has been undertaken. The effects of direction of measurement, deformation, constraint due to ply orientation, and variation in deformation constraint resulting from damage have been considered.[3] From related efforts, strong evidence has been provided to suggest that for the composite material laminates examined, a major portion of the AU disturbance becomes dispersed throughout the object by means of plate modes of stress wave propagation. This propagation is affected by the properties of the plate which are influenced by the condition of the material of the plate.

As the basis for this study, the following hypothesis was set: damage development occurs, subsequent to initiation, if and only if the material neighboring the point of initiation is unable to transfer the elastic strain energy released at a sufficient rate to avoid exceeding the critical level of strain energy to cause failure of any point in this neighboring region. The extent of this neighboring region would depend on the material constituency and the architecture, and their condition at the time of damage initiation. For example, suppose a fiber break occurs in an otherwise undamaged unidirectional fiber reinforced composite laminate. The fiber itself begins to unload and the surrounding matrix material begins to pick up the load that the fiber previously supported, and in turn transfers it to neighboring fibers. The magnitude and the rate of transfer of this load determine the transient state of stress that results. If the fiber unloads more rapidly than the load transfer in the matrix material, and if the magnitude is large enough, the load in the matrix can exceed its capacity and consequently it will also fail, further reducing the load bearing capability and exacerbating the situation. Alternatively, the fiber/matrix bond could fail or a nearby fiber. This may or may not continue, since in actual composite materials the strength is nonuniform. Furthermore, were damage present at the time of initiation it would influence the behavior, not necessarily in a detrimental way. (Damage has the effect of increasing quasi-static strength in notched laminates.)

The rate of load transfer in a material is characterized by the speed of sound propagation. Consequently, it is possible to assess the ability of a material to transfer load by examining the speed of sound propagation. For anisotropic material plates, the load transfer may occur by a variety of different modes--longitudinal, shear, surface, plate, etc. Figure 2 is a plot of the group velocity, the velocity of energy propagation, for a unidi-

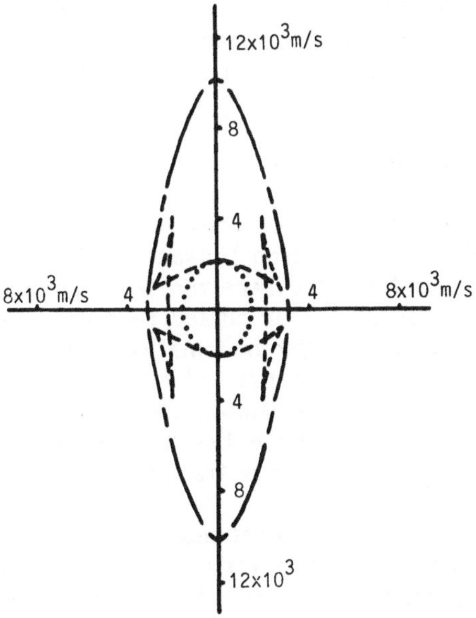

Fig. 2. Calculated group velocities for P, SH, and SV modes as a function of azimuthal angle for a unidirectionally graphite fiber reinforced epoxy lamina.

rectional graphite fiber reinforced epoxy lamina. The plot shows the variation of the velocity of longitudinal, horizontally polarized shear and vertically polarized shear modes with in-plane angle; zero degrees is parallel to the fiber direction. Since in general that state of stress in composite material is triaxial, unloading due to damage would result in a complex combination of modes endeavoring to transfer the load to the neighboring material. The task of determining the partitioning into various modes of load transfer resulting for each damage event under the myriad of possible conditions is indeed formidable, if not impossible. On the other hand, by using an analog of the transient load associated with damage events--the AU mechanical disturbance--and observing the manner in which the material, under various conditions, transfers this disturbance to the surrounding material, it should be possible to assess which damage mode/material condition combinations are critical.

The task of identifying and creating an AU input that is sufficiently analogous to the stress wave emission resulting from the damage mode of interest is by no means trivial. The typical experimental configuration has been motivated with the objective of observing the influence of the material condition on a portion of the AU input disturbance that has traveled to a point of observation that is in a direction, from the location of the AU input, that is parallel to an anticipated direction of service loading. For example, a typical configuration finds the AU sender and AU receiver along the axis of a specimen that will eventually be loaded in uniaxial tension. Often, values of the quantification of this AU response are found to correlate well with mechanical performance of specimens, as long as all specimens considered have experienced similar treatment prior to the scenario of evaluation.[4] Since the AU behavior changes with loading, prior variations in load history can negatively effect such correlative approaches. In the study to be reported here, a direct interpretation of the AU behavior is attempted, avoiding the necessity of considering the previous history which is often unknown in practical application. Although ultimately it is desirable to use such an approach to predict failure loads, the first step has been to predict patterns of damage development based on observed AU behavior.

EXPERIMENTAL CONSIDERATIONS

Both the AU sender and receiver used were piezoelectric transducers. A pulse generator, Model 1010 Accutron Pulser/Receiver, was used to excite the AU sender. The signals from the receiver were amplified when necessary by an amplifier and the amplifier stage of pulser/receiver, and digitally sampled by means of a data acquisition system installed in an IBM PC. Data records up to 4096 points with 8 bit resolution and real-time sampling rates of 25 MHz could be made. Software developed for comprehensively analyzing digital records[5] was utilized to calculate the fast Fourier transform (FFT), alter the spectrum, and calculate the energy transmitted at various frequencies.

Now, although typically longitudinal mode cylindrical elements are used as AU senders, the cylindrical symmetric nature of the disturbance is often ignored, with only a particular direction being considered. In this study, the analogy between this form of disturbance and impact loading of the composite material was exploited. The AU response for a $[0]_4$ and a $[90/0]_s$ laminate was recorded by placing the AU sender in a particular location and moving the AU receiver in a circle about the sender. In some instances the specimens were then impacted by means of a drop weight impact fixture (impact energy of 1,75 ft-lbs, 2.37 joules); the point of impact was the same as the location of the AU sender. In other instances the specimens were subjected to cyclic loading so as to introduce damage prior to again examining the AU response. After AU examination, these specimens were subjected to impact as described above.

To examine the influence of damage on the AU behavior and subsequent impact damage development, several specimens were cyclically loaded ($R = 0.1$, $\sigma_{max} = 5$ Kips) for 120K cycles. An MTS 442 controller and MTS 309 50 kip load frame, equipped with hydraulic grips, operating at 10 Hz, was used for the load cycling; tabs were bonded with FM 300 Adhesive to the ends of the specimens before loading.

Radiographs enhanced with zinc iodide penetrant were made of the specimens in order to provide evidence of the damage resulting from impact and cyclic loading. Preliminary findings indicated that AU evaluation after application of the penetrant was significantly influenced. Consequently, no radiographs were made following cyclic loading until after AU evaluation following impact. In each instance the zinc iodide solution was allowed to penetrate for at least 24 hours prior to radiographic exposure.

AU QUANTIFICATION

The AU signal has been quantified by scientists in several different ways.[6] For this study, a procedure adapted from narrow band analysis has been used to estimate the energy transfer occurring in various directions and at various different frequencies. The procedure involves calculating weighted areas of regions under the power spectral density function curve:

$$M_n(f_1,f_2) = \int_{f_1}^{f_2} f^n S(f)df$$

n is the order of the moment, f is the frequency, and S(f) is the power spectral density.

For the digital representation, the integral becomes a summation of discrete values. For example, the zeroeth moment for a digitally recorded signal is:

$$M_o = \sum_{i=1}^{m} \Delta f(S_{i+1} + S_i)/2$$

in which

S_i - power at the (i)(Δf) frequency
Δf - is ($f_s/2m$)
m - is the number of frequency samples (1/2 the number of time samples)
f_s - digitization rate.

M_o is directly proportional to the energy of the signal; the summation over a portion of the range of frequencies is proportional to the energy received in that range of frequencies. The first moment is the centroid of the area under the power spectral density curve.

Since this is a generic signal analysis procedure, we have used the notation

$$A_1(f_1,f_2) = M_o(f_1,f_2) \qquad\qquad A_2 = M_1/M_o$$

to distinguish our situation in which the signal being considered is a result of an acousto-ultrasonic evaluation. The bracketed parameters are omitted if the entire spectrum is used in the calculation.

RESULTS AND DISCUSSION

Figure 3 a) and b) are plots of the AU behavior as a function of azimuthal angle for a $[0]_4$ specimen before and after impact; the radial distance is proportional to the A1, Fig. 3a), A2, Fig. 3b) values. A1 is the area under the power spectrum curve, which is proportional to the energy of the signal, while A2 is the frequency value that corresponds to the centroid of this area. Figure 4 is a print of a radiography of the same specimen; damage resulting from the impact is visible.

The AU evaluation suggests that more energy flows in the direction of the fibers than any other with the amount in the direction perpendicular to the fibers being the smallest. These values reflect the rate of strain energy transfer from the input excitation. Consequently if the magnitude of the input is increased, the previously proposed hypothesis would suggest damage development in the 90° direction first. This is consistent with the damage observed as a result of impact.

Figure 5a) and b) are plots of the AU behavior for a $[90/0]_s$ specimen before and after impact; a) A1, and b) A2, are shown as a function of azi-

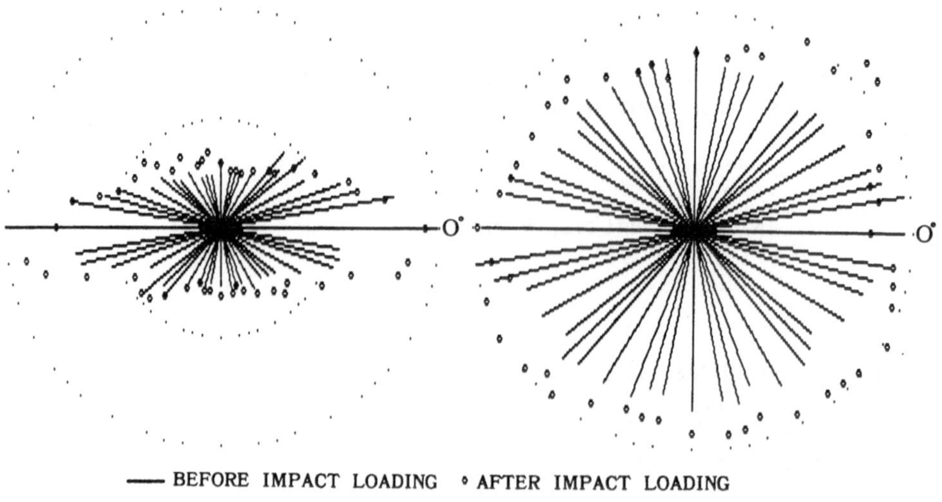

Fig. 3. a) A1 as a function of azimuthal angle for a $[0]_4$ laminate before and after impact, b) A2 as a function of azimuthal angle before and after impact.

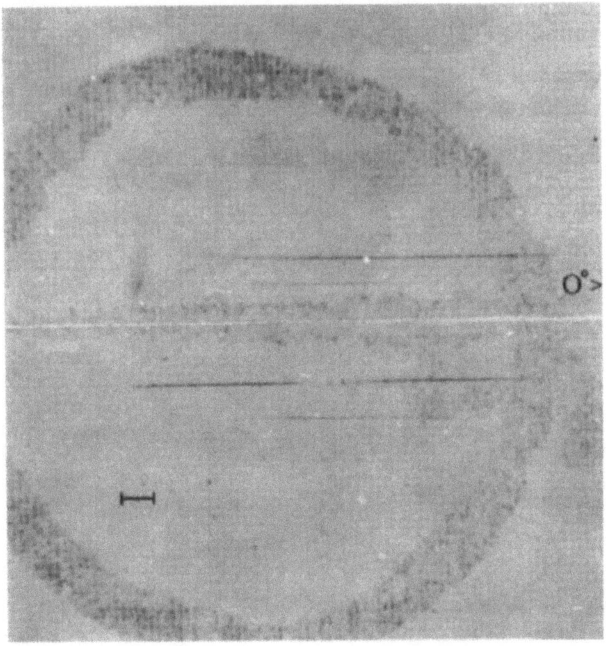

Fig. 4. Print of a penetrant enhanced X-ray radiograph of the specimen referred to in Fig. 3. The dark regions are areas where penetrant has entered the specimen; this damage is due to impact.

muthal angle. Figure 6 is a print of a radiograph of the same specimen; damage resulting from the impact is visible.

The before impact AU behavior is perhaps as might have been anticipated considering Fig. 3a); large energy transfer in the fiber directions, 0° and

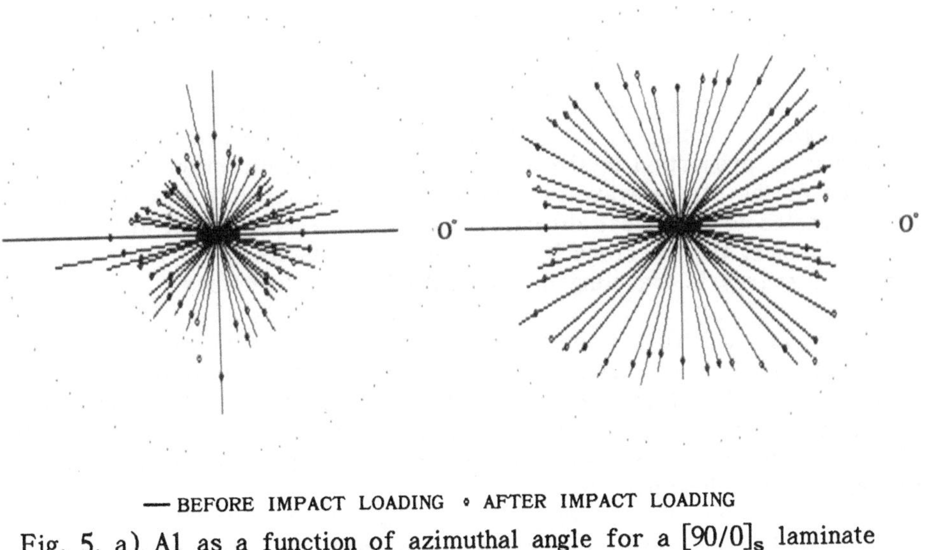

Fig. 5. a) A1 as a function of azimuthal angle for a $[90/0]_s$ laminate before and after impact, b) A2 as a function of azimuthal angle before and after impact.

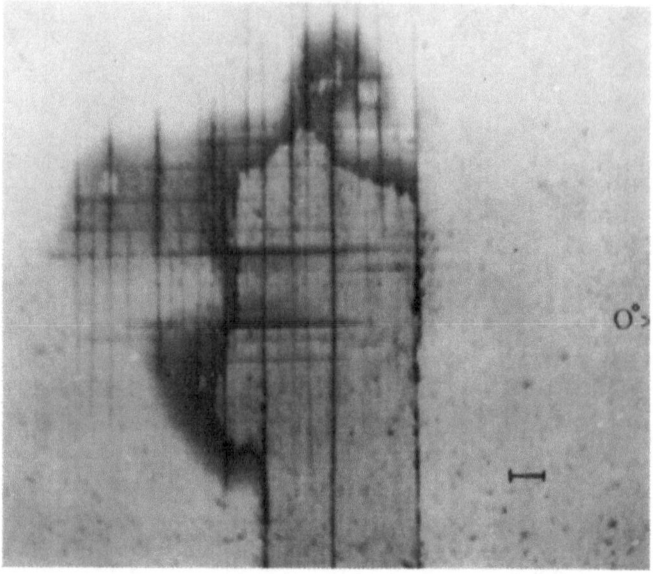

Fig. 6. Print of a penetrant enhanced X-ray radiograph of the specimen referred to in Fig. 5. The dark regions are areas where penetrant has entered the specimen; this damage is due to impact.

90°, and small in others. It also follows that with increased amplitude of the input, as would occur with impact, damage development would occur in directions other than 0° and 90°.

At this point the dilemma, of course, is that matrix cracks are observed in both the 0° and 90° directions as well as delamination. This suggests that any procedure for predicting damage development must reflect the manner in which the damage is manifest, that is, the various damage mechanisms. In fiber reinforced laminates this would include matrix cracks, delamination, fiber damage, and interlaminar and fiber/matrix bond degradation. In addition, the material varies not only with azimuthal direction but through the thickness as well.

Considering together the AU behavior, Fig. 3a) and Fig. 5a) and the damage apparent in Fig. 6, a consistent scenario may be proposed: energy transfer is in proportion to the amount of fiber in a direction. As the amplitude of the stress wave increases, matrix cracking is likely to occur in plies at angles to this fiber direction beginning with those that are orthogonal. Further, the interface between such plies is likely to fail or degrade if energy is traveling in nonparalllel fiber directions.

Figure 7a) and b) are plots of the AU behavior for a $[90/0]_s$ specimen after cyclic loading and after impact; 7a) A1, 7b) A2 are shown as a function of azimuthal angle. Figure 8 is a print of a radiograph of the same specimen; damage resulting from cyclic loading and impact is visible. The matrix cracking in the 90° plies away from the central region is typical of damage induced by cyclic loading.

The AU behavior after cyclic loading suggests that energy due to an impact will be most efficiently transferred in the 90° direction. At large

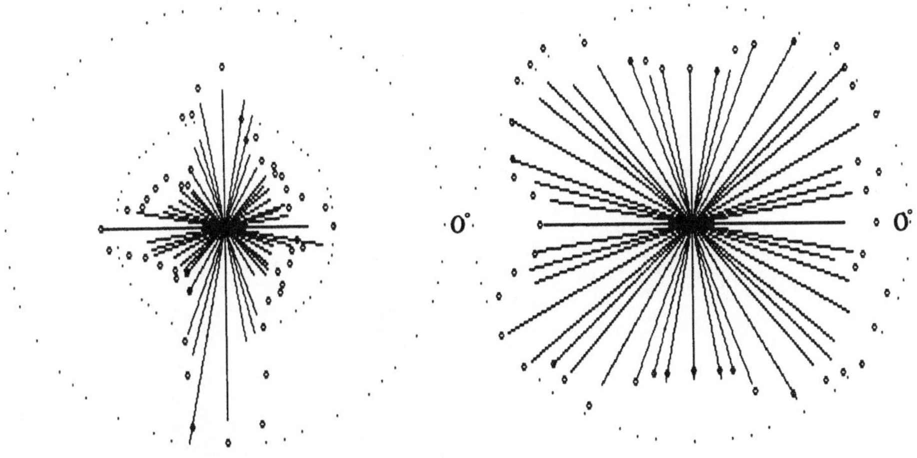

Fig. 7. a) A1 as a function of azimuthal angle for a $[90/0]_s$ laminate after cyclic, but before impact, and after impact, b) A2 as a function of azimuthal angle after cyclic loading, but before impact, and after impact.

amplitudes this might lead to matrix cracking in the neighboring 0° plies. In the immediate vicinity of the impact there is likely to be efficient energy transfer in the fibers of both the 0° and 90° plies. This transverse mismatch, however, is suspected as the source of delamination.

CONCLUSION

Although analytical predictive approaches are capable of handling the situations considered here, their potential for extension to situations involving complex fiber architecture (weave/wind, braided, etc.) is questionable. Through the use of advanced AU methodology, involving different frequency and different size input devices, a more in depth understanding of the dynamic stress-strain behavior is anticipated. An aspect which poses concern at this time involves establishing the level of input that will cause damage initiation. Perhaps by examining fundamental fiber/matrix constraint configurations, an indication of the critical level(s) may be established.

It is the conclusion that AU behavior can be used to predict damage development. The method used to measure the AU behavior should be selected based on the anticipated service loading conditions being considered. More detailed study is necessary before a routine procedure for interpreting AU behavior in terms of load induced damage will be available. The application of this approach to situations involving existing damage appears equally promising.

ACKNOWLEDGEMENTS

This work has been supported by NASA/Lewis Research Center through NAG 3-172; Alex Vary has been the technical monitor.

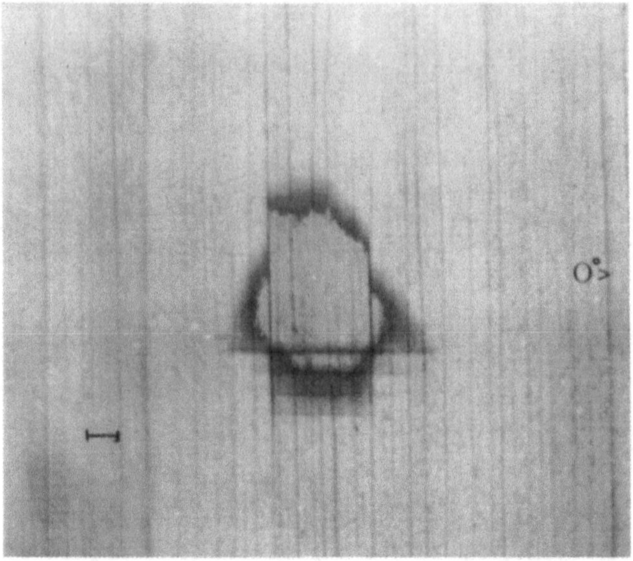

Fig. 8. Print of a penetrant enhanced X-ray radiograph of the specimen referred to in Fig. 7. The dark regions are areas where penetrant has entered the specimen subsequent to impact. The dark lines that are present away from the central region are typical of the damage due to cyclic tensile loading (matrix cracking); the other dark regions indicate damage due to impact.

REFERENCES

1. A. Vary and K. J. Bowles, An Ultrasonic-Acoustic Technique for Nondestructive Evaluation of Fiber Composite Quality, Poly Eng & Sc. 19:373 (1979).
2. A. Vary, Concepts and Techniques for Ultrasonic Evaluation of Material Mechanical Properties in: "Mechanics of Nondestructive Testing," W. W. Stinchcomb, ed., Plenum, New York (1980).
3. M. T. Kiernan and J. C. Duke, Jr., Acousto-Ultrasonics as a Monitor of Material Anisotropy, Matls Eval. 46 (1988).
4. J. C. Duke, Jr., E. G. Henneke, II, W. W. Stinchcomb, and K. L. Reifsnider, Characterization of Composite Materials by Means of the Ultrasonic Stress Wave Factor, in: "Composite Structures 2," I. H. Marshall, ed., Applied Scientific Publications, London (1983).
5. M. T. Kiernan and J. C. Duke, Jr., PC Software for Digital Signal Analysis, Matls Eval. 46, (1988).
6. A. Vary, The Origin and Development of the Acousto-Ultrasonic Approach, in: "Acousto-Ultrasonics: Theory and Application," J. C. Duke, Jr. ed., Plenum, New York (1988).
7. S. O. Rice, Mathematical Analysis of Random Noise, in: "Selected Papers on Noise and Stochastic Processes," N. Wax, ed., Dover, New York (1954).

LASER GENERATED ULTRASOUND

A. Sarrafzadeh, R. J. Churchill, and M.G. Niimura

American Research Corporation of Virginia
Radford, Virginia

ABSTRACT

This paper describes a laser-based ultrasonic system using high-energy, short-pulse trains for laser generation. Ongoing research includes the theoretical extension of laser-solid interactions related to laser generation of ultrasound in complex structures and advanced materials and the investigation of methods to achieve a high speed, high resolution laser excitation system for integration with a highly sensitive ultrasonic signal detector. Experiments are being conducted for the flaw inspection/characterization of rails using laser-based photoacoustic excitation techniques including ultrasonic detection by both piezoelectric and noncontacting laser interferometric means. This work was supported in part by the U.S. Department of Transportation under Contract No. DTRS-57-85-C-00132

INTRODUCTION

The use of advanced materials such as ceramics, ceramic composites and high temperature alloys in critical heat aerospace and energy-related applications has led to improved performance of critical aerospace and energy-related systems and components. However, the complex geometry, inhomogeneity, reduced critical flaw size and adverse inspection environments have increased the requirements for nondestructive evaluation (NDE) of these materials, have prolonged inspection times and have reduced confidence in estimating component lifetime. One class of NDE sensors, based on the laser generation and detection of ultrasound, offers several important advantages over conventional NDE techniques.

Ultrasonic characterization of flaws using piezoelectric devices for generating and detecting ultrasonic waves has widespread application. However, piezoelectric devices suffer several disadvantages, including bond variability, inability to operate on irregularly shaped objects and limitations in high temperature applications. Laser generated acoustic waves are, in part, an answer to the drawbacks of piezoelectric devices. Laser techniques are noncontacting, thus exhibiting no couplant problems; they can be used for rapid scanning, are amenable to use in hostile environments and can operate on irregularly shaped objects. Laser-based ultrasound is particularly appropriate for ceramic and composite materials. The laser/material interaction generates a high frequency pulse which can be used to detect defects in the 5-20 micron range, evaluate near surface microstructures necessary for inspection of coatings and impart a controlled, high-energy excitation signal necessary for NDE of dispersive composite media.

LASER/SOLID INTERACTIONS

The application of laser-based ultrasonics to the NDE of aerospace and energy related systems requires an understanding of the wave fields generated in and modified by structures having nonplanar boundaries or multiple boundaries. Ultrasonic waves can be generated by three different mechanisms depending upon the power density of the incident laser beam. For low power densities, ultrasonic waves are caused by the rapid thermal expansion of the material being irradiated. At intermediate power densities radiation pressure contributes to wave generation. At high energy densities, the surface of the material is ablated and ultrasonic waves are produced by the momentum transfer of the ejected material.

Pulsed lasers emit bursts of coherent electromagnetic radiation. For low power density the electric and magnetic fields of the light wave induce currents in the conduction band near the surface of the material. Part of the incident energy is absorbed and part is re-radiated as a reflective pulse. Most of the absorption and reflection take place within the skin depth. The absorption of laser energy causes thermoelastic strains equivalent to a sudden insertion of a volumetric change of material immediately below the surface, proportional to the incident energy. In the far radiation field close to the normal to the surface, this source can be approximated by a set of dipolar forces parallel to the surface. This thermoelectric source of ultrasound is different from that of a piezoelectric compressional transducer which generates a stress normal to the surface. For small beam diameters of incident laser radiation, the source of ultrasound may be treated as a point source. At intermediate power densities, a second mechanism contributing to the generation of ultrasound in both conductors and insulators is the transfer of momentum caused by the change in direction of the Poynting vector. For laser systems with higher power density, ablation of the surface occurs and a plasma forms at the surface of the material. However, for the purposes of ultrasonic generation, this is usually considered as producing an impulsive recoil force by transfer of momentum from the surface.

In both the thermoelastic and plasma regimes, the rapidly changing stress and strain fields produced just beneath the surface of the material act as sources of elastic waves. The strain energy is dissipated by both bulk and surface waves, and the irradiated area may be considered as a point source of ultrasound with time dependence related to that of the laser pulse. Calculations of bulk and surface wave directivities for both thermoelastic and plasma sources are available in the literature.[1,2] For insulating composite or ceramic materials the absorption of the laser pulse is a strong function of the wavelength. In the infrared, absorption arises from vibrational modes of the crystal lattice or from intermolecular vibrations of organic solids.[3] In the visible wavelength regions absorption may result from impurities or from the extension of strong ultraviolet absorption bands. Application of

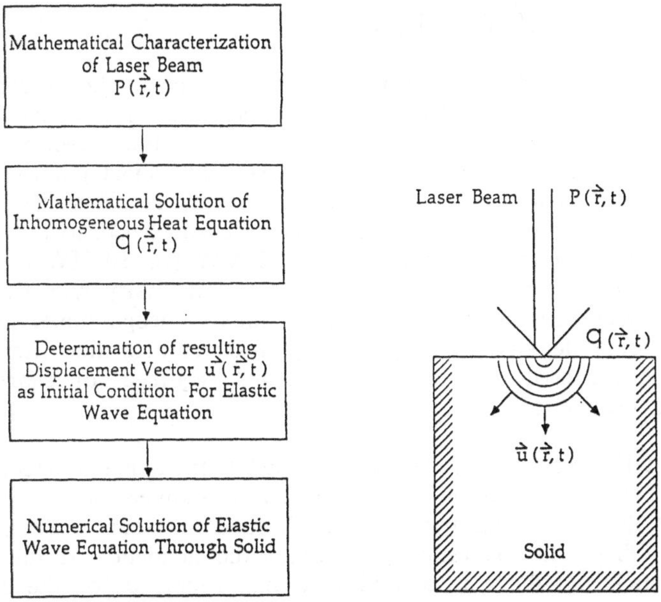

Fig. 1. General Approach to the Modeling of Laser Generated Ultrasound

these processes to laser generation of ultrasound may be accomplished through the approach shown in Fig. 1.

LASER EXCITATION SYSTEM

Laser generated ultrasonics (LGU) is a counterpart of piezoelectric transducer generated ultrasound (TGU) for nondestructive evaluation. The inspection speed of TGU technology is slow, however, because of the limited frequency response of PZT crystals and limited scanning speed of the transducer whose surface has to be in direct contact with couplers and/or the specimen. On the other hand, LGU technology has the potential to improve inspection speed dramatically. Because of the narrow laser pulses obtainable from N_2-lasers and from Q-switching or mode-locking of solid-state lasers, these pulses are intrinsically high-frequency sources of ultrasound excitation. Furthermore, recent technological development of light beam steering allows one to scan a large area in a very short time.

Suppose a high-power laser-pulse excites the surface of a specimen and the generated signals are detected at the point of excitation. Naturally this configuration allows the highest scanning speed, because one does not need to wait until the sonic wave propagates to the site of the detector. The time required is T_s, the round trip time of the wave across the thickness of the specimen. The existence of subsurface flaws can be detected by the elementary waves which travel back and forth between the flaw and material surface within the period T_s. Such a high frequency modulation over the slow wave structure of the surface undulation is a typical signature of material defects, as observed by the present authors as well as by AINDOW et al.[4] Since T_s is a measure of thickness, it can be very short when the specimen is very thin. Thus the first condition required for the laser pulse duration t_w is that $t_w \ll T_s$. A more precise expression may be

$t_w \leq T_f \qquad (T_f \leq T_s)$

Here, T_f is the time of flight from the wave launch to the return to the surface after being reflected by the subsurface flaw. The above condition indicates that a short duration pulse is essential for the detection of shallow defects. The N_2-laser produces reasonably short pulses (~5 ns), but the peak power available from commercial systems is not sufficient for metallic specimens. Experiments at the American Research Corporation of Virginia showed there is a minimum requirement of 1 MW (or 10 MW/cm^2) to generate clear signals from aluminum plates. The signal intensity was linearly proportional to power density and was neither linearly proportional to power density nor energy density. The pulse repetition rate is also limited for the N_2-laser because of the slow time constant required for charging the energy storage capacitor.

Recent LGU experimental results obtained from a metal specimen using a double-pulsed ruby laser as an excitation source showed two interesting features: (1) there is a finite time delay before the sonic wave lifts off (or launches) from the excitation point after the incidence of the laser energy (~100μs); (2) there is a minimum time required for the second pulse in order to reproduce the same waveforms as obtained from the first pulse (10-400 μs). The first phenomenon may relate to the time of energy conversion from radiation to thermal energies, thereby depending on the type of material. The second phenomenon depends primarily on the damping or resonance frequency of the material. Further study of the physical origin of these phenomena is needed since it limits the available repetition rate of excitation pulses.

The interval t_p of the repetitive excitation pulse should, therefore, satisfy the condition:

$$t_p \geq \gamma^{-1} \qquad \gamma = 2e^2\omega_o^2/(3mc^3)$$

Here, γ is the damping constant and ω_o is the characteristic resonance frequency of the specimen. Naturally, $t_p \gg t_d$ is also necessary, where t_d is the delay of the wave launching after $t = 0$, the laser incident time. Although small, the absolute value of t_d is important for precise measurement of the time-of-flight.

Although the ruby laser is an excellent high-power excitation source (3.3 MW, 30 ns), its high-speed repetitive operation is limited. Since continuous wave operation is impossible for the ruby crystal at the temperature, a constant amplitude pulse train cannot be expected even after a multiple Q-switching was successful within the flashlamp operating time. This poses a problem for through-transmission (C-scan) experiments. Further, present day optical fibers are less attenuative for the glass laser wavelength (1.06 μm) than for the ruby laser wavelength (0.69 μm). Nevertheless, the 3.3 MW excitation laser beam was successfully delivered by a flexible optical guide. The ruby laser beam was focused by a lens into the multimode multifiber optical guide whose transmission cross section was 3 mm in diameter. Upon incidence, the input surface of the cable suffered no damage.

A microscope objective (10x, 0.25 NA) was used to focus the laser beam into a spot-sized area. This success ensures high-speed steering and scanning of the excitation beam without using a computer-controlled polygon

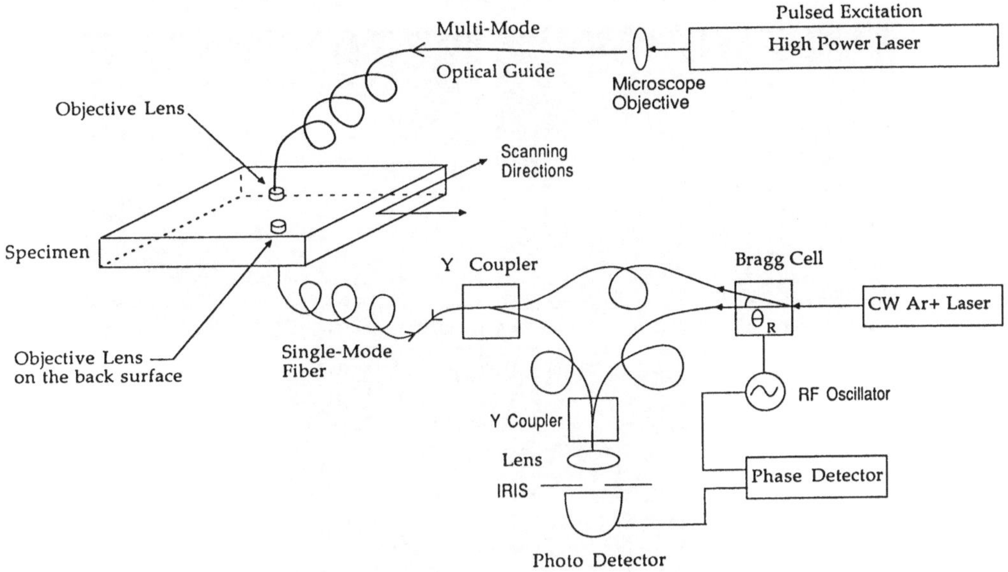

Fig. 2. Two-Fiber System for Through-Transmission Experiments

reflector. The experimental arrangement used is similar to that in Fig. 2, except that an He-Ne laser replaces the Ar ion laser. It incorporates two types of fiber guides: multimode fiber for the excitation laser and single-mode fiber for the interferometric detection laser.

An experiment was performed with an Nd-YAG laser as the excitation source for ultrasound in metals. The laser peak power corresponding to an average power of 9 W at a repetitive pulse frequency (RPF) of 1 kHz was 150 kW. Since nominal pulse width at this RPF is 60 nsec, the laser energy per pulse was 9 mJ. Unexpectedly, this small power/energy pulse (150 kW/9 mJ) excited well-defined acoustic signals in metals, although the signal had to be amplified 40 dB (using a Panametric Ultrasonic Preamplifier). In previous experiments with a ruby laser, a single pulse at a typical peak power/energy of 1 MW/30 mJ was necessary to obtain clear acoustic signals without an amplifier. This demonstrates that the signal intensity is still in the linear regime (or proportional) when the input laser power is as low as 150 kW. The metallic absorption of laser beam energy is supposedly less for the Nd-YAG laser (1.06 μm) than for the ruby laser (0.6943 μm). The photon energy of the Nd-YAG laser is smaller because it has a longer wavelength than the ruby laser. This drawback of the Nd-YAG laser is compensated for by its superior beam quality.

Experiments were also conducted on the surface of a 4-foot railroad beam containing several large surface defects (spall/blemish areas). A dedicated fiber optic interferometric sensor was used for noncontacting flaw inspection. The presence and absence of defects were faithfully identified by scanning the area with two laser spots. The photographically recorded signals (Fig. 3) were compared with the PZT-transducter detected signals. Both noncontacting fiber optic interferometer sensing and contact PZT-transducer detection methods showed very similar signatures; high frequency waves were heavily damped when defects were between the two laser spots. Since the

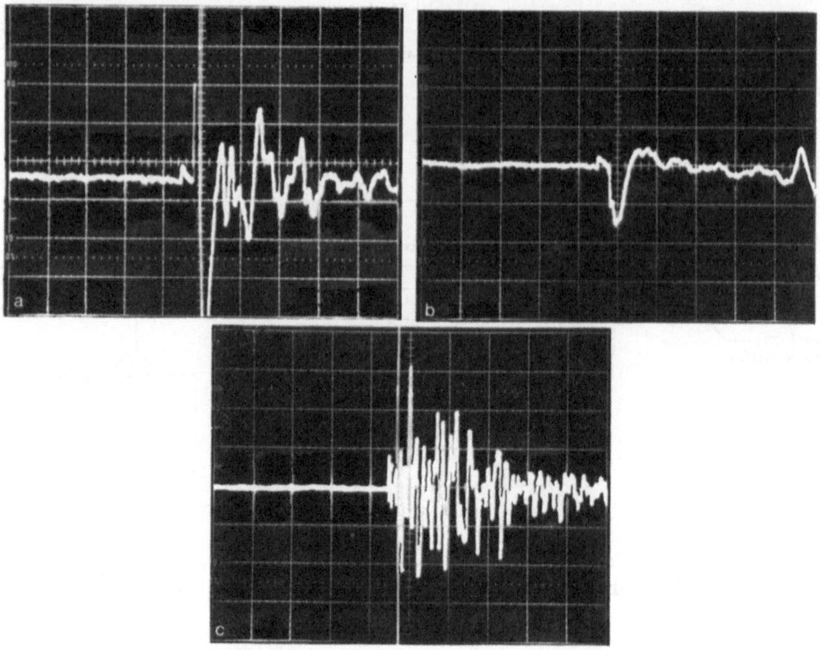

Fig. 3. Laser generated ultrasonic responses in a rail with surface defects located between the excitation and the detection points on the rail head. Detected fiber optic interferometric signals for a no defect area (a) and for a defect area (b), detected piezoelectric signal for the no defect area (c).

depth of penetration of surface acoustic waves is a function of ultrasonic frequency content, the damping of high-frequency components for signals propagated across the shallow defect surface was expected.

CONCLUSIONS

The technical approach to this ongoing research consists of the extension of laser/material interaction theory to the generation of ultrasound in coupler structures and composites, the investigation of methods to achieve a high-speed, high-resolution laser excitation system and the acquisition of laser interferometric detection systems. The overall goals are to demonstrate the feasibility of a research prototype system employing a practical, field-usable concept of source/detector integration for laser-based ultrasonics and to determine optimal ultrasonic wave generation and detection parameters for rapid area scanning and signal interpretation.

REFERENCES

1. C. B. Scruby, R. J. Dewhurst, D. A. Hutchins, and S. B. Palmer, Laser Generation of Ultrasound in Metals, in: "Research Techniques in Nondestructive Testing," R.S. Sharpe, ed., Academic Press, New York (1982).

2. C. Birnbaum and G. S. White, Laser Techniques in NDE, in: "Research Techniques in Nondestructive Testing, Vol. 7," R. S. Sharpe, ed., Academic Press, New York (1984).
3. W. W. Duley, "Laser Processing and Analysis of Materials," Plenum Press, New York (1983).
4. A. M. Aindow, R. J. Dewhurst, and S. B. Palmer, Laser-Generation of Directional Acoustic Surface Wave Pulses in Metals, Optics Comm. 42:116 (1982).

ELECTROMAGNETIC TRANSDUCERS FOR GENERATION AND DETECTION OF ULTRASONIC WAVES

P. J. Latimer and H. L. Whaley

Nondestructive Methods and Diagnostics Section
Babcock and Wilcox/Lynchburg Research Center
Lynchburg, VA 24506-1165

ABSTRACT

Electromagnetic Acoustic Transducers (EMATs) provide a noncontact method of generating ultrasound in metals. In its simplest form, an EMAT is a coil of wire and a magnet. An RF signal applied to the coil induces surface currents in the metal. These surface currents are acted upon by a Lorentz force due to the presence of the static magnetic field. This disturbance is transferred to the lattice of the solid by collisions resulting in body forces which drive the elastic wave. EMATs can be used to produce all of the wave modes produced by conventional piezoelectric and magnetostrictive transducers.

There are many advantages to EMATs which include operation without couplant, flexibility of wave modes, repeatability in sensor fabrication, and high temperature applications. The most significant disadvantage to EMATs is the low conversion efficiency; the loss is at least 50 dB, one way. However, achieving even this level of conversion efficiency requires special instrumentation, including carefully designed high-power RF drivers and very-low-noise preamplifiers. This means that EMAT applications should be chosen carefully. Inspite of this, EMAT technology is successfully making the transition from the laboratory to a versatile field inspection method.

INTRODUCTION

The usefulness of ultrasonic techniques in NDE applications is well established. The generation of ultrasonic waves is achieved primarily by means of some form of electromechanical conversion, usually piezoelectricity. This highly efficient method of generating ultrasound has a primary disadvantage: it requires the use of a fluid to mechanically couple the sound into the component being tested. The necessity of using a couplant means that the object being tested must either be immersed or be covered with a thin layer of fluid. This requirement complicates the testing procedures and, often, reduces inspection rates. In some cases, the test is not possible at all, because of this requirement. Couplant cleanup is also a significant problem in certain applications. It, therefore, becomes evident that an ultrasonic technique that does not require a couplant would have many advantages to offer in practical applications.

A technique possessing the attribute of not needing a fluid couplant has been developed within approximately the last twenty years. The technique depends upon electromagnetic acoustic interaction for elastic wave generation. An inductive element placed close to a metal surface induces eddy currents that experience a Lorentz force. This force then generates a mechanical disturbance by coupling to the atomic lattice by means of a complicated scattering process.[1] In electromagnetic acoustic generation, the electromechanical conversion takes place directly within the eddy current skin depth. Therefore, no mechanical coupling to the body is needed. The metal surface is its own transducer. Similarly, the reception takes place in a reciprocal way. The total class of transducers and receivers is referred to as Electromagnetic Acoustic Transducers (EMATs).

EMATs are capable of producing surface waves,[2] longitudinal waves, and vertically and horizontally polarized shear waves. The absence of a couplant makes it possible to design transducers that operate at elevated temperatures[1,3] and allows the use of rapid scanning. In addition, the operating characteristics of EMATs can be reproduced from one unit to another quite easily which makes them potentially useful as ultrasonic standards.

There is one major disadvantage to EMATs when compared to piezoelectric sensors. The insertion loss of EMATs can be as much as 50 dB[1] or more when compared to piezoelectric sensors. This makes the design of the instrumentation crucial. EMAT pulsers are low impedance devices. EMAT receivers are specially designed units that exhibit low noise input when operated in conjunction with low impedance loads. It, therefore, becomes important to choose applications where EMATs offer distinct advantages over piezoelectric sensors.

THEORETICAL CONSIDERATIONS

Nonferromagnetic Conductors

In its simplest form, an EMAT is a coil of wire and a magnet. If the coil of wire is excited with an RF signal and placed close to the surface of the conductor, then eddy currents will be induced in the conductor. These currents will be acted upon by a Lorentz force due to the presence of the static magnetic field. This is similar to the operation of an electric motor. The disturbance is transferred to the lattice of the solid by collisions, and this is the source of the elastic wave.

It should be noted that the electromechanical[1] conversion takes place directly within the eddy current skin depth. It should also be noted that the Lorentz process is reciprocal. Therefore, when EMATs are used as receivers, the elastic wave causes the metal surface to vibrate in the presence of the static magnetic field. This in turn produces an electric current that is sensed by the coil near the surface. This is similar to an electric generator.

Referring to Fig. 1,[2] it will be noted that for bulk waves the mode of the wave generated depends upon the relative direction of the static magnetic field with the conductor surface. If the magnetic field is perpendicular to the surface, then a shear wave is favored over longitudinal wave. If the field is parallel to the surface, then the longitudinal wave is favored. Finally, if the magnetic field vector makes some angle with the conductor surface, such as with a fringing field,[3] then both ultrasonic modes can be excited. A comprehensive bibliography of EMAT principles is covered by Frost[4] and Maxfield et. al.[5]

Ferromagnetic Conductors

The Lorentz theory of electromagnetic-acoustic generation generally applies to nonferromagnetic conductors. For ferromagnetic conductors,[2] the Lorentz force alone cannot account for the electromagnetic/acoustic generation/reception processes. The influence of magnetostriction also must be taken into account. In some magnetic materials, the time dependent magnetic field can modulate the magnetization which produces periodic magnetostrictive stresses that exist in addition to the stresses produced by the Lorentz mechanism.

A plot of the signal amplitude as a function of magnetic field is shown in Fig. 2. Note that the magnetostrictive contribution is generally signifi-

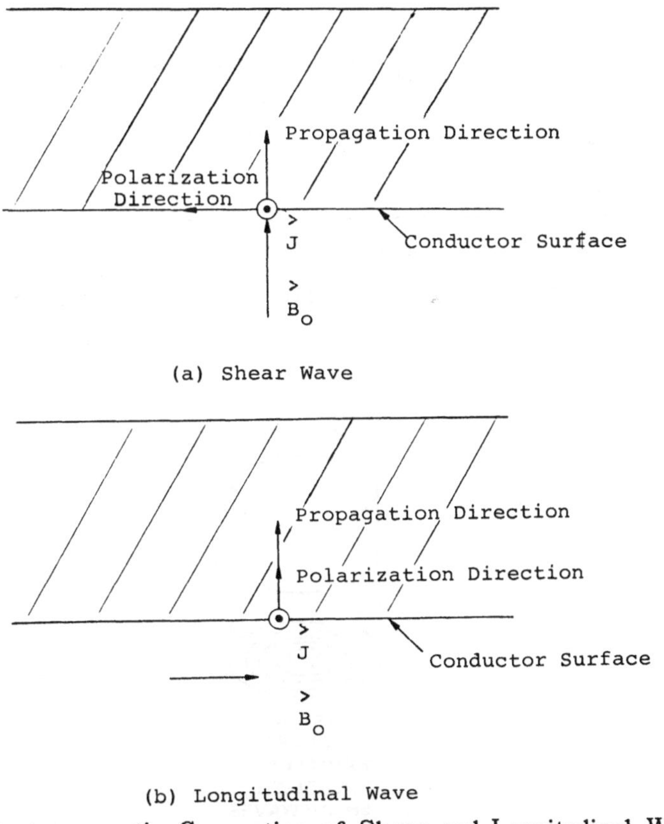

Fig. 1. Electromagnetic Generation of Shear and Longitudinal Waves (J is out of the page (from Reference 2)

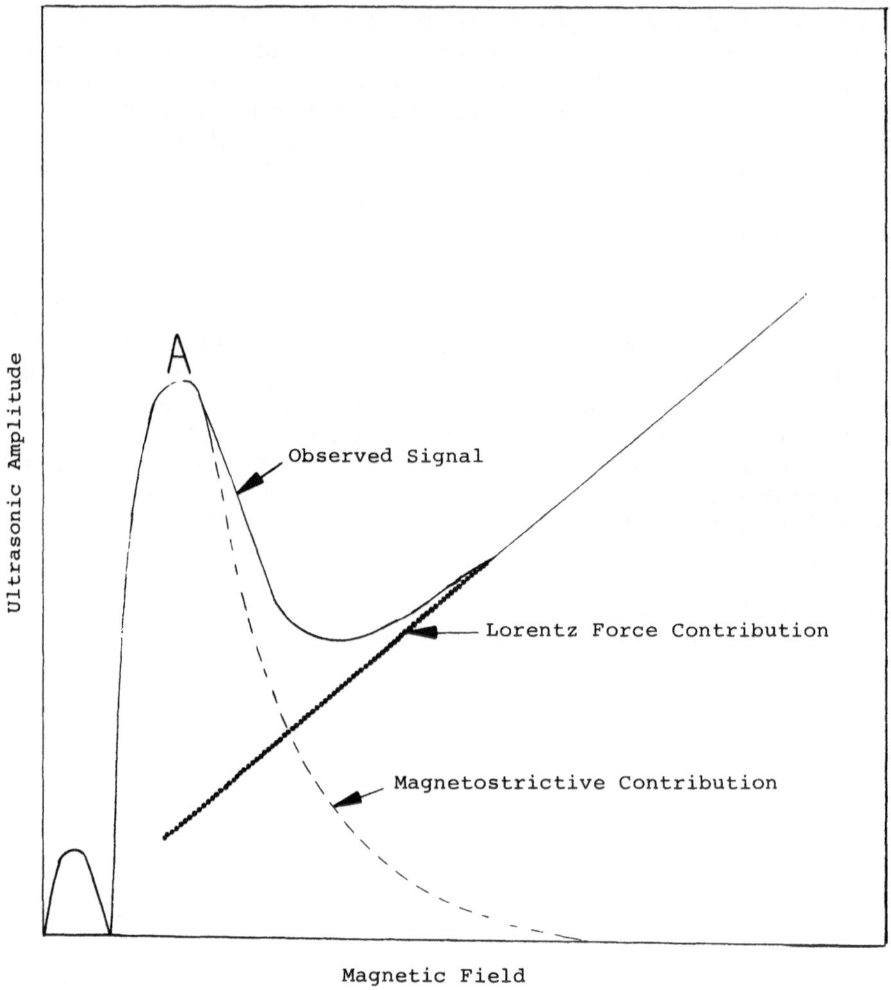

Fig. 2. The Magnetostrictive Effect in Ferromagnetic Solids

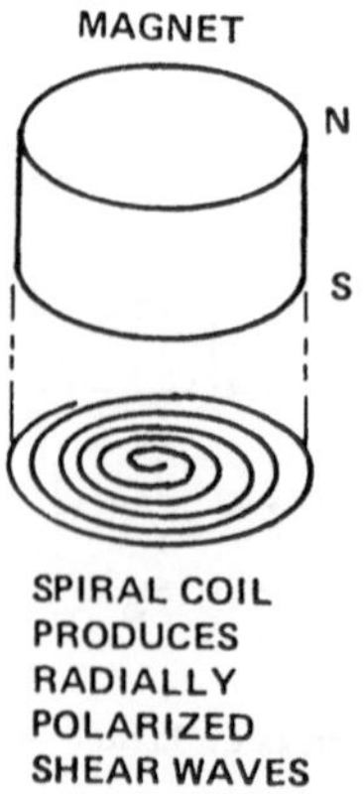

Fig. 3. The Spiral Coil EMAT

cant for lower fields. This actually enables one to achieve a much higher signal amplitude than would be possible with the Lorentz interaction alone. This is very useful, particularly when using small magnets for the bias field. At much higher fields which are well above the saturation value, it can be observed from Fig. 2 that the Lorentz force contribution dominates.

EMAT SENSORS

The Spiral Coil EMAT

The spiral-coil EMAT is illustrated in Fig. 3. The static magnetic field may be produced by either an electromagnet or a permanent magnet. The permanent magnets used until very recently were Samariam Cobalt; however, a new material called Neodymium-Iron-Boron has recently been introduced. This new material produces 30% higher fields than Samarian-Cobalt magnets.

Assuming that there is no fringing of the field, parallel to the coil, a radially polarized shear wave is produced. Since there is always a small gradient of the field lines parallel to the coil, there will always be a small amplitude longitudinal wave also present. By careful design considerations, however, the longitudinal component can be made almost negligible.[3]

A typical representation of a directivity pattern for the spiral EMAT[5] is shown in Fig. 4. It will be noted that there is always a small null in the center of the pattern due to phase cancellation, i.e., one side of the coil is 180° out of phase with the other side. This null will of course have to be considered when interpreting data from defect detection, etc. The received amplitude from EMAT decreases exponentially with liftoff.

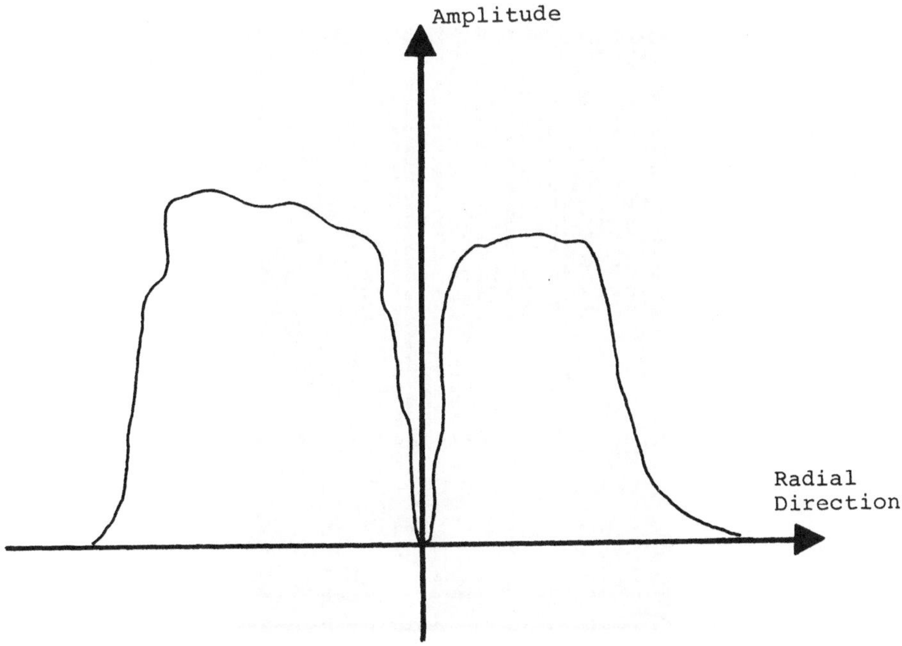

Fig. 4. Approximate Directivity Pattern for a Spiral Transducer

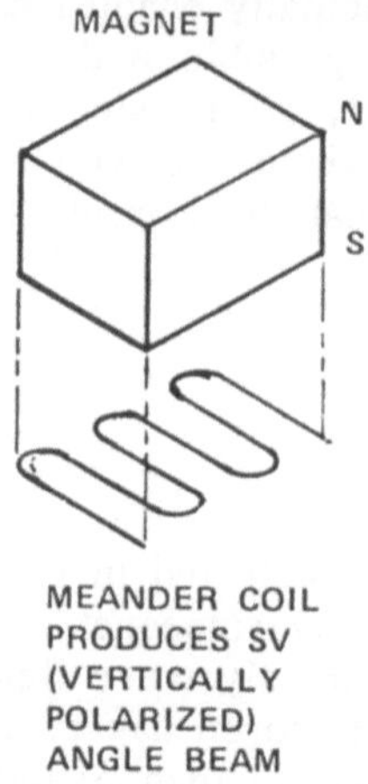

Fig. 5. The Meander Coil EMAT

The spiral coil EMAT is a highly efficient EMAT. When it is used with a permanent magnet, it externally resembles a conventional piezoelectric contact transducer.

The Meander Coil EMAT

The meander coil EMAT is shown in Fig. 5. The direction of the field can either be normal as shown in Fig. 5 or tangential. Referring to Fig. 6, let the distance D between adjacent conductors be $\lambda_s/2$ where λ_s is the wavelength for Rayleigh or surface waves. If the coil is excited with a signal of wavelength $\lambda = 2D$, then there will be periodic stresses on the surface of the

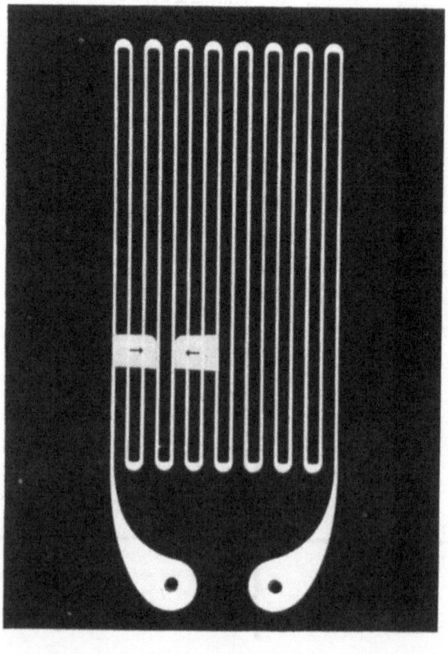

Fig. 6. The Separation of Adjacent Conductor in the Meander Coil is D

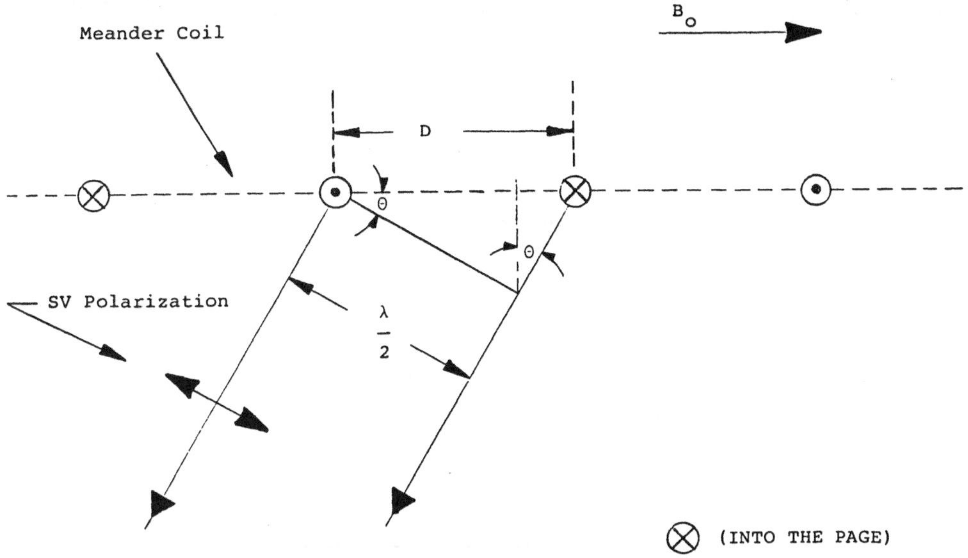

Fig. 7. Angle Beam Produced by Meander Coil

conductor which will produce a Rayleigh or Lamb wave depending upon the ratio of wave length to thickness of the conductor. If the wave length is comparable to the thickness, Lamb waves will be produced. If it is much smaller, a Rayleigh wave is produced.

If $\lambda < 2D$ then phase matching can occur at some angle θ. Referring to Fig. 7[6,7] $\lambda = 2D \sin \theta$ or in general $m\lambda = 2D\sin\theta$. This results in coupling to bulk modes. In general, there will be angle beams produced for both SV shear waves and longitudinal waves. It should also be noted that the meander coil is bidirectional, that is, it generates waves traveling in opposite directions.

In order to get some insight into the meaning of SV and SH polarization, refer to Fig. 8. Polarization in the plane defined by the incident and reflected rays is referred to as SV polarization. Polarization perpendicular to the plane defined by the incident and reflected rays is referred to as SH polarization.

SH Shear Wave EMATs

Shear waves polarized in the SH direction use a periodic array of permanent magnets with an encircling coil as shown in Fig. 9.[8-10] These EMATs have the capability of producing angle beam SH shear waves. This cannot be done conveniently with conventional ultrasonics. Only SV angle beam shear waves can be produced by a conventional piezoelectric transducer and wedge.

Shear waves with SH polarization are particularly useful since there is no mode conversion upon reflection from a plane interface. This greatly simplifies data interpretation in applications such as weld inspection. In addition, SH waves can propagate at grazing incidence. This is an important property. The primary beam angle is given by $\theta = \sin^{-1}(\lambda/2D)$ where D is the distance between adjacent magnets.

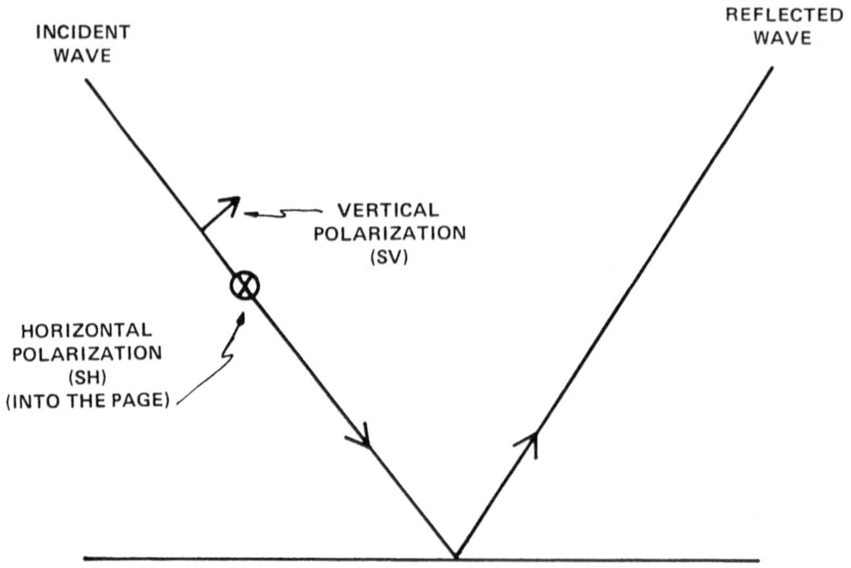

Fig. 8. Illustration of Horizontal (SH) Polarization and Vertical Polarization (SV) for Shear Waves.

This relationship also demonstrates another unique property of this EMAT. The beam angle can be scanned through almost 90° by merely changing the frequency. It should also be noted that this is also a bidirectional beam.

Thus far, the SH EMATs are limited to reasonably low frequencies (less than 2 MHz) due to the difficulty in machining thin sections of magnet materials.

APPLICATIONS

At the present time, more than 60% of the fossil fired utility boilers are reaching the end of their design life which is about 30 years. However, due to the increased emphasis upon maintaining fossil utilities as opposed to nuclear facilities, there is a demand to extend the operational life of these boilers another 20 to 30 years.

The most common cause of forced outages in fossil units is boiler tube failure. Of the failure modes, perhaps the most serious is hydrogen damage. The end result of hydrogen damage can be a sudden blowout. Also in some units, a very large number of welds can be damaged resulting in an almost continual chain of forced outages. Unscheduled outages of this type can be very expensive, up to $250,000/day in lost power revenues.

At the Babcock and Wilcox Lynchburg Research Center (B&W/LRC), a technique using conventional ultrasonics was developed for the detection of hydrogen damage. This technique proved to be very effective and led eventually to a 100% inspection of the furnace walls in a number of units. In one instance, this amounted to 16 miles of tubing and 1100 welds. The job required 250 gallons of ultrasonic couplant at a cost of about $30/gallon. This proved to be an ideal application for EMATs. An outside vendor was used

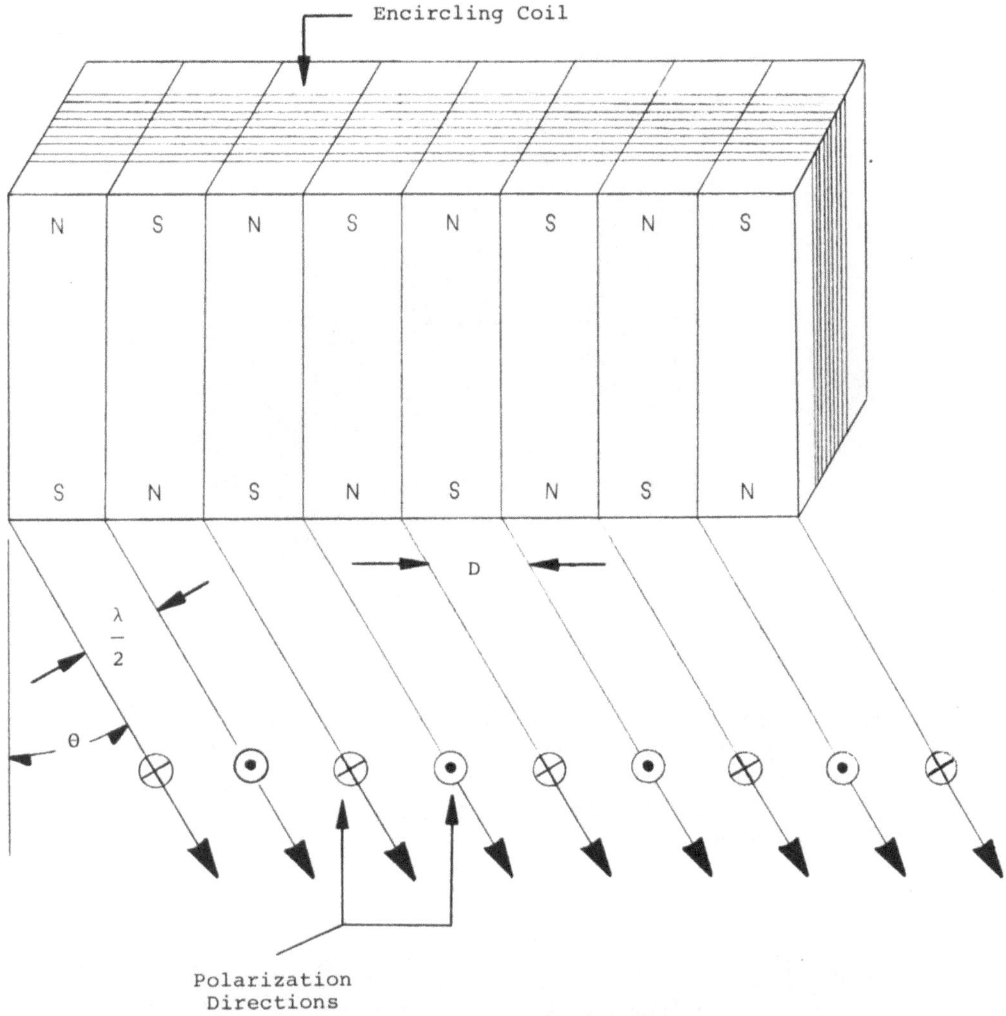

Fig. 9. SH Shear Wave EMAT

to develop the electronics for a portable system to meet the specifications for the hydrogen damage test.

The portable EMAT inspection system for hydrogen damage, shown in Figs. 10 and 11, has just been completed. Figure 10 shows the sensor used for scanning straight sections of tubing. The scan rate is extremely rapid, and the sensor is highly sensitive to hydrogen damage. The weld probe shown in Fig. 11 can be used to scan a typical weld in approximately 1 to 2 seconds. An automated data acquisition system will be implemented in the near future.

Another application for EMATs which is currently in progress is the use of surface waves to detect seams and laps in bars. These seams and laps cannot be detected by conventional techniques such as flux leakage due to their tightness. In order to accomplish this, it was necessary to develop a highly advanced AC magnet to be used with the EMAT. This system is shown in the first stages of field testing in Fig. 12. EMATs are particularly suited for this application because they are highly efficient at generating surface waves.

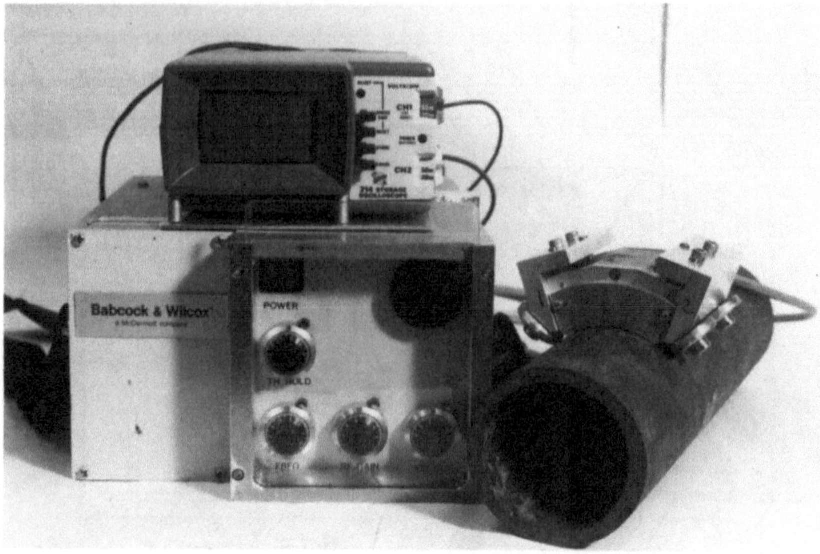

Fig. 10. EMAT System for Detection of Hydrogen Damage in Straight Tube Sections.

B&W/LRC has for a number of years been very much interested in EMAT research, development, and applications. Other applications in the fossil, nuclear, and tubular products areas are currently in the planning stage. It has become apparent that the number of practical applications of EMATs in NDE is rapidly accelerating.

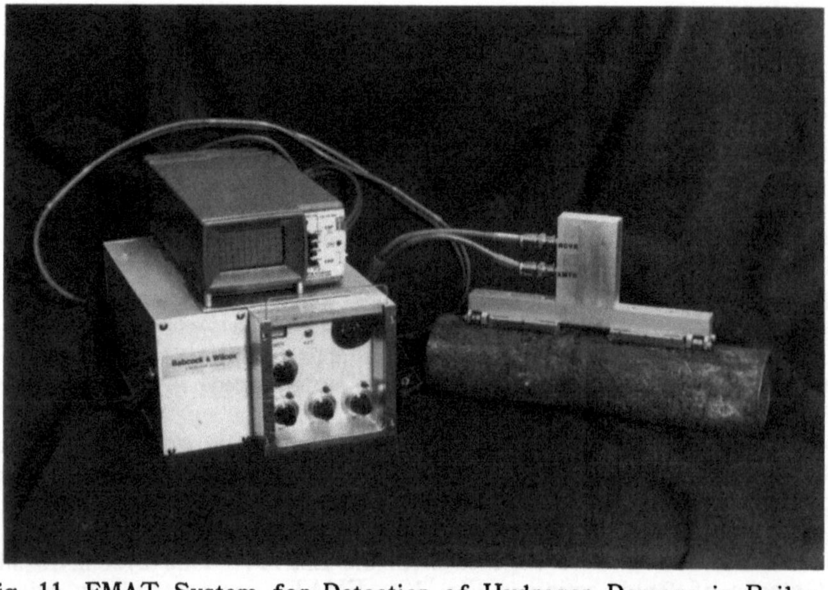

Fig. 11. EMAT System for Detection of Hydrogen Damage in Boiler Tube Welds

Fig. 12. EMAT System for Detection of Seams and Laps in Bars

ACKNOWLEDGEMENTS

The authors gratefully acknowledge the technical contributions made by J. H. Flora, K. C. Henderson, and B. I. Cantor.

REFERENCES

1. B. W. Maxfield and C. M. Fortunko, Design and Use of Electromagnetic Acoustic Wave Transducers (EMATs), Matls Eval. 41:1399 (1983).
2. R. E. Beissner, "Electromagnetic Acoustic Transducers: A Survey of the State-of-the-Art, NTIAC-76-1," Nondestructive Testing Information Analysis Center, San Antonio (1976).
3. B. W. Maxfield, M. Light, W. W. McConnaughey and J. Hulbert, Design of Permanent Magnet Electromagnetic Acoustic Wave Transducers (EMATs), in: "IEEE Ultrasonic Symposium Proceedings" (1976).
4. H. M. Frost, Electromagnetic Ultrasonic Transducers: Principles, Practice, and Applications, in: "Physical Acoustics, Vol. XIV," W. P. Mason and R. N. Thurston, eds., Academic Press, New York (1977).
5. B. W. Maxfield and T. K. Hulbert, Electromagnetic Acoustic Wave Transducers (EMATs) Their Operations and Mode Pattern, in: "Proc. 10th Symposium on NDE," San Antonio, (1975).
6. M. G. Silk, "Ultrasonic Transducers for Nondestructive Testing", Adam Hilger Ltd Tech. House, Bristol (1984).
7. T. J. Moran and P. M. Panes, Electromagnetic Generation of Electrically Steered Ultrasonic Bulk Waves, J. of Appl Phys 47:2225 (1976).
8. G. F. Vasile and R. B. Thompson, Periodic Magnet Non-Contact Electromagnetic Acoustic Wave Transducer - Theory and Application, in: "IEEE Ultrasonic Symposium Proceedings Collected Papers on Nondestructive Evaluation and Industrial Applications,".

9. C. M. Fortunko and R. E. Schram, An Analysis of Electromagnetic Acoustic Transfer Arrays for Nondestructive Evaluation of Thick Metal Sections and Weldments, in: "Review of Progress in Quantitative Nondestructive Evaluation," Volume 2A," D. O. Thompson and D. E. Chimenti, eds. Plenum Press, New York (1983).
10. J. F. Martin and R. B. Thompson, The Twin Magnet Configuration for Exciting Horizontally Polarized Shear Waves, in: "Proceedings of the IEEE Ultrasonics Symposium", (1981).

APPLICATIONS AND ADVANTAGES OF DRY COUPLING ULTRASONIC TRANSDUCERS FOR MATERIALS CHARACTERIZATION AND INSPECTION

J. A. Brunk

Ultran Laboratories, Inc.
Overland Park, KS

C. J. Valenza and M. C. Bhardwaj

Ultran Laboratories, Inc.
State College, PA

ABSTRACT

Practical self-coupling (dry coupling) piezoelectric ultrasonic transducer designs have been optimized for center frequencies up to at least 25MHz for longitudinal wave generation and at least 20MHz for zero degree incidence shear wave generation. While originally developed for examination of materials which would be damaged by liquid contact, they have been found to have practical and technical advantages for many materials characterization, discontinuity evaluation and gaging requirements.

Saving the time which would otherwise be required for the application and removal of liquid couplant is the most obvious advantage. Liquid intrusion is likely to cause unreliable data from porous materials and other materials which absorb water. Self-coupling facilitates the examination of assemblies with components such as electronic circuitry which could be damaged by liquids. Proper application of couplant is quite tedious for zero degree incidence shear wave examinations. Verification of complete removal of couplant is required where residue could interfere with subsequent processes such as chemical milling. For examination of low acoustic impedance materials, self-coupling provides increased transmission of ultrasonic energy into the test object and also improved near-surface resolution. Examples of several types of advantageous applications will be presented.

INTRODUCTION

The transducers discussed herein are, in most respects, quite similar to conventional direct contact and delay line piezoelectric transducers. They

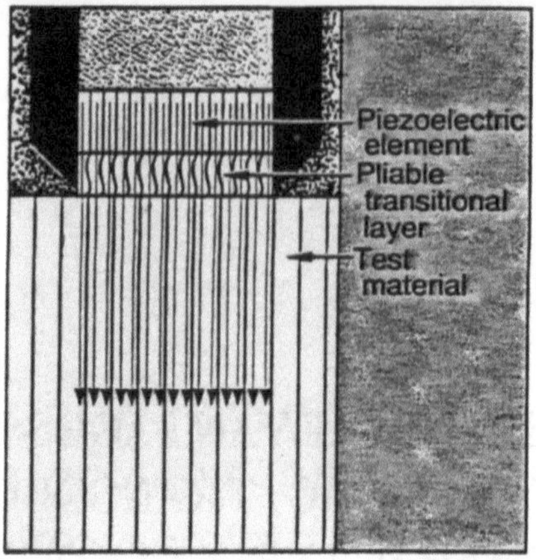

Fig. 1. Cross-section of contact self-coupling transducer

have been found to be direct replacements for their liquid-coupled counterparts for many applications.[1-3] The most obvious physical difference is between standard and self-coupling direct contact transducers. The usual hard surface wear plate is replaced by an acoustically active solid compliant layer which serves as the coupling medium, as shown in Fig. 1. Self-coupling delay lines are rigid like standard delay lines, except for a thin transitional layer on the contact surface.

Self-coupling transducers are used as if they were standard transducers with the couplant built-on or permanently attached. This is an oversimplification, because the entire transducer is designed especially to enhance self-coupled performance.

The acoustically active solid compliant layer is a proprietary material developed to produce good performance under a wide variety of testing conditions. It serves to reduce what would otherwise be a gross acoustic impedance mismatch between a hard transducer face and a test surface. It must also fill in surface irregularities on the test object so that there will be no significant air gaps at the interface. The material can be modified to meet special requirements. However, the same material is currently used for transducers with center frequencies from less than 200kHz to over 25MHz. All of the self-coupling data and oscilloscope photographs presented herein were produced using standardized self-coupling designs.

Self-coupling transducers can be used with liquid couplants without damage to the contact surface. There would normally be no reason to do so in actual materials examination operations. Some of the dry/wet coupling comparisons reported were made using the same transducer with and without couplant. This was done so that any changes in material response would be due entirely to coupling and not to variations in any other transducer characteristics. Repeatability studies, where absolute values were of less interest than statistical distributions, have generally been made comparing self-coupling with standard transducers of similar frequency range and sensitivity.

BACKGROUND

Transducers of this type were first made as direct contact longitudinal wave devices. They were developed for use on test surfaces where any liquid contact was strictly forbidden. They were made with center frequencies from 2 to 10MHz. The initial goal of producing transducers which would meet minimum performance requirements for this application was considered fairly ambitious, and it was a delightful surprise to discover that they performed extremely well. This led to several investigations of other possible applications since 1982, and to the extension of the operating frequency range as well as the creation of a complete "family" of self-coupling transducer types.

Direct contact styles now include dual element and zero degree incidence shear wave self-coupling transducers. Delay line longitudinal wave and zero degree incidence shear wave types have also become standard products.

Self-coupling angle-beam probes utilizing plastic wedges have also been made. These include longitudinal wave, shear wave and surface wave probes. They are not practical for the typical angle-beam application of weld inspection which requires "scrubbing" the probe over a relatively rough surface, but they have been used for a number of applications where "scrubbing" is not required. Dry coupling of the transducer to the replaceable wedge has been found to be quite advantageous as it provides stable performance for very long periods of time and eliminates the nuisance of replenishing transducer-wedge couplant periodically.

Direct contact longitudinal wave dry-coupled transducers can be constructed to also be very good transmitters and receivers of surface waves. This allows the possibility of some interesting test techniques using both wave types simultaneously. Construction which greatly suppresses surface wave transmission is also possible.

Self-coupling transducers are by no means a laboratory curiosity of limited practical value. They have undergone extensive field tests over periods ranging from weeks to years, and their design has been refined to make them useful and cost-effective for a very wide range of applications.[1-3] But they are by no means universally applicable. While it appears safe to say that virtually any contact ultrasonic examination could be performed using a self-coupling transducer, there are many instances where this would not be the sensible choice.

INHERENT CHARACTERISTICS

The materials which have been found to function most effectively as self-coupling transitional layer over a wide frequency range are low in acoustic impedance. This makes the transducers inherently more efficient for use with low acoustic impedance test objects than with high impedance materials. With most metals, for example, there is less energy transmission into the test object when a self-coupling transducer is used. This places some limitations on the use of self-coupling transducers with metals and gives some advantages for the examination of many nonmetals.[3] In many situations the differences are relatively minor, and the same results can be obtained with either type of device through minor adjustments to instrument settings.

To replace a standard transducer for an existing setup, it may sometimes be necessary to use a higher sensitivity or higher resolution design for the self-coupled device.

To be compliant, the transitional layer obviously cannot be as rugged and durable as a hard surface wear plate. Practical applications are for the most part limited to those where "scrubbing" the transducer over the test surface is not necessary. They have been used in robotic and other automated point-to-point contact systems. This can in effect give the same scan plan produced with many automated immersion testing systems where the ultrasonic instru-mentation is turned on after each step of the mechanical system.

Effective sliding contact is possible on very smooth surfaces, but there is often some difficulty in maintaining uniform contact pressure while in motion.

Sliding contact is normally not an issue in material characterization studies, acousto-ultrasonic examinations, robotic applications and thickness gaging. It is in these applications where self-coupling transducers are most likely to be particularly advantageous and/or cost-effective.

Damaged transitional layers can be replaced if there has been no damage to the underlying materials. However, it is not necessarily cost-effective to do this. In several studies of repetitive testing operations, it was found that a self-coupled transducer would in effect pay for itself in less than one day by saving the time that would normally be spent applying and removing couplant.

Self-coupling delay lines are much closer to their standard counterparts in durability and in performance. Since delay lines are much less expensive than transducers, economical as well as technical considerations favor self-coupling even more often when delay line devices are to be used.

The transitional layer is also compressible. This makes the ultrasonic signals somewhat sensitive to contact pressure, and spring loaded holders are advisable for some applications. The transitional layer thickness is matched to the transducer operating frequency range and is thinner for higher frequencies. Pressure effects are, therefore, inversely proportional to frequency and for most purposes negligible above 5MHz. It should be noted that pressure effects also occur when liquid couplants are used, and problems are not necessarily more pronounced with self-coupling.

The thinning of the transitional layer for higher frequencies leads to a practical upper limit for direct contact transducers of about 10MHz because of strength and durability. Delay line construction makes higher frequencies practical up to at least 25MHz. For transducers designed for much higher operating frequencies, there is a significant shift in response due to attenuation of higher frequencies by the transitional layer. For example, a transducer construction which provides a center frequency of 50MHz with glycerine couplant gives about 35MHz when completed as a self-coupling type.

Because self-coupling transducers are an excellent acoustic impedance match for many plastic and other nonmetallic materials, the entry surface reflection is often quite small in amplitude and short in time duration. This allows excellent resolution of near-surface discontinuities and the measurement of very thin sections.[4] At the same time sensitivity is improved with self-coupling. This is in contrast to the results typically observed with standard transducers where there is usually a tradeoff between sensitivity and resolution. The opposite is true when examining high acoustic impedance materials with direct contact self-coupling transducers. Both near-surface resolution and sensitivity are diminished compared to standard contact types[3]. However, these disadvantages are minimal or nonexistent when comparing self-coupling delay line and standard delay line transducers.

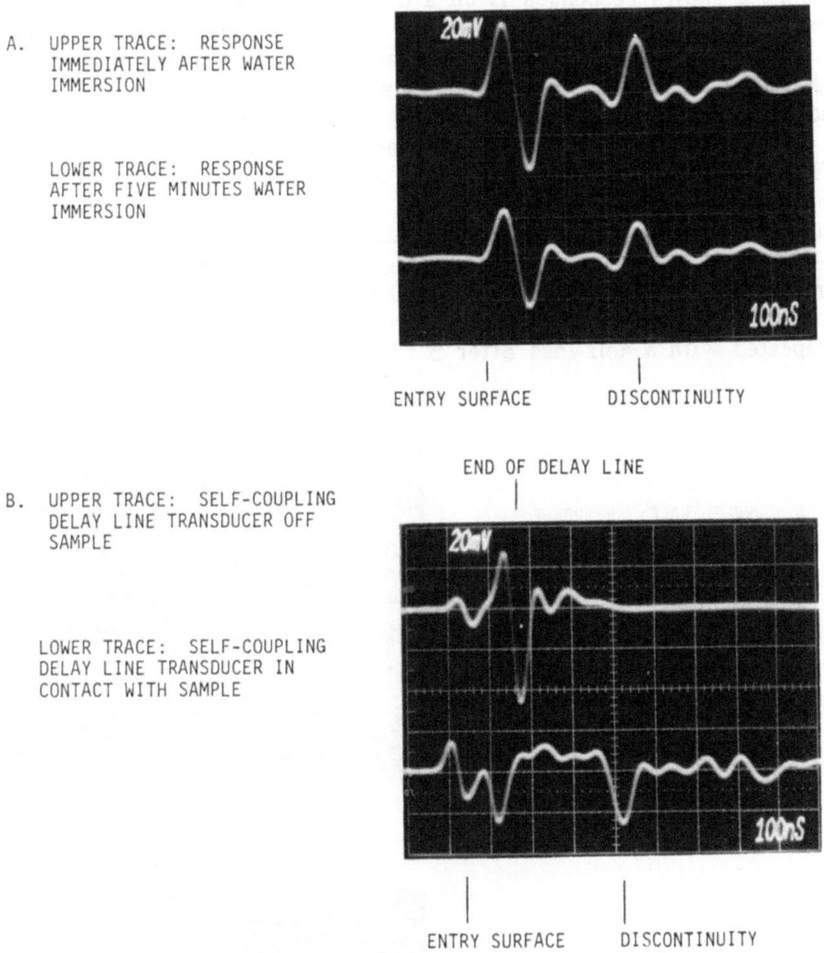

Fig. 2. Effects of brief water immersion on ultrasonic responses from a fiberglass/epoxy sample.

Self-coupling transducers require good alignment of the transducer with respect to the test surface for adequate transmission of ultrasonic energy. This makes it more difficult to perform a test incorrectly on curved surfaces where a liquid or gel could fill in the space with a considerably greater degree of misalignment.[4]

EXAMPLES

Porous Materials

Whenever a test specimen absorbs some of a liquid couplant, it becomes in effect a different material. The properties of the new material will change as more liquid is absorbed and penetrates deeper, and also as the material dries out after exposure to the liquid is ended. This can lead to variations in indicated acoustic properties and in responses to reference defects and natural discontinuities.

Figure 2 shows responses from a discontinuity in a fiberglass-epoxy composite, located 3.7mm below the test surface. The top photograph shows the response obtained with a 10MHz broad bandwidth immersion transducer immediately after placing the sample in water, and again after five minutes. No changes were made in the setup, and care was taken to assure that there were no air bubbles on the transducer face or sample surface. There was a significant loss of amplitude of both the entry surface and discontinuity reflections after five minutes in water. The initial amplitude of the discontinuity signal was 36mV. It was reduced by 33% (3.5dB) by water intrusion. Such a defect in a medium or large-sized structure could easily be immersed for a much longer time period before the scan path brought it into the ultrasound field. After the specimen was oven-dried overnight the test was repeated with a 4dB loss after 5 minutes.

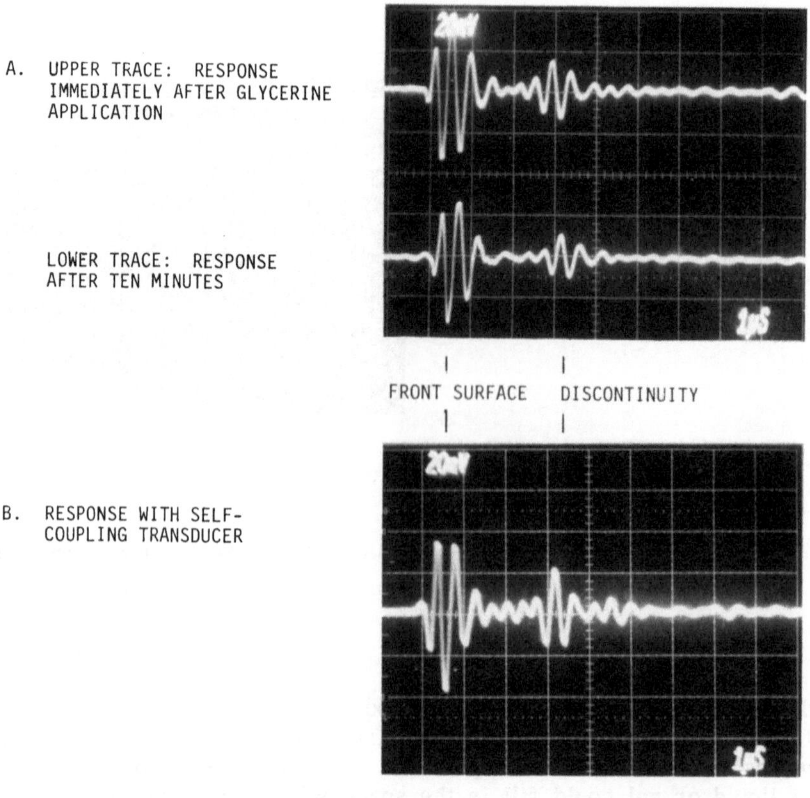

Fig. 3. Effects of couplant absorption on ultrasonic responses from a graphite/epoxy sample.

The bottom photograph was made using a self-coupling 10MHz broad bandwidth transducer. The upper trace shows the reflection from the end of the delay line not in contact with the specimen. The lower trace shows the response from the same discontinuity used for the top photograph. Using the self-coupling transducer the response was very stable and repeatable within less than 0.5dB as long as the specimen was not returned to the immersion tank. It is also notable that the time required for recovery of the entry surface reflection to the baseline level is about 50% less with the self-coupling transducer, while the discontinuity signal amplitude is almost the same as the initial value with the immersion setup.

Figure 3 shows similar results with a graphite-epoxy composite specimen examined with 2MHz direct contact transducers. The standard transducer was used with glycerine as a couplant. After 10 minutes exposure to the couplant, the discontinuity signal was reduced in amplitude by 3.5dB. With the specimen dry, the response with the self-coupling transducer was again stable and repeatable within less than 0.5dB.

Figure 4 shows the effects of water intrusion on the response of a core sample taken during a drilling operation. The first back reflection from the sample is gated to display its frequency spectrum. The top photograph was made using a self-coupling transducer. The transmit time to the peak (negative) of the back reflection was 7.3 microseconds, and the center frequency was 720kHz. The second photograph was made with a water-immersion setup using the same transducer so that probe characteristics would not influence the results. After less than 1 minute exposure to the water, the transmit time to the peak (now positive) of the back reflection was reduced to 6.9 microseconds, and the center frequency was increased to 840kHz. This was apparently due to the fact that water has higher ultrasonic velocity and less frequency attenuation than the air it displaced. The bottom photograph shows that after prolonged drying the material did not return to its pre-immersion state. Apparently the sample was permanently altered by erosion and/or reaction during its immersion in the water.

These results are typical of porous test specimens of a wide range of materials which have been examined by self-coupling and standard techniques. Dry coupling is necessary to give stable and repeatable results. Self-coupling transducers have been found to be suitable for characterization of a wide variety of porous ceramic material samples[5,6] and for discontinuity detection in several porous composites.[3]

Difficult Surface Contours

When the test surface is curved, the use of a liquid couplant permits transmission of ultrasonic energy between the transducer and test object over some range of alignment/misalignment of the transducer. This can lead to errors in thickness gaging or velocity measurement, and uncertainties about the actual sound path which can lead to errors in reflector location and evaluation. A properly chosen self-coupling transducer can virtually force correct alignment in many such situations by not allowing significant transmission until the transducer is properly positioned. Self-coupling transducers have been made with element diameters smaller than 1.5mm for limited access areas, and with shapes to match surface contours.

Figure 5 shows one example of the reduced possibility of error through the use of a self-coupling transducer on a curved surface. The test object

EXAMINED WITH SELF-COUPLING TRANSDUCER

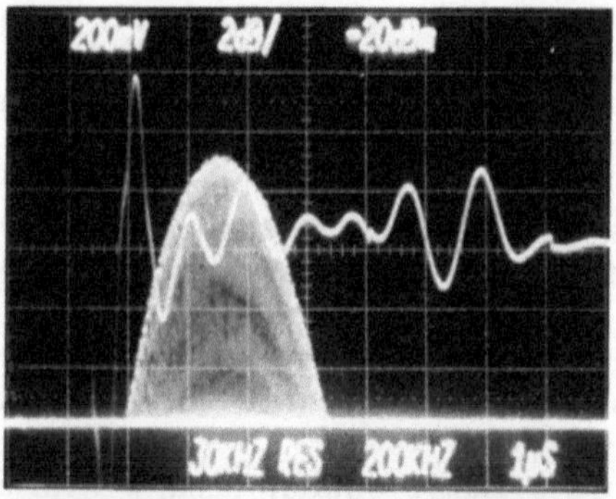

EXAMINED WITH WATER IMMERSION

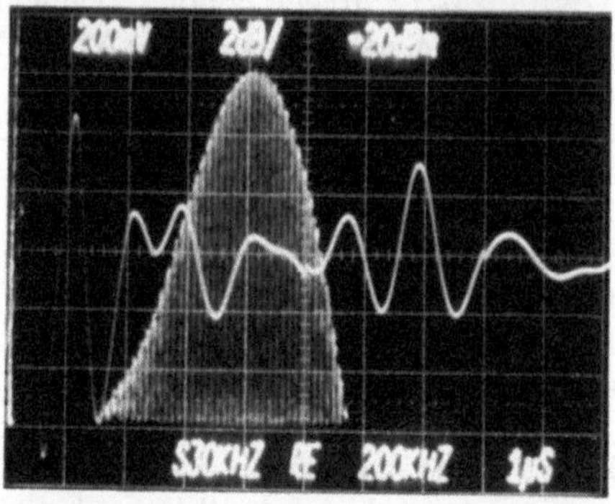

RE-EXAMINED WITH SELF-COUPLING AFTER IMMERSION TEST AND DRYING

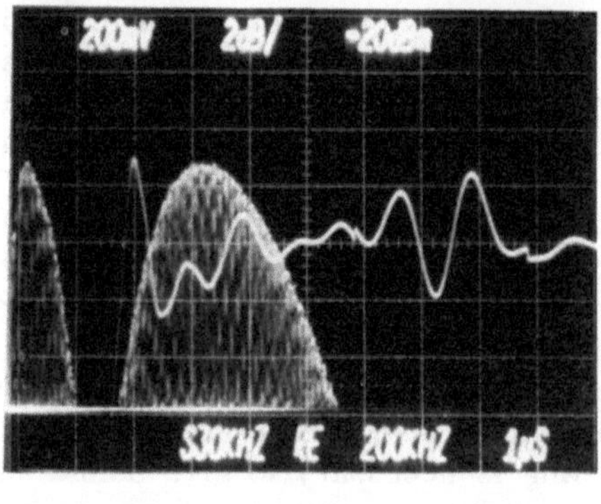

Fig. 4. Effects on Water Immersion on Porous Rock Core Sample

is a steel pipe, 74.3mm outside diameter. The same digital thickness gage was used with both standard and self-coupling 5MHz, 6mm diameter longitudinal wave single element direct contact transducers. The standard transducer normally used with the gage was coupled to the pipe surface with a commercial gel couplant. First, the self-coupling device was placed in contact with the pipe and moved with a rocking motion. This changed the alignment of the transducer axis with respect to the contact surface. The digital readout of the gage was observed while the first back reflection from the pipe ID surface was monitored with an oscilloscope. The indicated value of the nominal 7.87mm wall thickness ranged from 7.87 to 7.90mm as the transducer was rocked. Rocking to any greater extent resulted in no reading at all. The CRT traces corresponding to maximum, nominal, and minimum readings were photographed by multiple exposures on a single film. The photograph appears to show only two distinct traces with no observable difference in the leading edges of the signals.

STANDARD 5MHz, 6mm DIAMETER TRANSDUCER WITH GEL COUPLANT.

INDICATED WALL THICKNESS: 7.88 to 7.98mm

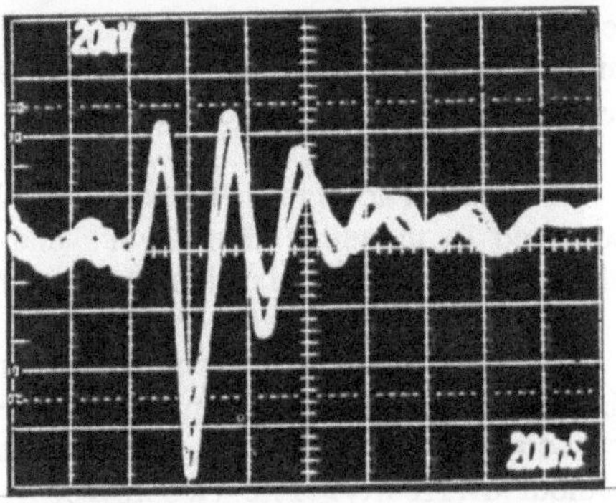

SELF-COUPLING 5MHz, 6mm DIAMETER TRANSDUCER

INDICATED WALL THICKNESS: 7.87 to 7.90mm

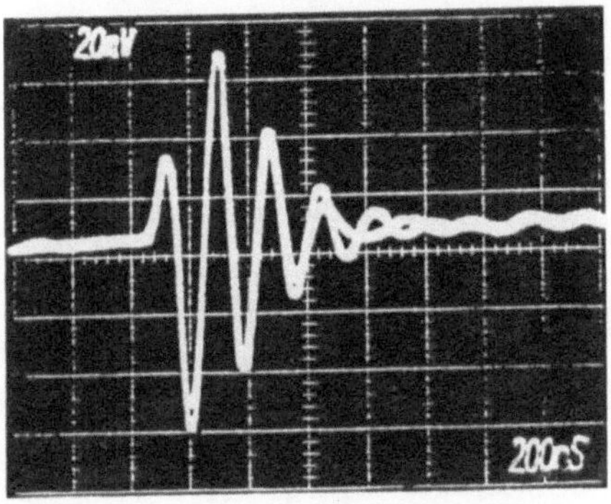

Fig. 5. Comparison of response variability of standard contact and self-coupling direct contact transducers on a curved surface of a alloy steel pipe, 74.30mm OD, 7.87mm nominal wall thickness.

The same process was repeated with the standard transducer and gel couplant. There was much more variation in the digital readout of the gage, from 7.88 to 7.98mm, before misalignment caused loss of display. The corresponding oscilloscope traces clearly show the increased variation.

It is true that for a gaging situation with parallel front and back surfaces the "best" reading will be the lowest one, and the operator could be instructed to "hunt" for the minimum and record this value. But this is often not so easy to accomplish under actual working conditions. This is especially true when difficult physical situations or long periods of repetitive measurements produce operator fatigue.

The same argument in favor of self-coupling applies equally well to automated or robotic testing systems. These may not be as effective as a skilled inspector in determining when the transducer position is slightly out of alignment because of positioning tolerances or an out-of-round test object.

Low Acoustic Impedance Test Objects

When low impedance surfaces are contacted by self-coupling transducers, there is very efficient energy transmission across the interface. This produces a relatively small entry-surface reflection which in turn allows resolution of near-surface discontinuities and the measurement of relatively thin sections. Figure 6 shows how interface impedance match/mismatch affects the entry surface reflection.

This characteristic of self-coupling delay line transducers can be exploited to measure thin layers and coatings which are difficult to gage otherwise, such as plastics and paint on nonmetals.

Figure 7 shows the responses from relatively thin sections of low and high acoustic impedance materials, using a 10MHz self-coupling delay line transducer. The upper trace shows that 0.10mm of polyethylene is resolved without difficulty. The two lower traces are both 0.50mm sections. The plastic drafting template is a very good acoustic impedance match for the

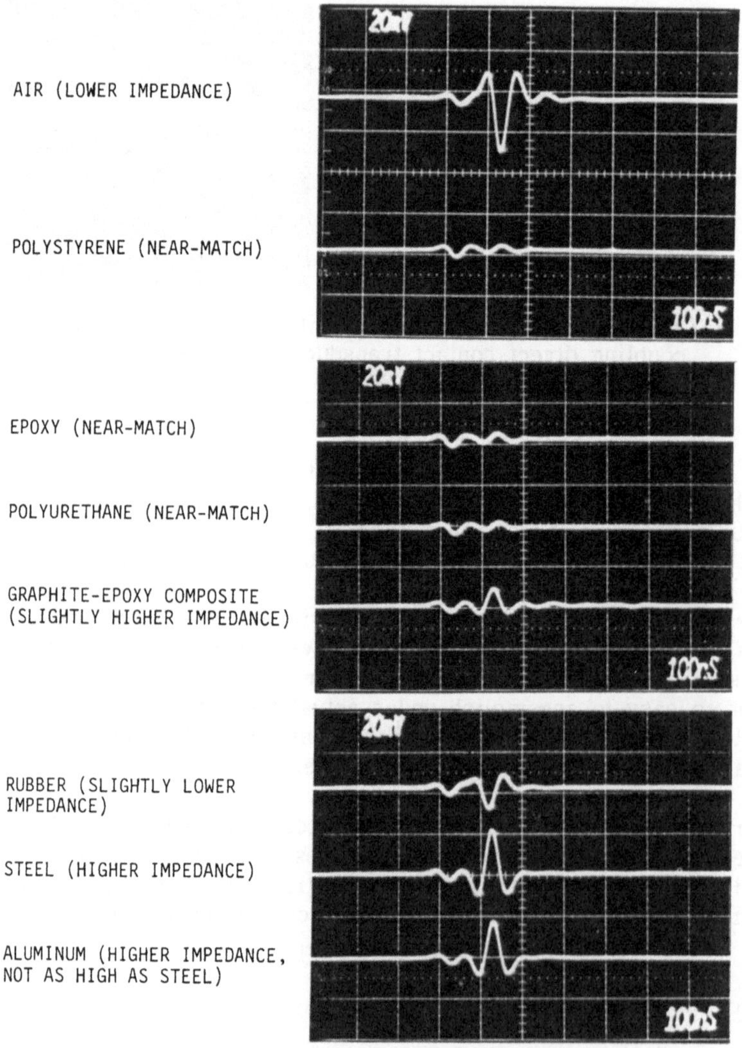

Fig. 6. Acoustic impedance match/mismatch as indicated by interface reflections observed with a self-coupled delay line 10MHz transducer in contact with various material specimens.

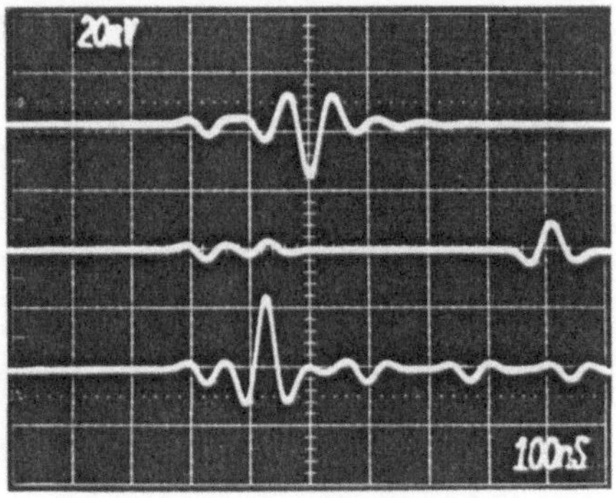

Fig. 7. Measurement resolution of relatively thin material sections with a self-coupled delay line 10MHz transducer.

self-coupling delay tip, and the longitudinal wave velocity in this material is approximately 1/3 of the value of the steel sample used to make the bottom trace. Both of the factors contribute to the ability to measure much thinner plastic sections compared to metal sections.

Figure 8 is an example of how good impedance matching can be utilized to measure thin coatings. The top photograph shows the responses from uncoated

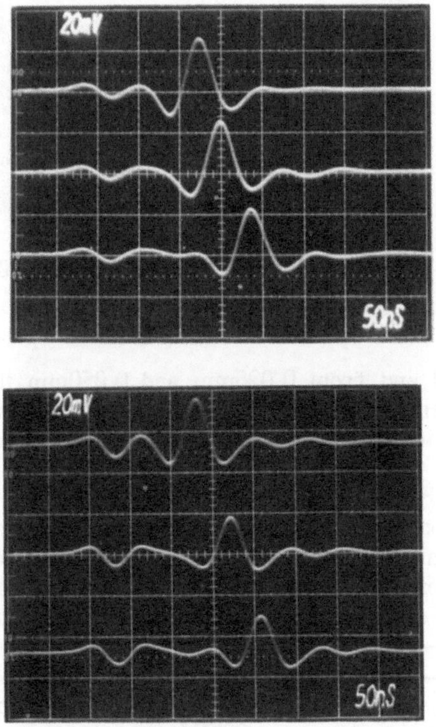

Fig. 8. Measurement resolution of a thin polyethylene layer.

A. RESPONSE FROM UNCOATED GLASS

B. UPPER TRACE: WELL-BONDED TEFLON COATING 0.05mm THICK

LOWER TRACE: DISBONDED AREA IN SAME COATING

C. WELL-BONDED TEFLON COATING 0.10mm THICK

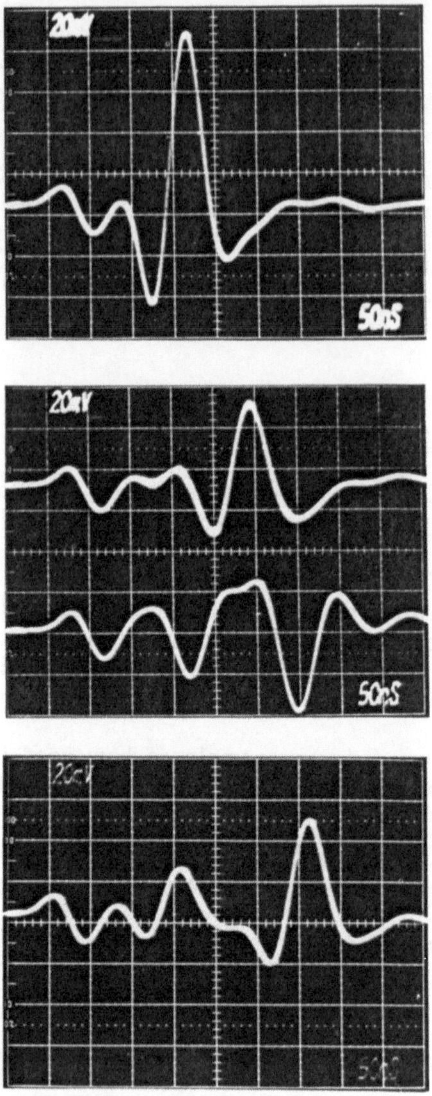

Fig. 9. Measurement of bond testing of teflon coating on glass using a 15MHz self-coupling delay line transducer.

steel and from 0.025mm and 0.050mm thick polyethylene coatings on the same material. The bottom photograph shows bare graphite/epoxy composite and 0.025mm and 0.050mm thick polyethylene coatings on the same composite.

Figure 9 shows the responses from an uncoated glass lamp, from bonded teflon coatings 0.05mm and 0.10mm thick, and from a disbonded area of the 0.05mm coating. It is generally possible to identify coating/substrate disbonds while measuring the coating thickness. This is a significant advantage over most other coating thickness measurement techniques.

The samples used to produce Figs. 8 and 9 were also successfully measured with standard delay line transducers and liquid couplant. However, it was necessary to use a 30MHz standard transducer to obtain the same resolution as

1. SELF-COUPLED

A. INITIAL OBSERVATION

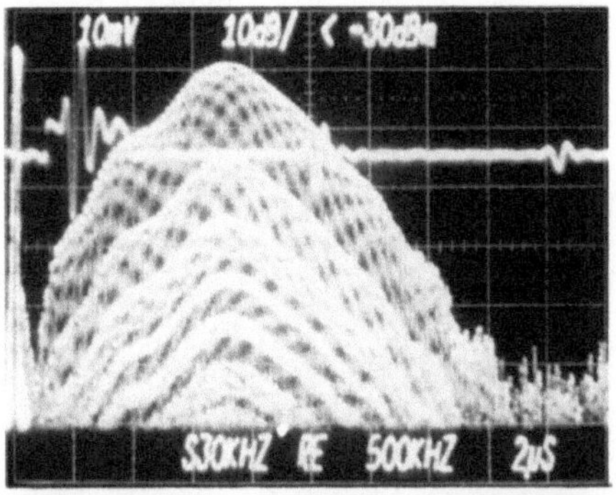

B. EXAMINATION REPEATED THE FOLLOWING DAY

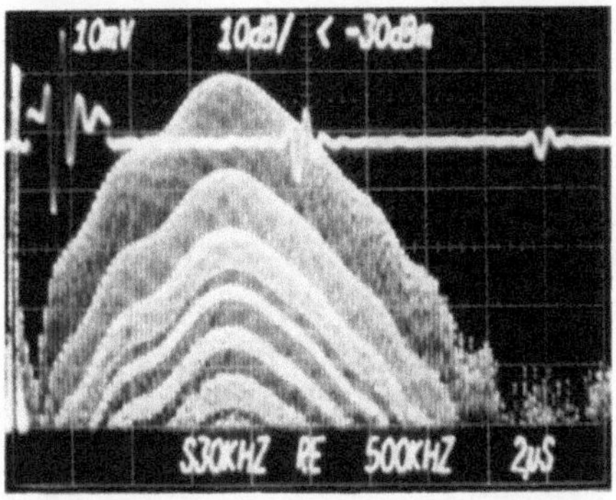

2. COUPLED WITH HONEY "A"

A. FRESHLY APPLIED

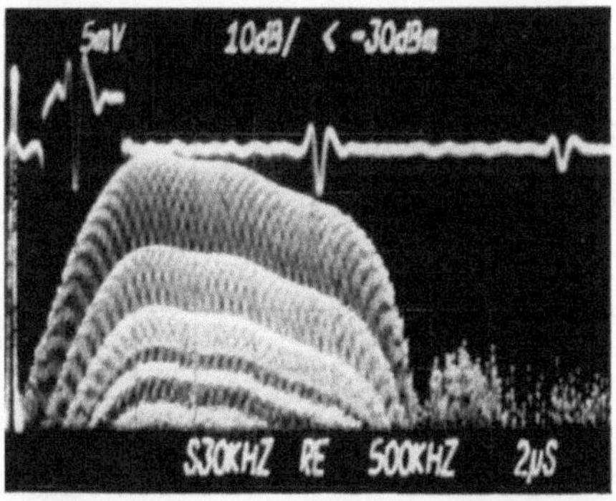

B. AFTER 5 MINUTES

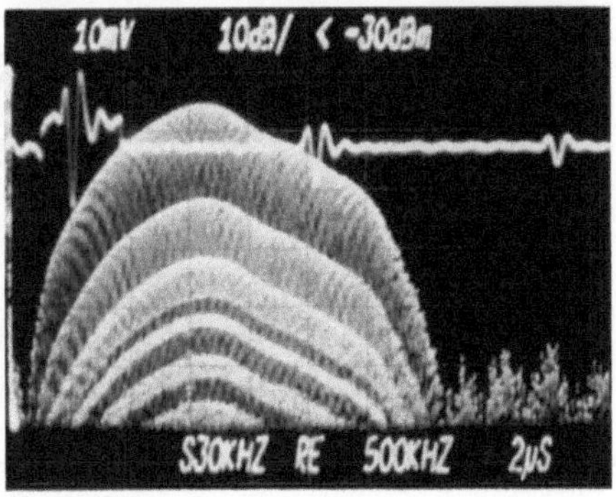

Fig. 10. (see page 235 for description)

3. COUPLED WITH HONEY "B"

A. FRESHLY APPLIED

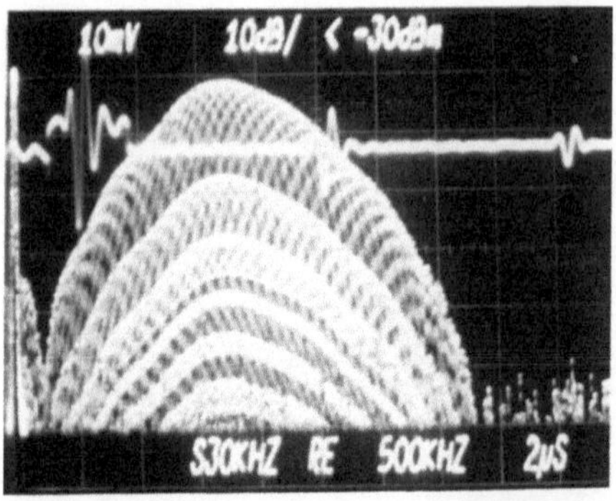

B. AFTER 5 MINUTES

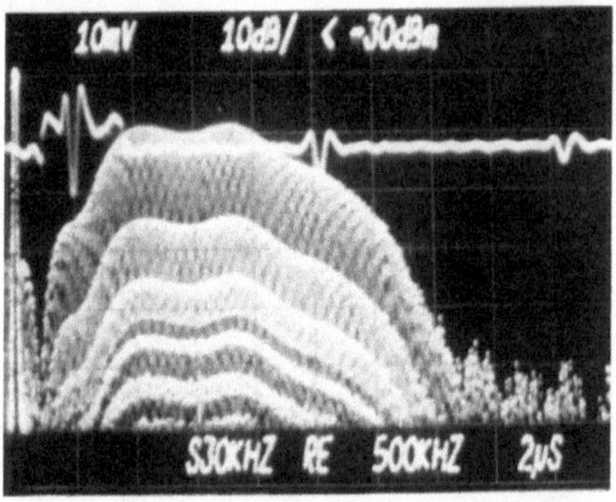

4. COUPLED WITH PETROLEUM JELLY

A. FRESHLY APPLIED

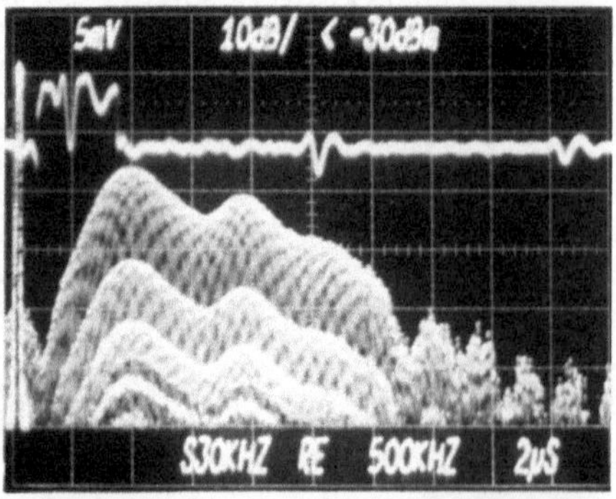

B. AFTER 5 MINUTES

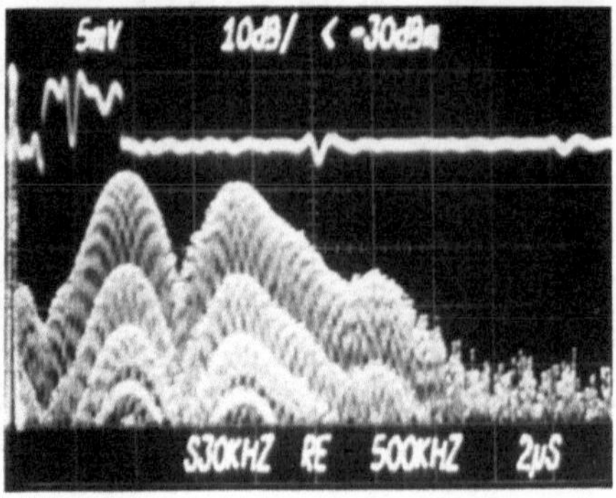

Fig. 10. continued (see page 235 for description)

with a 15 MHz self-coupled device. Substantial increases in test frequency result in more expensive transducers which may require more careful handling. The corresponding ultrasonic electronics are also often more expensive than instruments for use at lower frequencies. In any case, gaging is much faster when there is no need to apply and remove couplant.

Material Characterization - Longitudinal Wave Observations

The correlation of ultrasonic velocities to material elastic properties has been successful in a variety of applications.[7] Still more information about material structure and structural variations can be obtained through wide band ultrasonic spectroscopy.[5-8] For nonporous materials it has been found that longitudinal wave velocity and frequency data, obtained with water coupling, are in good agreement with self-coupling observations. Figure 4, however, is an example of how this is not the case when the sample absorbs liquid.

Material Characterization - Shear Wave Observations

Zero degree incidence shear wave transducers are typically used for shear wave velocity measurements for the determination of elastic properties. It has also been found that shear wave spectroscopy can be very sensitive to material variations.[6] The coupling of zero degree incidence shear wave transducers is accomplished by means of a highly viscous fluid such as honey when standard transducers are used. Slight variations in the amount of couplant or the contact pressure cause substantial variations in signal amplitude and frequency characteristics as well as slight variations in time measurements. Velocity variations due to coupling may require compensatory calculations or special measuring techniques, but coupling need not be a serious source of error in these measurements. For attenuation and frequency observations the customary coupling techniques are significant sources of uncertainty. It is difficult to obtain reproducible results because it is virtually impossible to repeat coupling conditions exactly. Self-coupling zero degree incidence shear wave transducers, both direct contact and delay line types, eliminate the major cause of such variations by making it easy to control and reproduce coupling conditions.

Figure 10 shows some typical response variations observed with changes in couplant material, couplant dwell time, and reapplication of couplant. These are compared to typical repeatability observations with self-coupling transducers. Time/velocity measurements were not significantly affected by couplant variables in this example. But amplitude/attenuation and frequency observations were stable and repeatable only with self-coupling. Two different brands of honey gave notably different results. Observations made with both honeys and with petroleum jelly changed significantly in a short time. The sample was nonporous PZT (lead zirconate titanate). The self-coupled examination consistently gave a first back-reflection peak frequency of 1.80MHz with a bandwidth (-10dB) of 1.2MHz. With brand "A" honey the

Fig. 10. Effects of transducer coupling on repeatability of observations made with zero degree incidence shear wave examination. Sample: PZT, 7.50mm thick; Method: pulse-reflection. The first and second back surface reflections are shown; the first is gated for the frequency-amplitude display (Transducer: 2MHz, 12mm dia.)

response appeared to stabilize at 1.1 MHz peak, 2.1MHz bandwidth. But after 5 minutes there was a change to 1.7MHz peak, 2.0MHz bandwidth. In fact, the response never really stabilized. With brand "B" honey the observations changed from 1.8MHz peak, 1.4MHz bandwidth to double peaks at 1.5 and 1.85MHz with 1.5MHz bandwidth after 5 minutes. Again, the observations never stabilized. Similar effects were observed, but with greater variations, repeating the same setup with the same honey at different room temperatures, and also using honey from a new jar and later from the same jar when it was less than half full. Petroleum jelly was also used as a couplant as part of this experiment. This produced a quite significant drop in peak frequency to 1.0MHz, and a dramatic change in the shape of the spectrum after 5 minutes. This never really stabilized. Likewise, overall attenuation as measured from the real-time trace and attenuation at a given frequency was found to be repeatable with self-coupling and erratic with viscous liquid couplants. A number of other viscous liquids and pastes were evaluated with similar results, including materials sold commercially as "shear wave couplant". Several examples of the sensitivity of shear wave spectroscopy to material property variations and to isolated discontinuities which were relatively difficult to detect by longitudinal wave/amplitude techniques have been reported.[6] It appears that self-coupling has created the possibility of using shear wave spectroscopy as an effective technique for material analysis.

Time Savings and Nuisance Avoidance

In many repetitive point-to-point testing operations a very large part of the operator's time is spent applying and removing couplant. Some couplants containing fluorescent dyes are available so that parts may be examined under black light to assure that all traces of couplant have been removed when any residue would interfere with subsequent processing. Any such situations are good candidates for cost savings through the use of self-coupling transducers.

Angle beam probes with wedges which self-couple to the test surface are not durable enough to be practical for most tasks which require "scrubbing" the probe over the test surface. But surface wave examinations are normally performed on rather smooth finishes, and excess surface couplant can be a serious nuisance to the operator in these types of examinations. The problem can be eliminated by the use of a self-coupling surface wave probe. Whether this is a cost-effective change depends upon the surface roughness and the amount of care taken by the operator. Probes in which the transducer self-couples into a replaceable wedge and liquid couplant is used between the wedge and the test surface are quite practical. Evaluations of 2.25 and 5MHz probes compared to their conventional counterparts have shown that transducer-wedge self-coupling produces stable performance for a much longer period of time (indefinitely in most cases), with negligible increase in "noise" and typically 2-6dB loss of sensitivity. The decreased sensitivity would very rarely create a problem.

SUMMARY AND CONCLUSIONS

Self-coupling transducers have been found to be not only effective but preferable to standard contact and delay line transducers for a wide variety of applications. They are particularly practical for repetitive point-to-point examinations, whether manual or automated. They are clearly technically superior for many examinations of porous and low acoustic impedance materials. They provide the most practical if not the only reliable means of performing zero degree incidence shear wave spectroscopy, which has been shown to be a promising technique for material characterization.

REFERENCES

1. J. A. Brunk, Applications of Dry Contact Ultrasonic Transducers, in: "Proceedings of the 11th World Conference on Nondestructive Testing, Vol. 2," The American Society for Nondestructive Testing, Columbus,(1985).
2. J. A. Brunk, "An Investigation of Dry-Contact Ultrasonic Gaging, BDX-613-3421," U. S. Dept. of Energy , Washington, DC (1986).
3. J. A. Brunk, Performance Comparisons of Dry Coupling with Conventional Contact Ultrasonic Transducers, 0.5 - 20MHz, in: "Proceedings of the 16th Symposium on Nondestructive Evaluation," NTIAC, San Antonio (1987).
4. J. A. Brunk, "An Investigation of Dry-Coupled Ultrasonic Thickness Gaging," paper presented at the American Society for Nondestructive Testing Spring Conference, Phoenix (1987).
5. M. C. Bhardwaj, "Advances in Ultrasound for Materials Characterization," paper presented at the 89th Meeting of the American Ceramic Society, Pittsburgh (1987).
6. J. A. Brunk, C. J. Valenza, and M. C. Bhardwaj, "Ultrasonic Characterization of Ceramics by Dry Coupling Techniques," presented at the 89th Meeting of the American Ceramic Society, Pittsburgh (1987).
7. M. C. Bhardwaj, Principles and Methods of Ultrasonic Characterization of Materials, Adv Ceram Matls 1 (1986).
8. M. C. Bhardwaj, An Industrialist's Perspective on Ultrasonic Characterization of Materials, J of Wave - Matl Inter 1:3 (1986).

FIBER WAVEGUIDE SENSORS OF STRESS WAVES IN SOLIDS

Richard O. Claus, J. A. Wiencko, and R. E. Rogers

Fiber and Electro-Optics Research Center
Department of Electrical Engineering
Virginia Polytechnic Institute and State University
Blacksburg, VA

ABSTRACT

Optical fiber sensor systems have been applied to the detection of stress waves in solids. The optical fiber sensing elements in such systems may be embedded within the solids or attached to the surfaces of the solids. Fiber sensor output is proportional to the strain integrated along the sensing length of the fiber. This paper reviews the operation of such sensors and sensor systems, the reported experimental measurements of stress waves using such techniques, and the potential implications of such sensors in acousto-ultrasonics.

INTRODUCTION

Optical Fibers as Sensors

Optical fiber waveguides, shown to be useful for practical long distance telecommunications nearly twenty years ago, have been applied to the sensing of a wide range of physical observables since 1977.[1] All optical fiber sensors operate on the principle that one of the measurable quantities associated with the propagation of an electromagnetic wave in an optical waveguide structure, namely, amplitude, phase, polarization, wavelength, frequency, or mode, may be modulated by perturbing the geometry and the materials of the waveguide. Two examples of practical optical fiber waveguide sensor systems are all-optical fiber gyroscopes and towed underwater acoustic sonar sensor arrays.

Optical Fiber Sensing in Materials

Optical fiber sensing in materials has been reported by several authors within the past five years.[2-6] Most of this work has been directed at the detection of point strain within materials, advanced composites in particular. Research in this area has been concentrated on the problems of 1) the embedding process, especially the chemical and mechanical boundary conditions between the glass-on-glass optical fiber waveguide, its buffer and jacketing materials, and the host material, 2) the interaction between the fiber sensor and the environment, and the development of different sensor geometries to achieve specific interactions, and 3) the detection and processing of output signals to infer material characteristics. Several recent applications of embedded fiber sensors include: the analysis of resin cure, the measurement of vibrational components of structures fabricated from materials containing embedded fiber sensors, as well as the nondestructive evaluation of materials, are reported on at a NASA workshop.[7]

This paper reviews the few publications which report the measurement of stress waves in solids using optical fiber sensors. Basic optical fiber principles are first reviewed. Next the analytical basis of all reported stress wave detection methods is considered. Experimental measurements performed by the authors are reported and observations concerning the limitations of such detection are presented.

OPTICAL FIBER MODAL SENSING OF STRESS WAVES

Optical Fiber Basics

Optical fibers are typically constructed as cylindrical inner layers of glass core and cladding, and external layers of polymer buffer coatings and jacketings. Important to the optical field waveguiding properties of the fiber are the indices of refraction of the core and cladding glasses. The core index of refraction is typically one percent higher than the cladding index, allowing optical fields coupled into one end of the fiber to propagate coupled into one end of the fiber to propagate as guided modes within the waveguide. The mismatch between core and cladding index also determines the numerical aperture, (NA), or the field of view of the input end of the fiber. The NA is given by

$$NA = ((n_1)^2 - (n_2)^2)^{0.5} \tag{1}$$

where n_1 is the index of the refraction of the core glass and n_2 is the index of refraction of the cladding glass. The NA, the wavelength of the light injected into the waveguide and the core and cladding geometry and indices determine the mode content of the waveguide. Specifically, the V-number of the fiber is defined as

$$V = (2\pi a/\lambda)((n_1)^2 - (n_2)^2)^{0.5} \tag{2}$$

where λ is the free space wavelength of the light propagating in the fiber and "a" is the radius of the core of the fiber. By solving the electromagnetic field propagation equations subject to cylindrical boundary conditions in the fiber, it may be shown that for $V<2.405$, only one degenerate electromagnetic field mode may exist within the fiber, and for $V>2.405$, more than one such mode may be supported. The number of modes in the fiber may be approximated by

$$M = V^2/2. \tag{3}$$

Typical fibers for telecommunications applications are designed for single mode operation ($V<2.4$ and one propagating mode) or multimode operation

(V equal to 40 or more, and as many as 1000 supported modes). If instead the parameters which influence the number of modes in the fiber given in eq. (2) and eq. (3) are chosen in such a way that the V-number is only slightly greater than the 2.4 cutoff limit, several modes are supported. Such "few-mode" optical fiber waveguides have been used exclusively in the detection of stress waves in solids as considered below.

Fiber Mode Perturbations

The measurable properties of optical fibers may be modulated by the application of a wide range of physical perturbations to the exposed sensing length of the fiber. Acoustic waves supported within a material containing a sensing fiber, in particular, interact with the fiber by causing time varying distributed strain along the fiber length. If the waveguide of the acoustic wave is long with respect to the diameter of the fiber, the fiber may be considered to be subjected instantaneously to a uniform hydrostatic pressure[8]. Such pressure causes a change in the phase of the optical field propagating in the fiber. This phase change may be detected using differential optical fiber interferometric techniques; such techniques are the basis of underwater acoustic hydrophone systems.

If the wavelength of the acoustic wave is not much larger than the diameter of the optical fiber and if the geometry of the acoustic field is such that several acoustic wave maxima and minima interact with the fiber simultaneously, then the assumption of uniform hydrostatic pressure is no longer valid. Thus, a simple model of optical field phase modulation is not appropriate in this case.

The resulting fiber modulation in this case, however, may be considered in terms of the distributed stress-induced interaction between different modes within the fiber. This distributed interaction has been analyzed by several authors and compared with initial experimental results.[9-11] For multimode telecommunications-grade optical fibers having many modes, such a modulation produces a complicated interaction among all of the propagating modes. This interaction may be observed visually by imaging the light from the output end of the fiber on a screen placed in the far field. Under static conditions, the observed optical intensity pattern is a complex speckle formation similar to that produced by a laser alone. If the fiber is perturbed, however, the speckles appear to randomly rearrange themselves. This rearrangement is the basis of the "Fiberdyne" sensing method reported by Kingsley and coauthors.[12,13] An exact mathematical treatment of this complicated pattern modulation is extremely difficult due to the existence of a large number of coupled modes, each of which interacts differently with the applied perturbation.

If instead the number of modes in the fiber is restricted to just a few, as suggested above, the analysis of the mode-mode interaction is significantly simplified, a closed-form solution is possible, and the calibrated detection of the applied field may be obtained. Detection of fiber modulation in few-mode fiber is based upon the difference in propagation constant between different modes. For example, if just two modes propagate within the fiber, applied strain causes the total length of fiber through which the modes propagate to change. If the propagation constants of the two modes differ slightly, the resulting phases of the mode fields at the far end of the fiber differ due to the applied strain.[14]

Detection of Mode-Mode Interference

A phase difference between two mode fields induces interference; this interference may be detected using several methods. First, the output field at the far end of the fiber may be projected onto a screen and the motion of the pattern on the screen may be monitored. If this screen is replaced by a two-dimensional electro-optical array, the pattern may be detected as a two-dimensional intensity function and changes in that function may be monitored as a function of time.[14] Second, mode-selective optical fiber components may be inserted into the fiber system to guide low-order modes into one output fiber and high-order modes into a second output fiber. Specifically, asymmetrical fiber optic couplers may perform this differential fiber mode separation task.[14] Using either of these two implementation methods, differential fiber mode measurements may be made to yield information concerning differential mode propagation constants and thus applied fiber strain.

EXPERIMENTAL RESULTS

Strain Measurement

Experimental measurements of stress waves in solids using optical fiber sensors have been made using few-mode sensing methods. A block diagram of a modal sensing system is shown in Fig. 1. Here, light from a monochromatic laser source is coupled into a few-mode optical fiber. The fiber is subjected to perturbation in a localized region and the output light from the fiber is interrogated using one of the measurement methods indicated in the above section. Output signals which yield information concerning differential mode modulation and thus applied strain are recorded and processed.

By controlling the launch conditions of the optical field into the input end of a fiber, the exact modal content of the fiber may be controlled. Specifically, polarized light from a helium-neon laser which operates at a wavelength of 633nm, may be injected into a fiber having a 0.2 NA and core and cladding diameters of 9 and 125 microns, respectively, to produce either HE_{11}, and HE_{21} or HE_{11} and HE_{31} mode combinations. The complete analytical

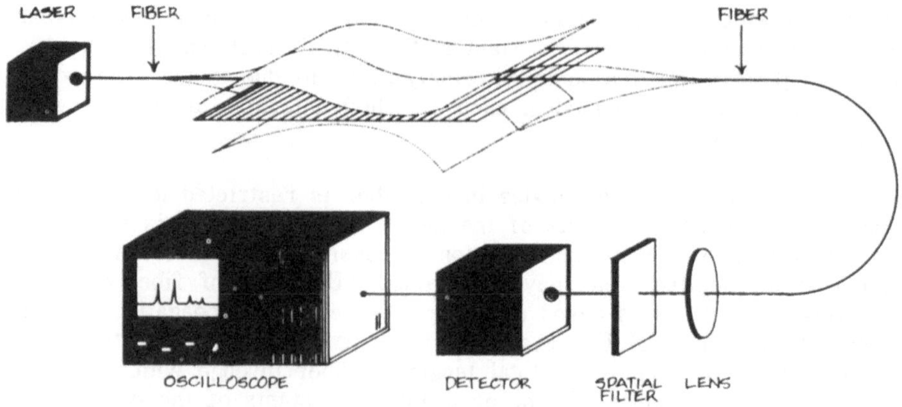

Fig. 1. Modal Domain Optical Fiber Detection System

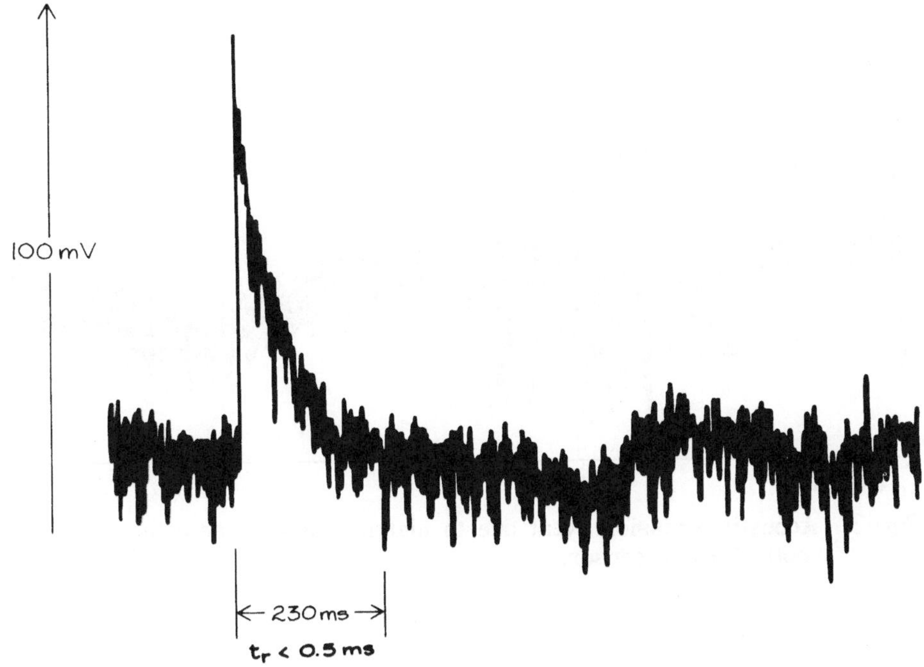

Fig. 2. Acoustic emission event due to graphite fiber break detected by optical fiber sensor.

derivation of the form of these fields and their dependence upon launch conditions is given by Ehrenfeuchter.[14] These combinations produce output intensity patterns characterized by either four or three circularly symmetric lobes, respectively. Axial strain applied to the fiber produces a difference in the propagation constants of the different modes, causing the lobed patterns to rotate about their axes. Alternatively, such an applied strain produces an increasing difference in the high- and low-order mode phases of optical fields in the output fiber.

Stress Wave Measurement

If instead the few-mode fiber embedded within a solid is subjected to a vibration or a propagating stress wave, the fiber is stressed in a nonuniform way along its length. Waveguide analysis indicates that such stress produces periodic changes in fiber strain and microbending,[15] and that the differential modal phase contributions may be interpreted to yield information concerning the frequency spectrum of the acoustic source.[16]

Measurements of acoustic emission in loaded graphite/epoxy laminate coupons obtained via differential mode-mode interference methods are shown in Figs. 2, 3, and 4. The specimens corresponding to each figure were 2.54 cm by 25.4 cm six-ply laminates having few-mode optical fibers embedded between the central two plies. Figures 2 and 3 show specific acoustic emission event waveforms detected using the embedded optical fiber. In each case, waveforms were verified as being induced by acoustic emission sources by attaching a piezoelectric acoustic emission sensor to the exterior surface of the coupon during loading and monitoring both the piezoelectric and optical fiber sensor outputs during excitation. A comparison of two such signals obtained for the same acoustic wave event is shown in Fig. 4.

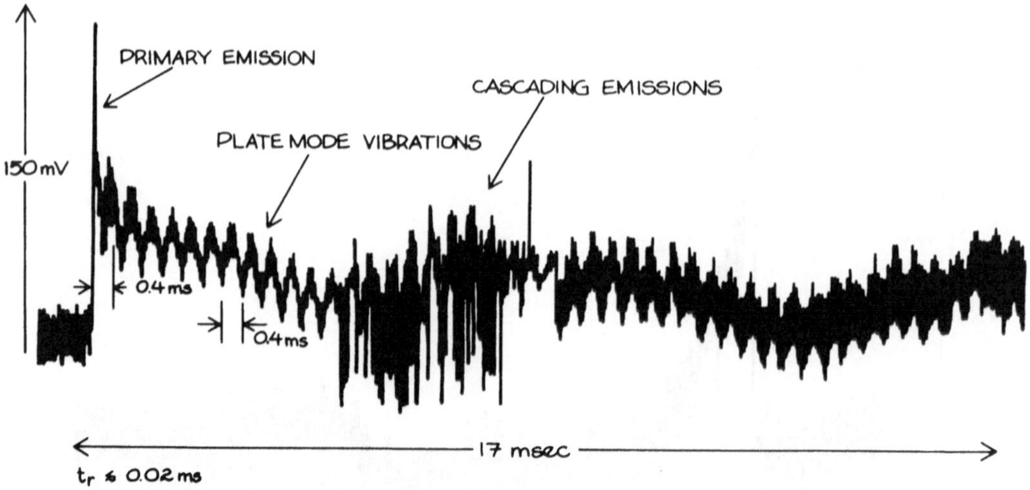

Fig. 3. Acoustic emission event due to matrix cracking detected by optical fiber sensor.

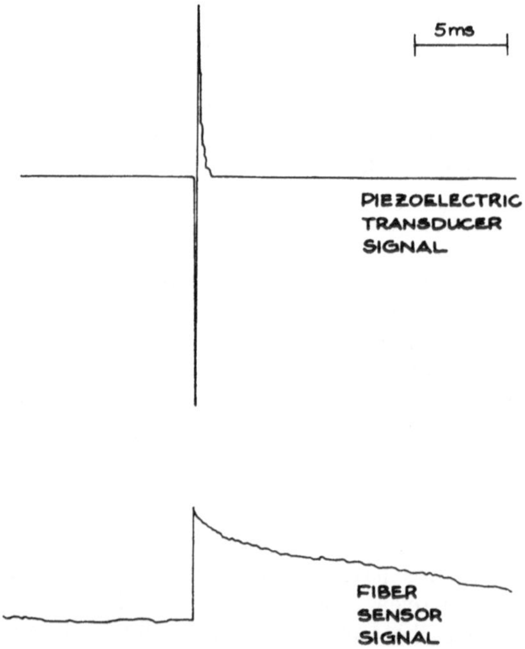

Fig. 4. Comparison between acoustic emission event signatures detected by piezoelectric transducer (top) and optical fiber sensor (bottom).

OBSERVATIONS AND CONCLUSIONS

Optical fiber waveguides embedded within materials may be used to detect acoustic waves which propagate within the materials. Certainly the response of the optical fiber sensor is different from that of an ultrasonic sensor with a relatively small surface area. First, the optical fiber output signal is proportional to the strain integrated along the gauge length of the sensor. Second, the optical fiber has a frequency response which is broader than that of tuned piezoelectric detectors. Thus, acoustic emission data analysis equipment, or acousto-ultrasonic systems, will in general not work using optical fiber signal outputs due to the incompatibility of piezoelectric and fiber detector signals.

Embedded optical fibers, however, offer several advantages for nondestructive structive evaluation. First, they may be embedded within advanced materials to monitor material processing and cure conditions without compromising material integrity. Second, they may be used to interrogate materials for strain, vibration or acoustic emission events during the use lifetime of the material. Finally, they are all-dielectric sensors and so are inherently immune to electromagnetic interference effects.

Applications of optical fibers as sensors in acousto-ultrasonic systems are suggested by this and previous work. Such applications could, for example, utilize piezoelectric source transducers and optical fibers as the receiving transducers. Issues which must be considered prior to the development of a system using such a transducer scheme include 1) the optimization of the acoustic wave sensitivity of the fiber, probably through fiber coating modification, and 2) analysis of the impulse response function of the fiber.

ACKNOWLEDGEMENTS

This work has been supported in part by the NASA Langley Research Center and Simmonds Precision. The authors thank A. Goette and T. Freeman for their assistance.

REFERENCES

1. C. D. Butter and G. B. Hocker, Fiber Optics Strain Gauge, Applied Optics, 17:2867 (1978).
2. G. Meltz and J. R. Dunphy, Proc. SPIE 566:159 (1985).
3. B. W. Brennan, W. B. Spillman, and J. R. Lord, Proc. 3rd Ann. SEM Conf. on Hostile Evv. and High Temp. Measurements, Cincinnati (1986).
4. R. O. Claus and J. H. Cantrell, Detection of Ultrasonic Waves in Solids by an Optical Interferometer, Proc. IEEE Ultrasonics Symp. 2:719 (1980).
5. W. J. Rowe, E. O. Raush, and P. D. Dean, Embedded Optical Fiber Sensor for Composite Structure Applications, in: "Proceedings of the SPIE Conference on Fiber Optics Sensors," SPIE, Cambridge (1986).
6. R. L. Crane, A. B. Macander, and J. Gagorik, Fiber Optics for a Damage Assessment System for Fiber Reinforced Plastic Composite

Structures, in: "Proceedings of the Review of Progress in Quantitative NDE," D.O. Thompson and D.Chimenti, eds., Plenum, New York (1982).

7. Proceedings of the NASA Workshop on Intelligent Materials, NASA, Hampton (1987).
8. J. A. Bucaro and E. F. Carome, Single Fiber Interferometric Acoustic Sensor, Applied Optics, 17:330 (1978).
9. K. D. Bennet, R. O. Claus, and M. J. Pindera, Internal Monitoring of Acoustic Emission in Graphite/Epoxy Composites Using Embedded Optical Fiber Sensors, in: "Review of Progress in Quantitative Nondestructive Evaluation," D. O. Thompson and D. Chimenti, eds., Plenum, New York (1986).
10. K. D. Bennet and R. O. Claus, "Proc. IEEE Region 3 Conference", IEEE, Richmond (1986).
11. N. K. Shankarnarayanan, Optical Fiber Modal Domain Detection of Stress Waves, in: "Proceedings, IEEE Ultrasonic Symposium," IEEE, Williamsburg (1986).
12. S. A. Kingley and D. E. N. Davies, Multimode Optical Fibre Phase Modulation and Discrimination I and II, Electronics Letters, 14:322 (1978).
13. M. R. Layton and J. A. Bucaro, Optical Fiber Acoustic Sensor Utilizing Mode-Mode Interference, Applied Optics 18:666 (1979).
14. P. A. Ehrenfeuchter, "Modal Domain Sensing of Vibration in Beams," M. S. Thesis, VPI&SU, Blacksburg (1986).
15. N. K. Shankaranarayanan, Mode-Mode Interference in Optical Fibers: Analysis and Experiment", M. S. Thesis, VPI&SU, Blacksburg (1987).
16. B. Zimmerman, K. Murphy, and R. O. Claus, Local Strain Measurements Using Optical Fiber Splices and Time Domain Reflectometry, in: "Review of Progress in Quantitative Nondestructive Evaluation," D. O. Thompson and Dale Chimenti, eds., Plenum, New York (1987).

CONSIDERATIONS FOR DEVELOPING CALIBRATION STANDARDS FOR ACOUSTO-ULTRASONIC INSPECTION

James D. Leaird and Matthew C. Kingdon

Acoustic Emission Technology Corporation
Sacramento, California 95815

ABSTRACT

The advent of a calibration standard for use in ultrasonic inspection practices meant that ultrasonic inspection could be applied with some degree of repeatability and reliability. This paper is intended to be a starting point for similar development, or at least the opening frame of discussion, of a calibration standard in the technology of acousto-ultrasonic inspection. This paper identifies and discusses parameters used to define the pulse that generates the simulated acoustic emission event. The purpose is to alert the operator/engineer to potential variability in the output of systems. This variability can exist due to slight changes in the characteristics of the pulse. Items that are discussed include pulse rate, height, width, and rise/fall time. It was discovered that slight changes in some of these characteristics can have a dramatic effect on the energy detected by the receiving sensor. This illustrates the need for at least two types of calibration standards. This paper discusses numerous considerations for calibration standards.

INTRODUCTION

The absence of a test standard "the flat bottom hole" in ultrasonic inspection kept that technology from advancing past the novelty stage for several years. Likewise the advancement of acousto-ultrasonics (AU) is being retarded by the lack of similar standards and practices. This paper is intended to form the basis for discussion leading to development of a standard to be used in acousto-ultrasonic inspection. It should be admitted that this technology is being applied to materials that are often difficult to inspect, mainly fiber reinforced composites. The difficulty mentioned is not limited to the nature of the construction of this material, but rather compounded by the nature of the defects requiring inspection. The defects can range from distributed problems such as porosity to localized problems such as impact damage. What this amounts to, in the opinion of the authors, is that at least two types of calibration standards must be developed. The first standard would allow for establishing a basis to compare and tune systems. The intent of this standard would be so that the engineer/operator can compare results between systems, labs, over time and between operators. The second test standard would be application dependent. This standard would be constructed by controlled engineering specification that allowed for the manufacture of a sample that mimicked the desired inspection requirement. That is, if the inspection problem is to detect or measure the aforementioned porosity, then, the sample would have to exhibit the porosity condition. Further, the sample would have to represent the geometry, nature, and variability of the part to be inspected. This paper is divided into two sections. The first section deals with items that the acousto-ultrasonic user should be aware of and possibly even take into account when developing a test standard and/or test procedure. The first section deals with variations in the pulse and how these variations effect the measured signal. The second section deals with putting together a test block to assure the user that the system is constant with time or across systems. Part of the second section contains some ideas on what should be included in a calibration block for a specific application.

CALIBRATION OF THE SYSTEM PULSE GENERATION CIRCUIT

The first step in a quality assurance level system check is to compare the system response against the system specification and note any variation. The acousto-ultrasonic system is comprised of three elements: A. the pulse generation circuit, B. the pulse sensor, and C. the signal measurement circuit. What is of particular interest is to find out how changes in the pulse that drives the pulsing sensor, or changes in the sensor affect the measurement performed by the system. The pulse generation circuit can be measured with a simple oscilloscope and the output signal with a true RMS meter.

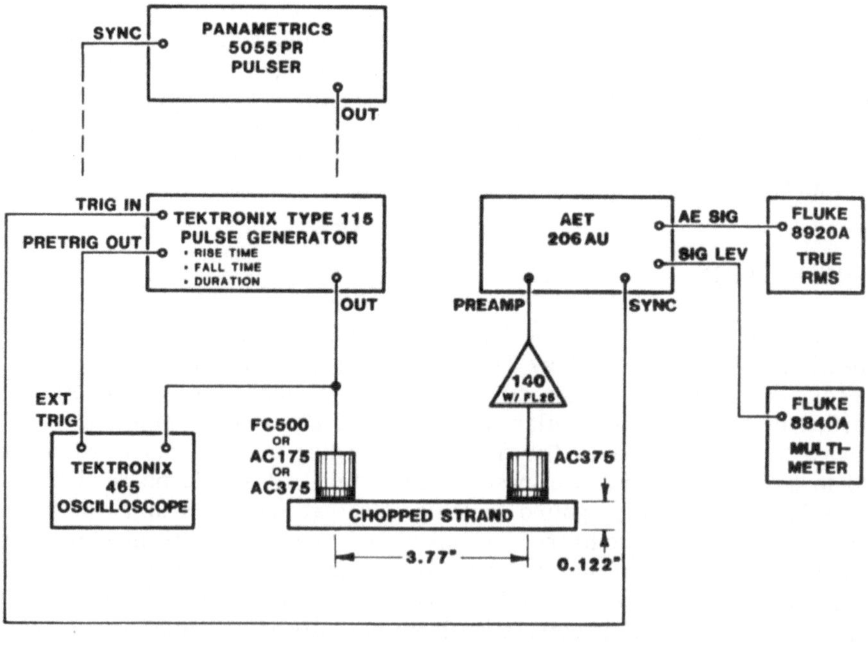

Fig. 1. Schematic of the equipment layout

Figure 1 is a schematic of the equipment layout used to make the necessary measurements. A pulse generator is included so that the authors may create pulses with specific characteristics. The points of interest in the pulse generation circuit are: 1. the rise time and decay time of the pulse, 2. the duration of the pulse, 3. the pulse voltage, and 4. the repetition rate of

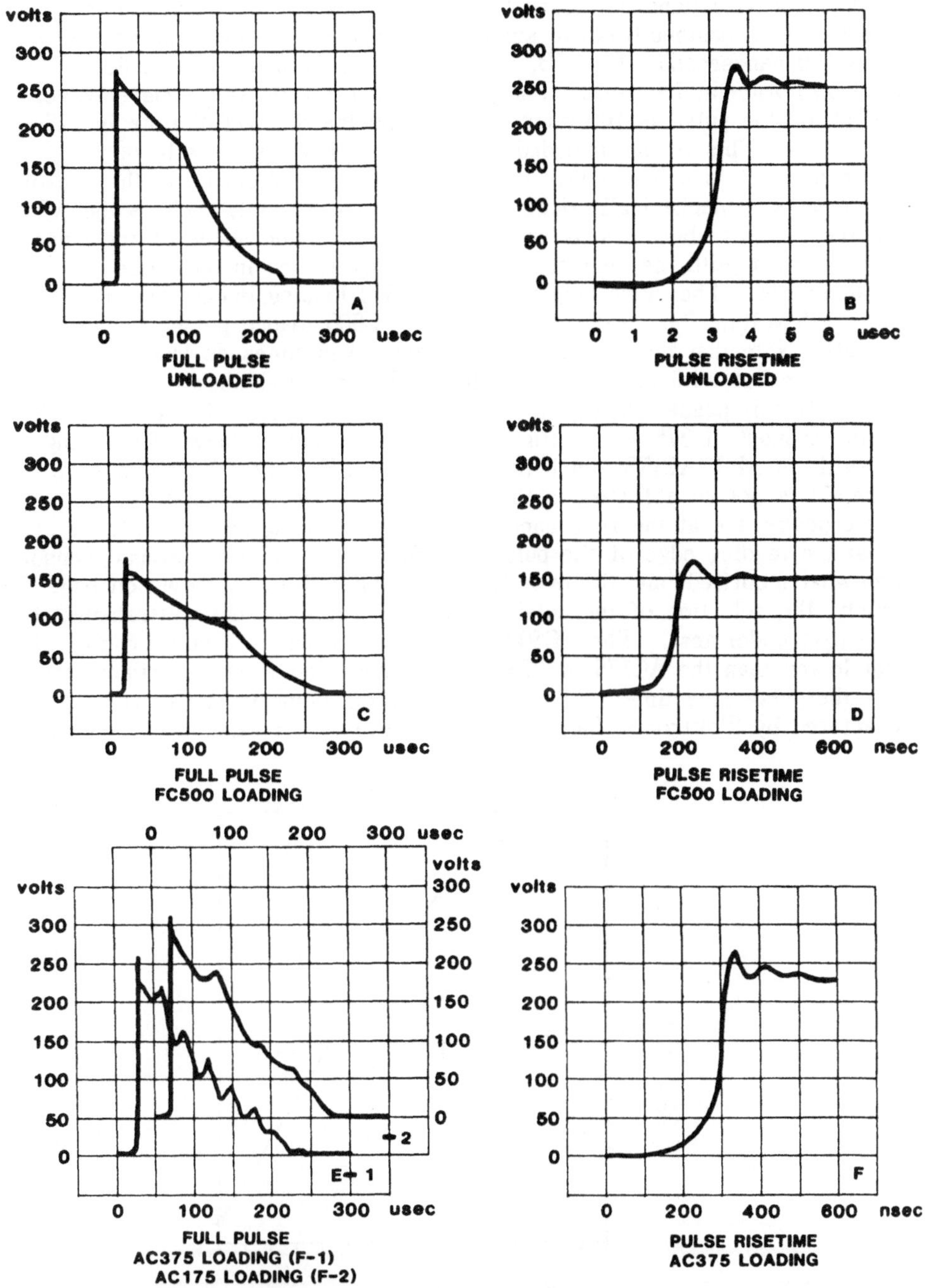

Fig. 2. Full pulse without and with transducer loading, an the corresponding pulse risetime plots.

the pulse. A series of simple experiments can show the effects of the variation of these factors on the received signal. The first endeavor is to determine what the pulse looks like from the system. Then there is assurance that the pulse that is generated in the test circuit is correlative to the pulse being generated by the an acousto-ultrasonic system.

Figure 2A shows the pulse/sync output of a AET 206 A/U system (AET is Acoustic Emission Technology Corporation, Sacramento, CA.). This system was designed to specifications developed by Mr. A. Vary for the first commercially available acousto-ultrasonic system. Note that the rise time of the signal is 100 nanoseconds, Fig. 2B, with a slight overshoot. The pulse height is 275 volts at its peak. The pulse is a nonsymmetric square wave. The width of the pulse is 100 microseconds and the decay/fall time is 125 microseconds. This is the unloaded pulse output. The output will vary according to what is being driven with the pulse. For instance if the sensor being driven is a mechanically damped piezoelectric crystal, such as the AET FC500, the response becomes as shown in Fig. 2C. The shape of the pulse remains constant with the biggest effect being witnessed in the pulse height/amplitude. The pulse height of the FC500 loading brings the peak height to 175 volts. The leading edge of the FC500 loaded pulse is as shown in Fig. 2D. Note that the leading edge is now seven times as long as the unloaded pulse but with the same overshoot. Figure 2E-1 shows the shape of the pulse after it passes through a resonant sensor such as the AET AC375 which is resonant at 375 kHz. Figure 2E-2 is the pulse shape after being loaded with an AET AC175 which is resonant at 175 kHz. Note that here the pulse height is not as affected as the overall shape of the signal. The harmonic peaks are at the resonance of the sensor being pulsed. Please also note that the leading edge of the pulse that is damped with a resonant sensor is somewhat modified, and the overshoot is amplified as is shown on Fig. 2F. Obviously, the selection of the pulsing sensor affects the energy transmitted into the part under test. The FC500 type sensor provides a pulse voltage 30 percent lower than the AC175 or 375 type sensor. The first experiment then is to vary the pulse rise time. The signal generator being used (Tektronix 115, manufactered by Tektronix, Inc. Beaverton, OR.) is not capable of generating

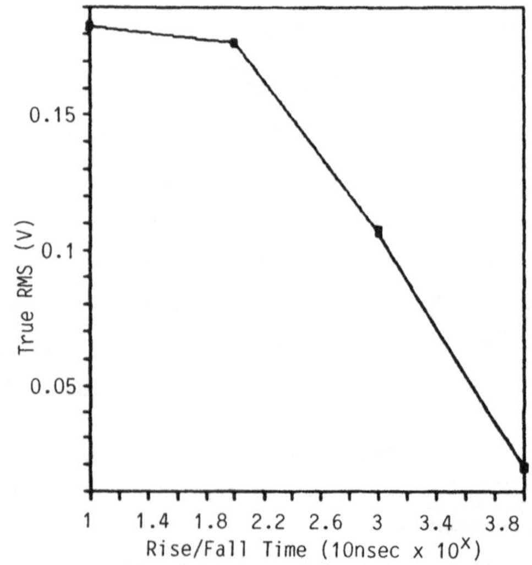

Fig. 3. The effect of large changes in Rise/Fall time

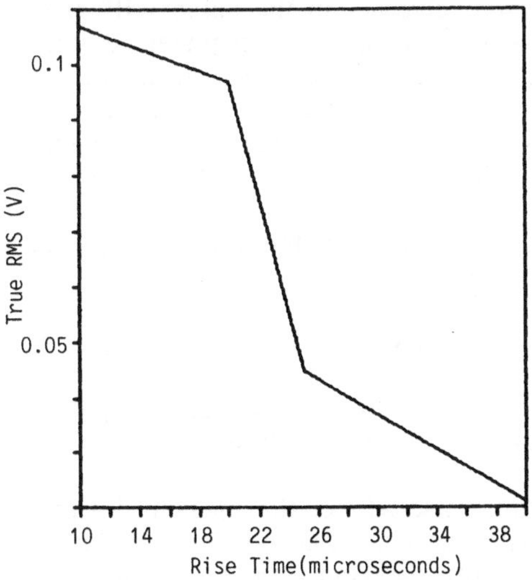

Fig. 4. The effect of small changes in Rise time

a 250 volt pulse, so a model is constructed that is approximately 20:1 scale. The peak pulse out of the generator is 9.6 volts; therefore, all the values will be scaled accordingly. Variations of the pulse generator will be over several orders of magnitude, then in smaller steps at close to the conditions that exist in the 206A/U system. Figure 3 is a record of the rise/fall time variation data. The values on the abscissa are powers of 10 nanoseconds. The first point is 1 ns x 10 to the first power, the second point is 1 ns x 10 to the second power and so on. The pulsing sensor in all the experiments is the FC500 flat frequency sensor. The receiving sensor is the AC375 resonant sensor. Major changes in the pulse signal will obviously change the

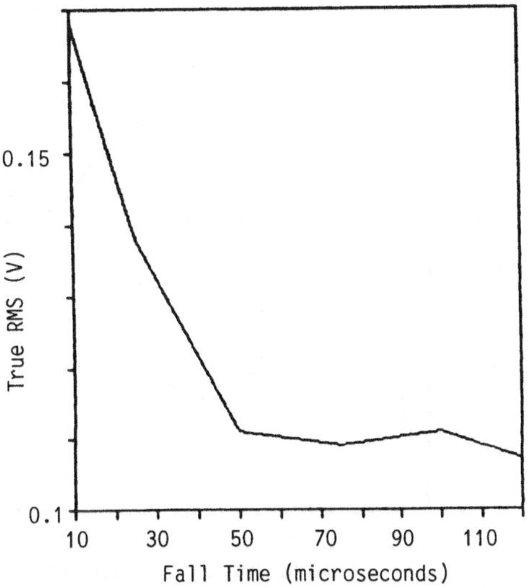

Fig. 5. The effect of small changes in Fall/Decay time

received signal but small changes in the pulse will also create changes. Therefore, it is wise for the operator/engineer to know what subtle changes in the pulse rise time will do to the received signal. Nominal in this experiment is 1 microsecond rise time. The actual rise time was varied with the fine adjustment on the Tektronix 115 Pulse Generator and measured on the scope. Figure 4 presents this rise time data. It is immediately apparent that in the range of greater than 20 microseconds the dropoff is substantial. This could represent a harmonic in the specimen geometry. The calculation for specimen geometry is:

wavelength (w) = velocity (v) / frequency (f).

In the case where v = 3070 m/s and f = 500 kHz, then w = 6.1 mm, which is roughly half the thickness of the specimen.

Fall time variations also affect the measured signal. Figure 5 is a plot of this data for small changes. The nominal signal and variations to it were the same as described for the rise time small changes discussed above. The dropoff from a very rapid fall time to a moderately rapid fall time (10 μs to 50 μs) is dramatic while the slow fall times (longer than 50 μs) are level to 120 μs. Pulse width variations fluctuate wildly as is shown on Fig. 6. The nominal pulse is 100 ns in rise/fall time and has a period of 100 μs. The first peak is at 3.5 μs pulse width which is close to the resonance of the receiving sensor. The second peak is at 9.1 μs.

Both pulse height and pulse rate effect the signal in an expected way. Increases in either cause an increase in the received signal RMS. Figure 7 shows the near linear increase with pulse rate for large pulse rate changes while Fig. 8 shows the same for small pulse rate changes. Figure 9 shows the effect of pulse height on the received signal.

In summary what is attempted here is to indicate the effects of changes in the pulse shape on the received signal. Changes in any of the four parameters used to define the pulse in this report resulted in changes in the

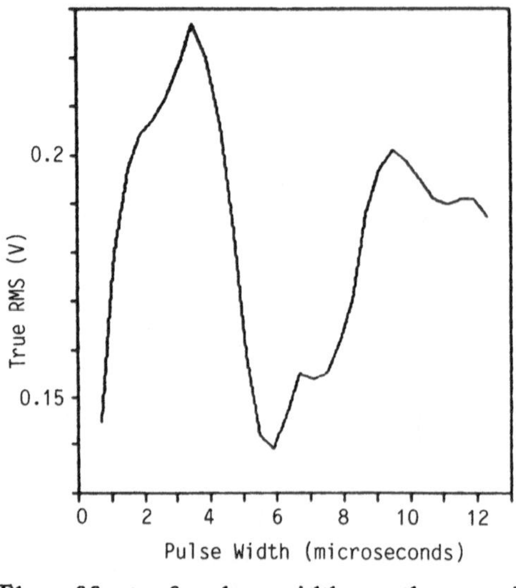

Fig. 6. The affect of pulse width on the received signal

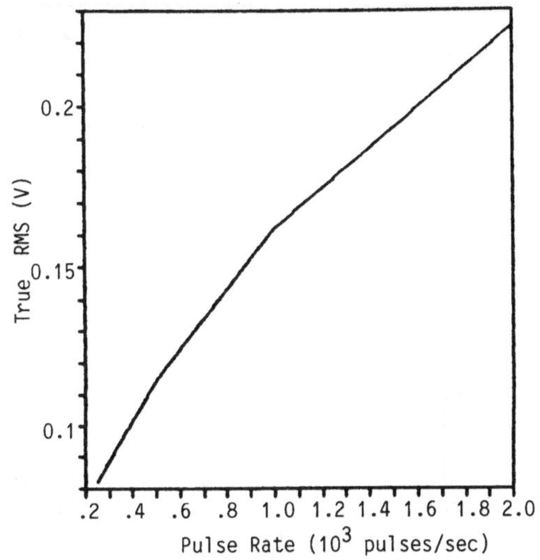

Fig. 7. The affect of pulse rate on the received signal

received signal. Some changes like pulse height or pulse rate had a linear effect as was anticipated. On the other hand, pulse width and pulse rise or fall time had a varying effect. The results of this investigation should make it clear that it may be difficult to make two instruments behave the same way. For instance if one system has a square wave pulse whose width is 3.5 μs and a second system has a square wave pulse width of 5.9 μs then the received sensor RMS signals would be 0.227 V and 0.139 V respectively, all other factors being the same. As long as the operator/engineer knows that this condition exists in the two units then it may be possible to make adjustments in amplification to correct for this effect.

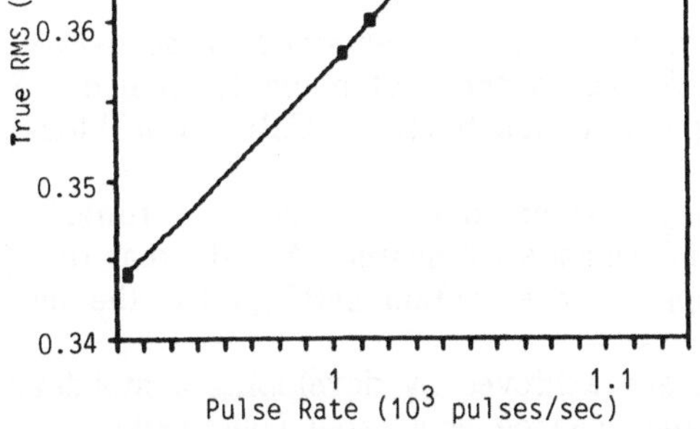

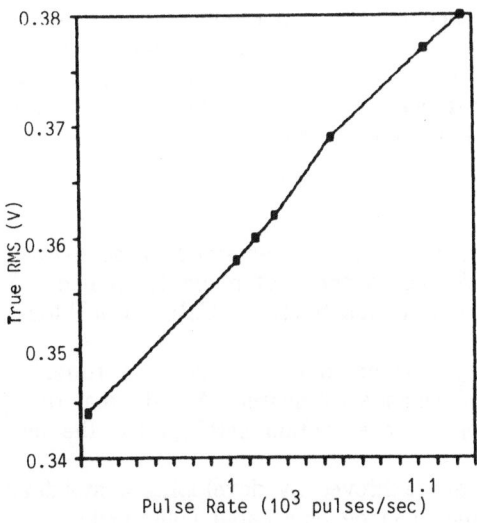

Fig. 8. The affect of small changes in pulse rate on the received signal

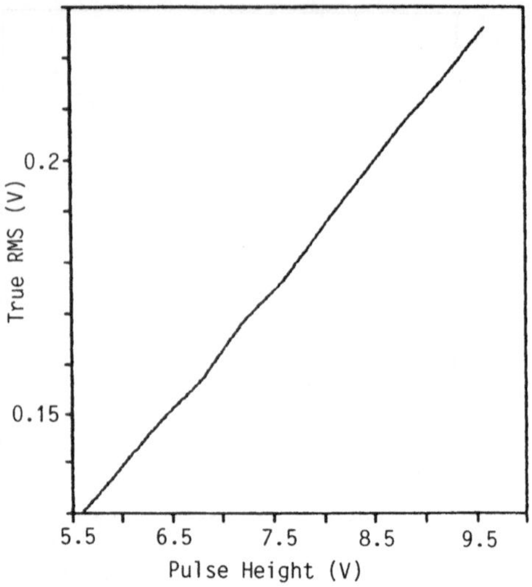

Fig. 9. The affect of pulse height on the received signal

SIGNAL MEASUREMENT CIRCUIT

The measurement side of the system must also be calibrated. The first step is to generate a signal with a source that can mimic a typical acoustic emission wave form. The device used here is the AES-2 (manufactered by Acoustic Emission Associates, San Juan Capistrano, CA). The AES-2 system can generate this wave shape on either of two channels. By combining these two wave shapes and feeding them into the acousto-ultrasonic system it is possible to measure the system response. This wave shape is first set up to appear as the classical AE wave form and then the second channel is overlayed on the first to determine the ability of the system to accurately measure counts. The system should be able to measure this compound signal regardless of the nature of the overlap (in terms of time). One obvious problem area is the timing network that establishes the period over which the signal is measured. A difference in timing between two systems can result in one system reporting larger stress wave factor reading than the second system. This, however, should be a uniform difference and a correction factor can easily (and even empirically through the use of a test block, such as type one described below) be determined.

Test Block 1

The measured results (data) determined by acousto-ultrasonic systems have been shown to be affected by the system performance. System performance may be verified by using calibration blocks. Calibration blocks allow the user to:

1. Check (verify) system performance over time.
2. Compare performance of system A with that of system B.
3. Determine appropriate system settings for the intended application.

Items 1 and 2 above are achieved by developing a standard test block. This standard test block then can be duplicated (obtained) world wide. This allows lab A to compare test results with lab B without sharing the same test block. Items 2 and 3 are accomplished with application specific test blocks appropriate for the task at hand.

The first type of standard that is required is one that assures the operator that the system is consistent in its output. This standard can be constructed of many (any?) materials since it does not have to have features similar to the material actually being tested. This standard may be manufactured from a block of aluminum, such as 7075-T6. The shape of the block is not important unless the sensor fixture that is being used with the acousto-ultrasonic system has special requirements. A 0.5 cm thick plate (ie. approximately four times the wave length at 500 kHz) that is at least 15 cm on a side would suffice. Place the plate horizontally. A damping material may be used to attenuate effects arising from the surface on which the plate is placed. The method developed is such that it is easily repeatable from site to site and over time.

The general idea that is being advanced here is the need to be sure that the system including sensors are behaving with consistency. The system is set into a predefined configuration and the sensors are placed a fixed distance apart and coupled with a repeatable technique. One means of achieving a fixed configuration is to predrill flat bottom holes that are the same size as the sensors being used.

In the following description the selections chosen are typical for the AET 206 AU system. For instance, the system may be placed in the pulse-sync mode and pulse at maximum or 250 volts per pulse and 2000 pulses per second. The receiving sensor signal is amplified 50 dB and the RMS and stress wave factor of the signal is measured. In the Stress Wave Factor the system is set up with predefined instrument selections:

1. the data acceptance gate is set, eg. 60 microseconds, from the time of the pulse.
2. A sample interval is selected, eg. 0.125 seconds.
3. The threshold is adjusted to a fixed amount, eg. 1.00 volts.
4. A scale factor is set at a specified value, eg. 10.
5. A pulse voltage is selected, eg. 250 volts.
6. A pulse rate is selected, eg. 2000 pulse per second.

The next requirement for this test block is a method of attaching the sensors. If standard sensors are used, that is, sensors whose shape is cylindrical with a ceramic wear plate and a side mounted connector (or otherwise but with a general shape similar to the one described herein) then the fixture is simple. The sensors are acoustically coupled with the couplant that is normally used in a test configuration. This couplant is applied in a manner that assures the operator that a sufficient quantity is placed. The couplant should be adequate to wet the surface but not so plentiful that it runs up the side of the sensor. If couplant runs up the side of the sensor it must be removed from the sides of the sensor before any readings are made.

A standard mass is placed on the sensor to mechanically couple the sensor to the plate. A kilogram (kg) on each sensor would be adequate for the mass. If the operator is using a dry coupled wheeled type fixture then coupling the sensor to the plate becomes much easier. The sensor is held in a fixed orientation to the plate, eg. vertical, and pressed down with a given force. This could also be accomplished with a mass placed on a vertically aligned wheel fixture. Care must be taken to insure that the load is evenly distributed to both wheels of the fixture. It may be advisable to manufacture a tricycle arrangement for the wheeled fixture. The advantage is that the tricycle arrangement has is that the wheels are held in a fixed orientation with respect to the plate. If a special sensor fixture is manufactured for a given application then the operator has to insure that the basic principles outlined above are considered.

Test Block 2

This block is used to develop the relationships for which the Acousto-ultrasonic system is to be used. This block will have typical defects manufactured, or otherwise generated, eg., impact damage, that are to be detected by the system in operation. A series of AU scans are to be made over the defect block until the optimum technique is developed. In some applications the best technique may be through transmission and in another application the best technique may be side-to-side. In some cases the technique is defined by the application. That is, it may not be possible to perform through transmission or one of the other possibilities due to geometry or part conformation.

Here must be kept in mind the theory of operation of the system. A pulse is generated in the material and received after it has passed through the material that contains the defect type. The important thing to remember is that the system is not generally used to detect a single defect but rather distributed type defects. The exception to this generalization is impact damage or other pervasive type damage. Since different materials and configurations are going to affect the signal, the strongest correlation will be derived from samples of the actual material in the configuration of interest. This implies that the test block be manufactured out of the same material that is to be inspected. Not only that but the configuration must be the same. That includes shape, thickness, ply orientation, number and density, and construction technique by which is meant all backing and interplies must be the same as the part to be inspected.

The calibration standard then becomes application dependent. If the application is to measure porosity on a production line then the standard developed should contain the range of porosity changes observed in the real parts. This will probably entail a complicated engineering controlled development, and independent verification. Keep in mind that the configuration of the part to be inspected must be duplicated by the standard test block. This is in terms of thickness, ply orientation, layup and any other consideration. For instance let us assume that the part is a tube whose wall thickness varies. Engineering will have to assure itself that the wall thickness variations do or do not affect the ability of the system to measure porosity. If thickness does affect this ability then the standard will have to take this into account. Another consideration is the backing on the part. Does it have any effect, does it change in dimension (or even presence) and most important does this affect the AU reading and how? If there is an effect then this will also have to be built into the standard.

CONCLUSION

Operator awareness of system performance is key to the successful implementation of acousto-ultrasonic inspection. The development of calibration blocks will help advance the technology of acousto-ultrasonics past the novelty stage.

Two calibration standards (blocks) were proposed. The first block allows the operator/engineer to compare one system through time or several systems to each other. This block is simply a plate of metal (aluminum for instance) that can be defined and obtained world wide. This choice is made so that lab A and lab B can compare results on a semi-independent basis. The paper suggests a set of operating conditions for the system check but these are not as important as reporting what is used so the test may be duplicated. This first calibration block is application independent.

The second test block should exhibit representative examples of all the dependent characteristics affecting the acousto-ultrasonic measurement. This means the second test block is application dependent. All the features of the part to be inspected must be determined by an engineering evaluation. If a feature is determined to affect the measurement then this feature must be accounted for in the test block. This would include such items as thickness variation, backing or the lack of it, geometry, material changes, ply orientation, adhesives or changes in them, coatings or changes in them, chemical composition, porosity, and density.

TEST CONDITIONS IN STRESS WAVE FACTOR MEASUREMENTS FOR FIBER REINFORCED COMPOSITES AND LAMINATES

M. Bhatt and P. J. Hogg

Queen Mary College
University of London
United Kingdom

ABSTRACT

The acousto-ultrasonic technique using Vary's definition of the stress wave factor has been evaluated (SWF = g.r.n.). It was found that the technique needed to be reappraised and conditions under which the stress wave factor is reproducible defined carefully. The Acoustic Emission Technology Model AU 206 was used for the study. Instrument variables were sequentially eliminated and reduced to the instrument gain and the background noise. Regions of validity were then defined in terms of the background noise levels, and of the instrument gain on various unreinforced and glass reinforced polyester resins, with and without introduced defects. The stress wave factor is found to be most reproducible when independent of background noise, and at high gains. The equivalence of readings taken using through transmission, and those taken by placing transducers on the same side is investigated. These were found to be equivalent only at high gain, with doubtful results for low gain. The stress wave factor is interpreted in light of the present study.

INTRODUCTION AND AIMS

A novel technique that has recently become available in the United Kingdom is the commercial equipment that allows stress wave factor measurements, as marketed by AET (USA), the model 206 AU.

In essence, the technique consists of exciting a transducer by using a pulse, allowing the transducer response to propagate through the system under test, and analyzing the signal using what are termed acoustic emission techniques. The latter consists of counting the number of oscillations that exist above a particular noise threshold within the transducer pulse as modified by the system. The number of oscillations present are characteristic of the energy spectrum that exists due to interaction with the system.

If a low energy pulse is generated at a pulse rate g, then the number of oscillations, (or "ringdown counts"), n, may be counted within a time window, r. The stress wave factor (SWF) is then defined as SWF = g.r.n. and is considered to be a material parameter, given that other conditions are constant, characteristic of the system under test. In work by the several groups that have used the technique so far[1-9, 11-14], the stress wave factor has been shown to be sensitive to deviations from perfection within a specimen, and has been considered to be capable of distinguishing inferior material correctly. In addition, the stress wave factor is variously claimed to provide information about a flaw population, regions of residual stress, local microcracking, and other failure phenomena. However, it should be emphasized that the stress wave factor is only a comparative number as cited within the literature, and it is uncertain that one specimen may be compared, and ranked, correctly with another made of the same material. In fact, a quick survey of the sources cited above indicate that, apart from being a comparative technique, rather than an absolute one, there appears to remain the problem of consistent reproducibility, from one reading on a single site to the next reading, and of the conditions under which the readings are taken.

Initial work done by Ono & DeSpain[4], and by Govada et al.[1], and Duke et al[3], indicates that the stress wave factor, and its interpretation is open to considerable ambiguity. This view was reinforced by work done in the U. K. previous to the present study, and indicated that a re-appraisal of the technique was in order.

EXPERIMENTAL

In order to evaluate the technique, it is necessary to identify all the contributions to the cause of the ambiguity in the results. The factors are many and varied, some central to the test technique, and some the result of actual material variation.

Those that affect the actual technique may be grouped into the following areas:

i. coupling
   a) contact pressure
   b) type of coupling

ii. emitted ultrasonic pulse
   a) emitter transducer type
   b) rate of transducer excitation (g)
   c) pulse energy
   d) pulse or burst mode

iii. received signal and processing
   a) receiving transducer type
   b) gain of amplified signal (G)
   c) time window of reading (r)
   d) background noise threshold (v)
   e) signal gating

iv. relative position of transducers

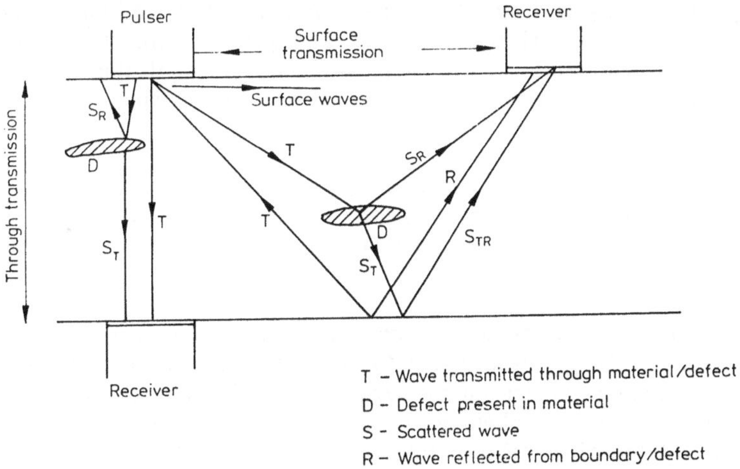

Fig. 1. Schematic of wave modulation within material.

If all of the above variables are kept constant, then the SWF should be a reflection of material variation in the specimens under test. Parameters that probably characterize the material are thought to be

v. structure of material
   a) processing history
   b) morphology
   c) internal interfaces within the structure
   d) defects (debonds, cracks, etc.)

vi) geometric effects
   a) extent of test specimen
   b) thickness of test specimen

vii) mechanical response of material

Conditions of test were narrowed down according to the following arguments, especially as far as instrument settings were concerned.

The 206 AU comes with a 1 MHz bandwidth, 0.5 MHz resonant frequency FC500 piezoelectric transducer, which is used as the emitter, and a 375 kHz resonant frequency transducer, used as a receiver. Most of the studies in the literature have concentrated on thin plates, such that the wave penetrates through the specimen under test, and therefore allows the emitted pulse to be transmitted through the body of the material, and also to be reflected off the opposing specimen boundaries, Fig. 1; this restricts the test to a continuous pulse, rather than a burst mode. It also allows for examination of both through transmission of the signal, and of propagation limited to the surface.

The thickness of the specimens under test was kept constant, and in through transmission, the transducers are therefore placed opposite each other at a constant distance, namely the through thickness. The extent of the specimens under test was also kept constant. This limited the center to center distance for transducer placement on the same surface. A constant distance of ten centimeters was chosen.

The pulse energy is limited to three values, and the middle value of -140 volts was chosen on an arbitrary basis. The gate, within which each pulse is read, was set at maximum in order to obtain as much of the signal as possible, including any contributions from constructive superposition of signals.

The product of the time window (r) for reading the continuous signal, and the rate of transducer excitation (g) was chosen as g.r. = 1000 as a referent, in order to facilitate comparison at varying gains, noise levels, and other variables,

One of the major difficulties was that of reproducibility. The couplant provided by AET (SC-6) was found to be messy, and required a set-up that provided a constant pressure on the coupling. This was eliminated by using a vinyl film with a rubber adhesive on both surfaces (double sided sticky tape), such that the film relaxed into a constant static state, without continuous application of pressure. The settling-in time of the order of seconds.

Given that all the above conditions were kept constant, this then left the gain (G) of the amplified signal, (from 40 to 100 dB), and the background noise threshold (v) as variables. Note that the 206 AU has found possible rates of transducer excitation (0.25, 0.5, 1, 2 kHz), with complementary values for r, providing four ways of obtaining the referent g.r = 100.

The approach was to characterize initially the noise levels that exist near and within the test environment, and thus to determine the region of validity for any readings of the SWF that may be indicated. The aim was to attempt to find regions within which the only signal that was picked up by the receiver was that of the emitter. This was done without any material being present between the transducers, and corresponded to one oscillation per excitation pulse, that is, the SWF equalled 1000.

Secondly, the equivalence of behavior using through transmission, and same-surface placement of the transducers was determined on a sheet of poly ether ether ketone (CFR-PEEK or APC-2), as a function of the background noise (v), and once suitable noise levels had been established, as a function of the instrument gain (G).

The third step was to characterize the response of the coupling agent, in terms of the background noise. This was done using through transmission, as the vinyl film is only several microns thick.

In order to cross compare between materials, a neat polyester resin (NR), a chopped glass reinforced polyester resin (CSMR), and a unidirectional glass fiber reinforced polyester resin (UGRP) were selected as systems under test. The response of these materials as a function of the background noise (v), and as a function of the gain (G) was now determined.

Finally, circular paper defects of one inch diameter were introduced into the latter three systems, designated NR-D, CSMR-D, and UGRP-D. The similar response using through transmission as a function of the background noise (v), and of the instrument gain was then determined. These results are presented and discussed below.

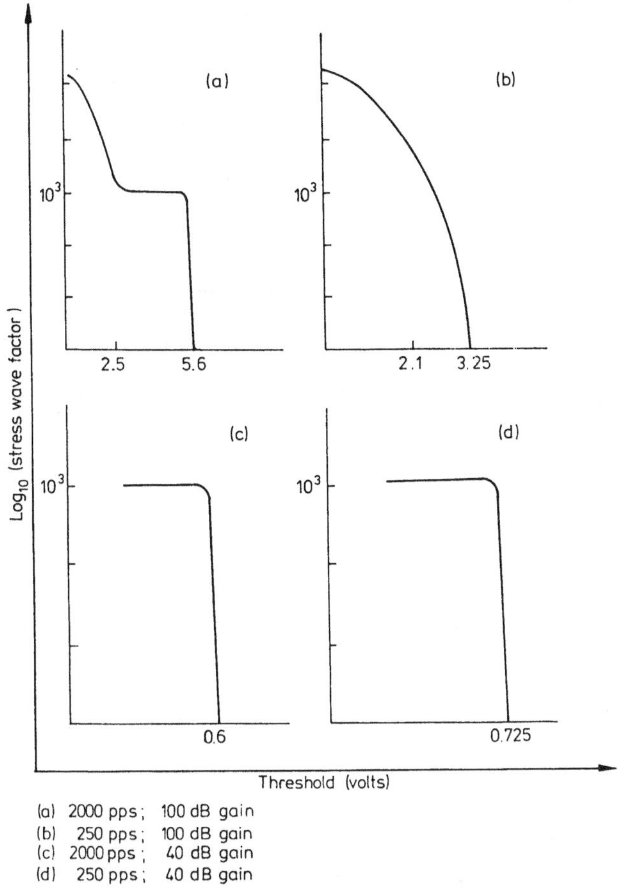

Fig. 2. Noise levels recorded at varying pulse rates, threshold levels and gain of system --no material between transducers.

RESULTS

SWF measurements were initially obtained from transducers separated by an air gap (with no material or couplant). The transducers were maintained a constant distance of ten centimeters apart. It was noted that the stress wave factor in this portion of the study was insensitive to the relative orientation of the transducers, and to their separation. The pulse excitation rate was set to 2 kHz or 0.25 kHz, with time windows (r) set to 0.5 and 4 seconds respectively. The gain of the total system was set to varying values, generally the two extremes of 40, and 100 dB. The results obtained from measuring the SWF under these conditions are fully reproducible. Figure 3 shows the individual data points obtained by progressively increasing the threshold voltage, at a gain of 100 dB, and an excitation rate of 2 kHz. The SWF measured at the minimum threshold level was found to be greater than 100000 but this fell progressively as the threshold was increased. A constant value of 1000 (the constant of referent), was recorded between the threshold voltage values of 2 and 5 volts, after which the slightest increase in threshold voltages caused the SWF to fall rapidly to zero.

The dependence of the SWF on gain was found to be different when gain and excitation pulse rate were varied (Fig. 2). At a gain of 100 dB, and an

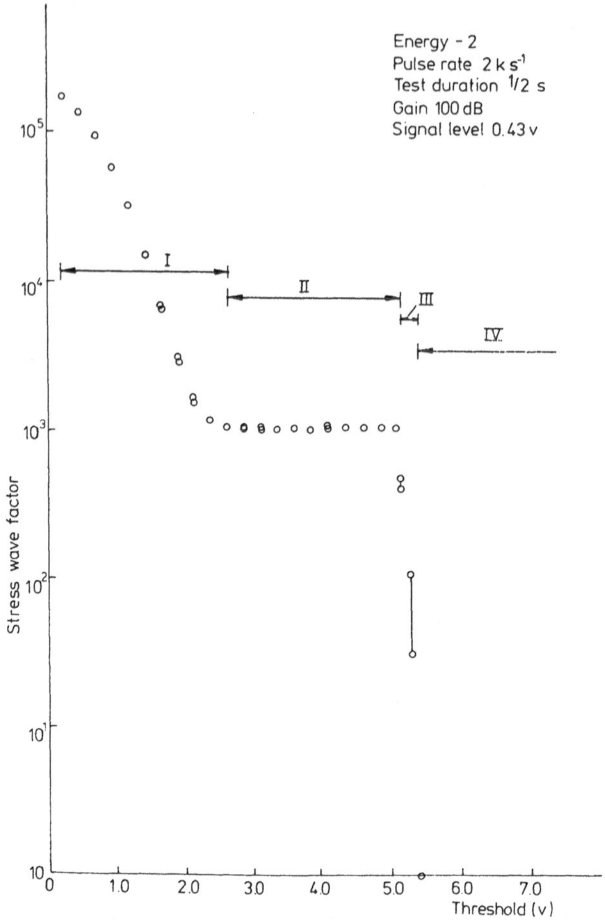

Fig. 3. Noise level at high gain and high pulse rate.

excitation rate of 0.25 kHz (Fig. 2b), the SWF did not exhibit a clear noise independent region, and progressively fell to zero with increasing noise threshold. At a 40 dB setting, the SWF maintained a constant value of 1000 at very low threshold voltages, but very rapidly fell to zero for values of the threshold voltage greater than 0.6 V, for both the 2 kHz and 0.25 kHz excitation rates (Figs. 2c,d). It should be noted that the actual threshold at which the SWF fell to zero differed for each instrument setting.

The effect of the couplant was checked by bonding the transducers together using the coupling medium (double sided sticky tape), thereby employing a through transmission mode of testing. The SWF was recorded for varying threshold voltages with a gain of 100 dB, and 40 dB signal gain, at a pulse excitation rate of 2 kHz, and a sampling time of 0.5 seconds (Fig. 4). A plateau region where the SWF is found to be independent of the background noise was clearly observed at the 100 dB setting, from the lowest threshold voltages values, up until the highest attainable value of about 6 V, when the SWF decreased rapidly to zero. At the 40 dB setting, the SWF versus threshold voltage curve is found to be similar in shape to that for an air gap at a 100 dB, although the noise independent region is found to be somewhat indistinct. Very similar behavior was observed when a sheet of neat polyester resin (4 mm thick) was interposed between the two transducers (Fig. 5). The

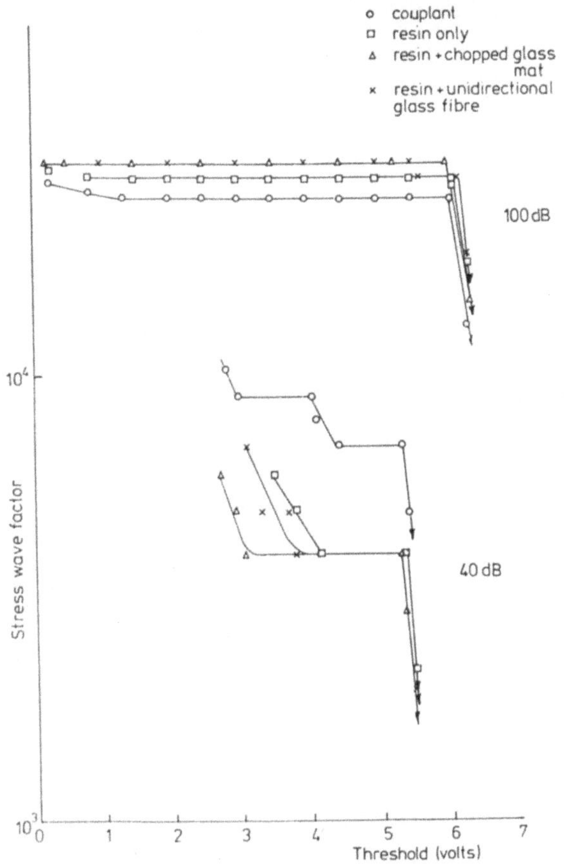

Fig. 4. To show differences between materials at varying gain and threshold level using through transmission.

system consisted of transducer - couplant - resin - couplant - transducer, tested in the through transmission mode, again with an excitation rate of 2 kHz, test duration of 0.5 seconds, and gains of both 40 and 100 dB. The SWF values for the resin and the couplant systems differed, as did the critical voltage values that bound the noise independent region for the SWF. The noise independent values measured at both 40 and the 100 dB gain values are considerably higher than the 1000 characteristic of the air gap, with 25000 for the couplant, and 28000 for the resin at 100 dB. Corresponding values for these systems at 40 dB were 8000 and 4000 for the couplant and resin respectively. Note that the couplant exhibited the higher value of the SWF at 40 dB gain, whereas the resin had the greater value of SWF at 100 dB gain.

Comparison of all the materials within the present study was facilitated within the noise independent region, as it provided a common reference region. Readings of the SWF for all the materials without a defect, and for the couplant are shown in Fig. 4, for gains of 40 and 100 dB. At 40 dB, although the couplant can be distinguished quite clearly as having had the highest value of the SWF (= 8000) in the noise independent region, there was no significant difference between the neat resin and the reinforced materials (SWF = 4000). The materials presented in this set of data were nominally kept defect free. At 100 dB, differences between the materials were appar-

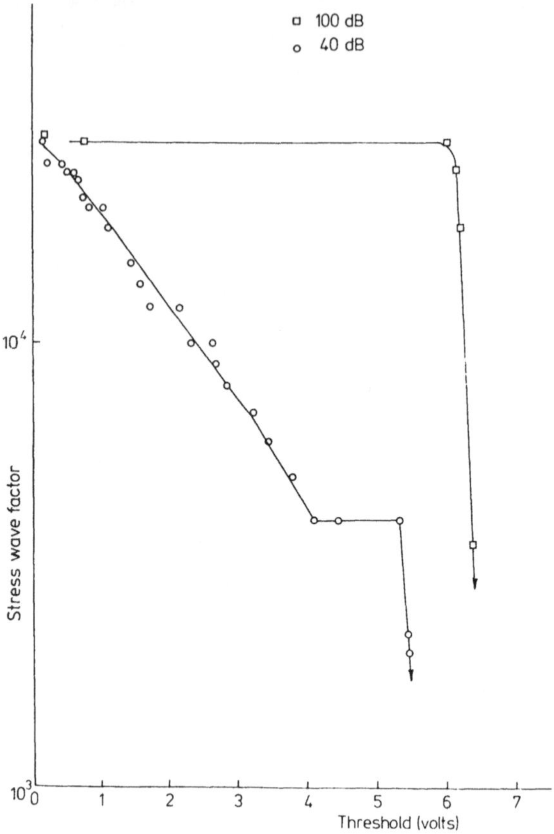

Fig. 5. Variation of stress wave factor with threshold level for polyester resin using through transmission.

ent. The couplant now had the lowest SWF wave factor (25000), within the noise independent region, the next lowest being the neat resin (SWF = 28000). The highest factor (SWF = 30000) belonged to both reinforced systems, with no apparent difference between them.

Within similar noise independent regions, the materials with introduced defects were studied and the results presented in Figs. 6 and 7. The factor readings for material without defect were taken within the same plate, and are therefore expected to be different from the absolute values presented in Fig. 4 for the neat resin. In effect, both the NR-D and the CSMR-D show significant differences in value for both high and low gains. The CSMR had a value of 26000, as compared to the CSMR-D, with a value of 28000. At 40 dB, the CSMR showed a noise independent value of 4000, whilst the CSMR-D had no noise independent region at all at this gain.

The neat resin showed similar behavior with respect to threshold voltage for introduced defects. At 100 dB, the NR showed a factor value of about 30000, whereas the NR-D showed a noise independent value of 26500. At 40 dB, the NR had a noise independent region of 4000, whilst the NR-D at 40 dB had two noise independent regions, the one at lower noise level having a value of 5000, the upper noise level having a SWF value of 3000, with the transition

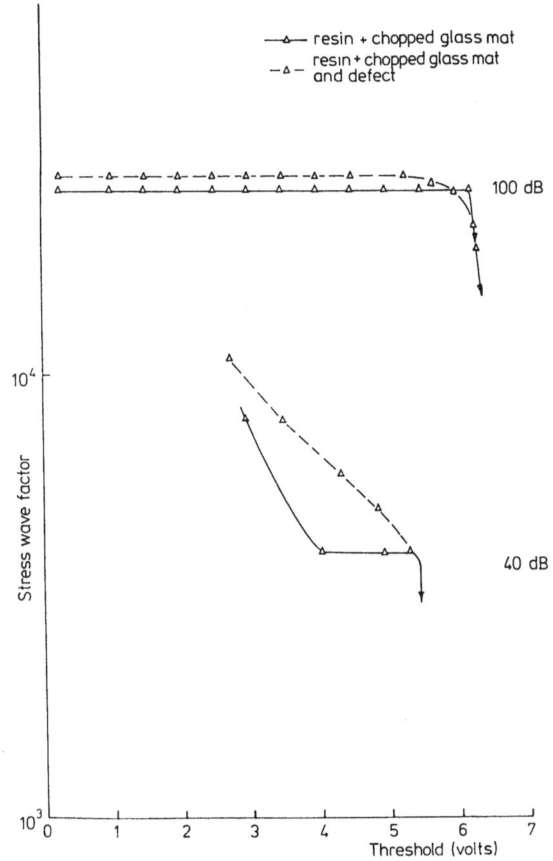

Fig. 6. To show the effect of an introduced defect on the stress wave factor for resin reinforced with chopped glass.

occurring at 3.6 volts. Although results are not presented here, the UGRP showed a higher SWF at higher gain, and a lower SWF at lower gain, than the UGRP-D.

Again, once noise independent regions had been identified for all the materials, the response of the SWF as a function of gain was studied, and results are presented in Fig. 8 for readings taken using through transmission. In general, the SWF was highly gain dependent until about 60 dB, after which it passed through a transition to reach a saturation level which varied with the material. Again, as the SWF is expected to be sensitive to minute variations in the material structure,[11,12] and to major defects,[9,10,12] it is not expected that the absolute stress wave factors would be identical to those presented previously. In terms of the saturation levels, the NR had a SWF value of 31100, the UGRP had a value of 30000, the CSMR 29100, the UGRP-D had a value of 28500, the CSMR-D 27900, the couplant 25100, and the NR-D a value of 24000. These results, along with the saturation values for CFR-PEEK are presented in Table 1. Table 1 also includes the initial slope of the stress wave factor versus gain curve in the 40 to 60 dB linear region.

Values of the SWF for same-surface and through transmission testing on CFR-PEEK sheet are shown in Fig. 9 for the extreme cases of 100 and 40 dB,

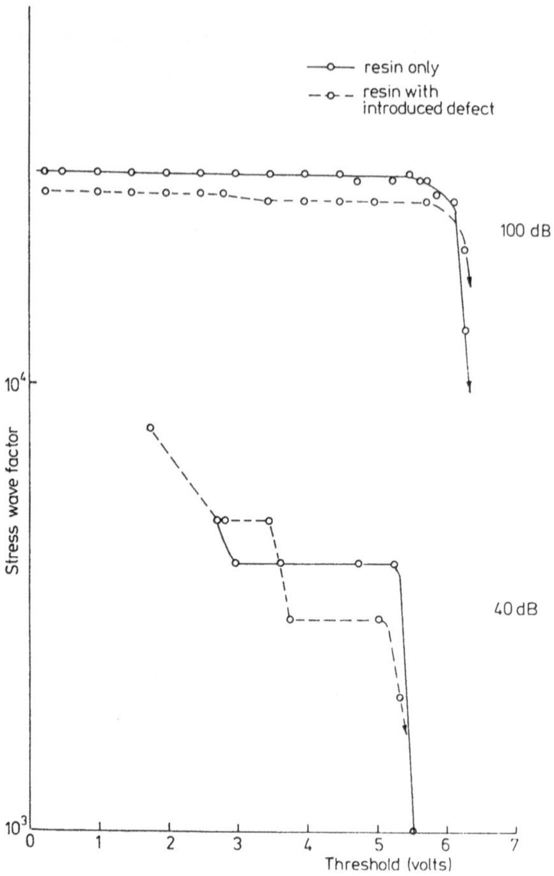

Fig. 7. To show the effect of an introduced defect on the stress wave factor for the polyester resin using through transmission.

when the threshold voltage is varied. The results of through transmission testing for the CFR-PEEK at both gain settings were found to be qualitatively similar to those obtained from testing the resin and couplant systems. The critical threshold voltages values above which the SWF fell to zero were also comparable. At the 100 dB level the same-surface transmission mode exhibited similar trends concerning the variation of the SWF, as in the through mode, albeit with slightly higher values of the SWF. At the 40 dB gain setting however, the behavior using the same-surface transmission mode was markedly different. The SWF fell very rapidly with increasing threshold voltage, showing a very restricted noise independent region. Furthermore, the SWF fell to zero at a critical upper noise threshold of 1.5 V for the same-surface mode compared to 5.5 V for the through transmission mode. It should be noted that the plateau observed in the same-surface configuration at 40 dB gain occurs at a value of 1000, which corresponds to the unmodified signal.

At gain settings greater than 40 dB, a well defined noise independent region existed in the SWF versus threshold voltage curve, which in every case encompassed the 4.5 V threshold level. The SWF within the noise independent region was measured from 40 to 100 dB, in steps of 10 dB, keeping the threshold voltage constant at 4.5 V, and is shown in Fig. 5. For both the through

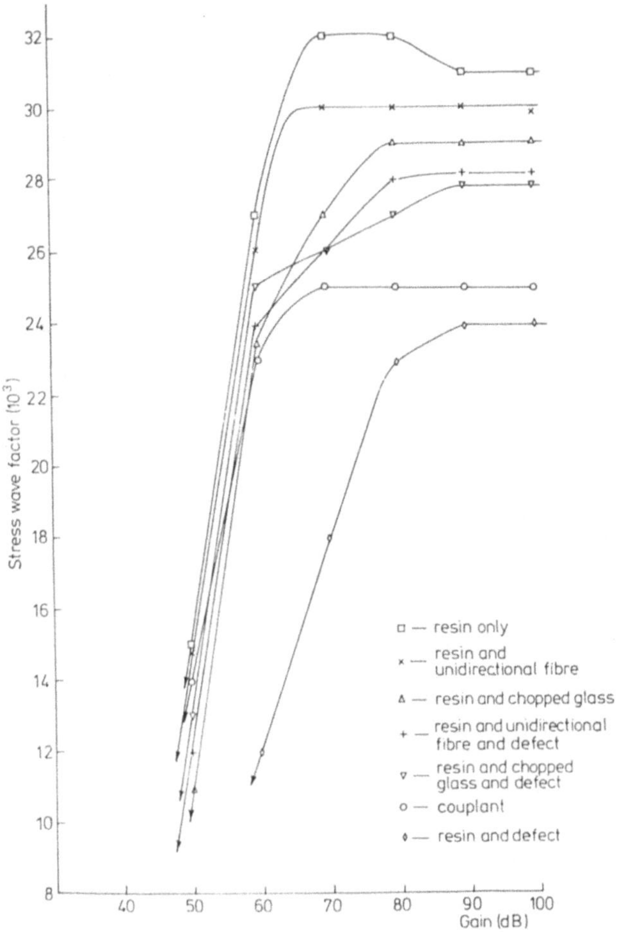

Fig. 8. To show the variation of stress wave factor with gain for materials with and without introduced defect.

mode of transmission, and the same-surface mode of transmission, a maximum threshold voltage independent value of the SWF may be determined, that is also independent of the gain. A critical gain setting must be exceeded in order to obtain this SWF, and the exact value of the noise and gain independent SWF for the CFR-PEEK, differs strongly for the same-surface and through transmission configurations.

DISCUSSION

The results that appear to show all the salient features of the response with respect to background noise level are to be seen at the highest excitation rate and at the highest gain. A schematic of this is shown in Fig. 3, which also divides the curve into four distinct regions. At present, it is suggested that region I is a reflection of true background noise, the instrument being at its most sensitive at these levels of threshold voltage. Within this region, the receiver also picks up signal noise from the emitter, such that there are a thousand pulses per second that can be attributed to the emitter, as well as the generally accepted background noise.

Table 1. Variations of SWF with different material systems.

System	Stress wave factor within gain and threshold independent regions (10^3)	SWF (60 - 40)dB ($10^2 dB^{-1}$)
Resin only	31.1	11.6
Resin and unidirectional fibre	30.0	11.1
Resin and chopped glass mat	29.1	9.75
Resin and unidirectional fibre and defect	28.25	10.0
Resin and chopped glass mat and defect	27.9	10.05
Couplant	25.1	7.6
Resin anddefect	24.0	6.0
APC 2 through	25.8	9.96
APC 2 surface	34.8	15.95

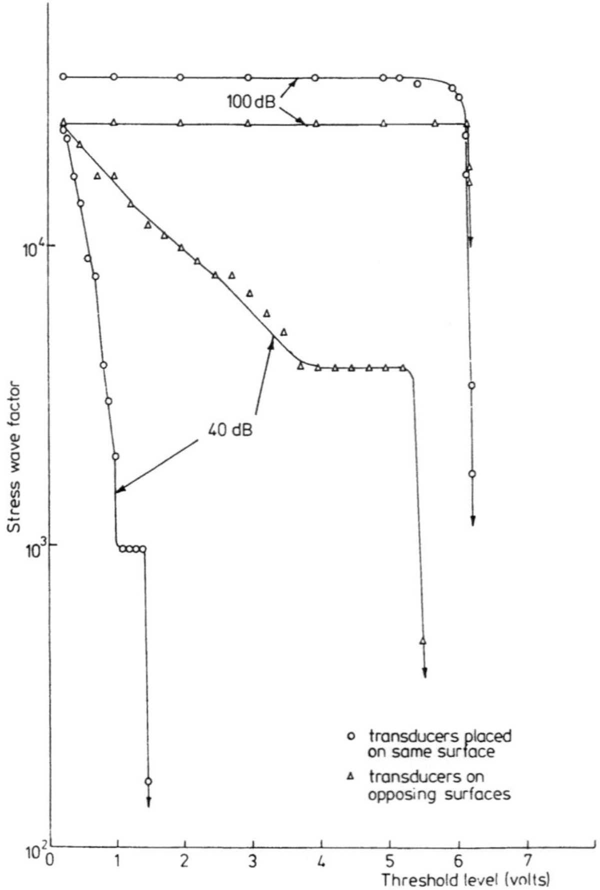

Fig. 9. Comparison of through and surface transmission for carbon fiber reinforced peek (APC-2).

Region II is the region where the threshold voltage signal has successfully masked out the true, random background noise. However, within this region, the emitter signal is still strong enough such that it is picked up by the receiver. Recognition of this is facilitated by the use of g.r. = 1000. The SWF has become independent of the background noise, and is responsive only to the emitter signal. If such a region exists when a material is being tested, then presumably this would correspond to that region within which readings of the SWF ought to be taken, as it would be free of noise interference.

As the threshold voltage is increased, even the signal from the emitter is masked such that the SWF drops rapidly to zero (region III in Fig. 3). Above this, a dead zone exists (region IV), where no readings of the stress wave factor exist, as all the signals have been masked.

Note that all the materials in the present study have shown all these four regions, in some combination. The indication appears to be that noise

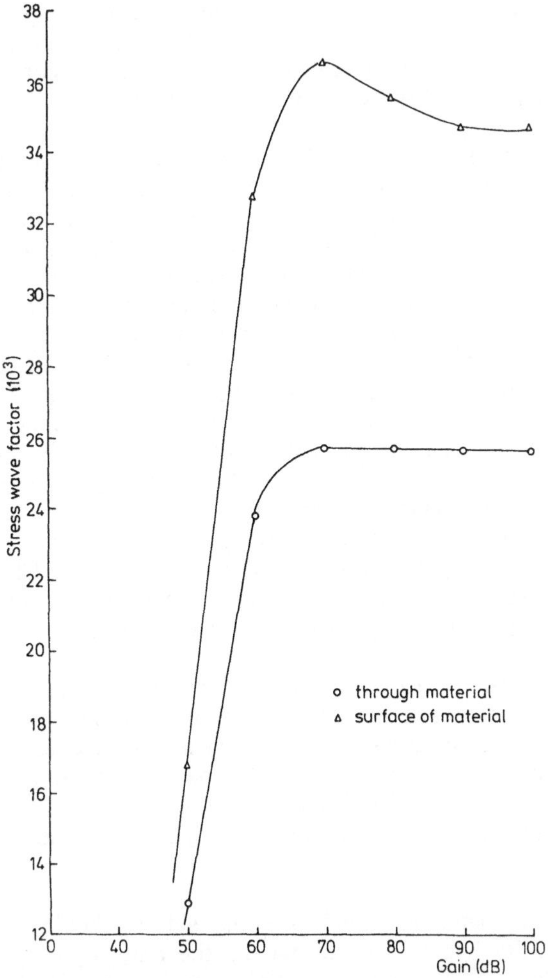

Fig. 10. Comparison of through and surface transmission for APC-2 at constant threshold level.

levels have to be carefully masked, as seen in the various figures referred to above. In particular, the region of validity (region II), where the receiver is confidently responding only to signals from the emitter that have been modified by the system, is one that shifts from material to material (Fig. 4), but this same range also shifts depending on what gain the readings are taken at (Figs. 5-7), and also according to what defects are present (Figs. 6 & 7), and is additionally dependent on what configuration is used for the test.

The least variation in the extent of the region of validity is found to be at high gains. It is seen that the noise independent SWF is reproducible at high gains over a much wider range of the background noise, regardless of the test configuration (Fig. 9), type of material (Fig. 4), or of changes in the material structure (Figs. 6 & 7).

High gain regions also provide the greatest sensitivity to variations in the stress wave factor due to variations in the type of material, or differences in the material structure (Fig. 8), or in the test configuration (Fig. 10). As the gain is changed, the SWF increases rapidly with the gain, and then reaches a saturation level. There appears to be a strong correlation between the slope of the initial linear portion of this curve, and of the saturation value of the SWF at high gains. The latter correlation is shown in Fig. 11; the origin of this behavior is unknown.

Qualitatively, values of the SWF, at high gain taken within the noise independent region provide information about the material strength, such that all the materials with defects have a lower stress wave factor than those with no defects (Fig. 8). This is consistent with the accepted view[1,3,9,11-14] that a lower stress wave factor indicates a lower strength, defective material. However, there is ambiguity when comparing the reinforced materials with the neat resin without any defects. The reinforced materials have a lower SWF than the neat resin, when there are no defects present, suggesting that glass fiber does not act as a reinforcement. However, when a defect is introduced, the reinforced materials with defects have a higher SWF than the neat resin with a defect, which is qualitatively correct. A possible explanation for this is that in the absence of any major defects (larger than the glass fibers), the fibers act as a modifying mechanism for the transducer pulses, thus giving rise to the observed inversion. The view is supported by the observation that the 16 ply CFR-PEEK is seen to have a lower stress wave factor than the resinous materials without a defect, when comparison is made using through transmission.

The link between the values of SWF and the mechanical and the physical properties of a material are not defined at present, and there is no justification for ascribing a change in SWF (increase or decrease) to the weakening or strengthening of the material. It would be prudent at this stage merely to note that a change in the SWF within a specific material system indicates a change in material structure without specifying the consequence of the change.

When the background threshold is being set, it should be manifest that the region of validity has to be that where the only signal that is modified and received is that of the emitter. This implies that if the noise level is set too high, or too low, such that the threshold voltage is found to be near or within regions I or III, the SWF is going to be difficult to define. This

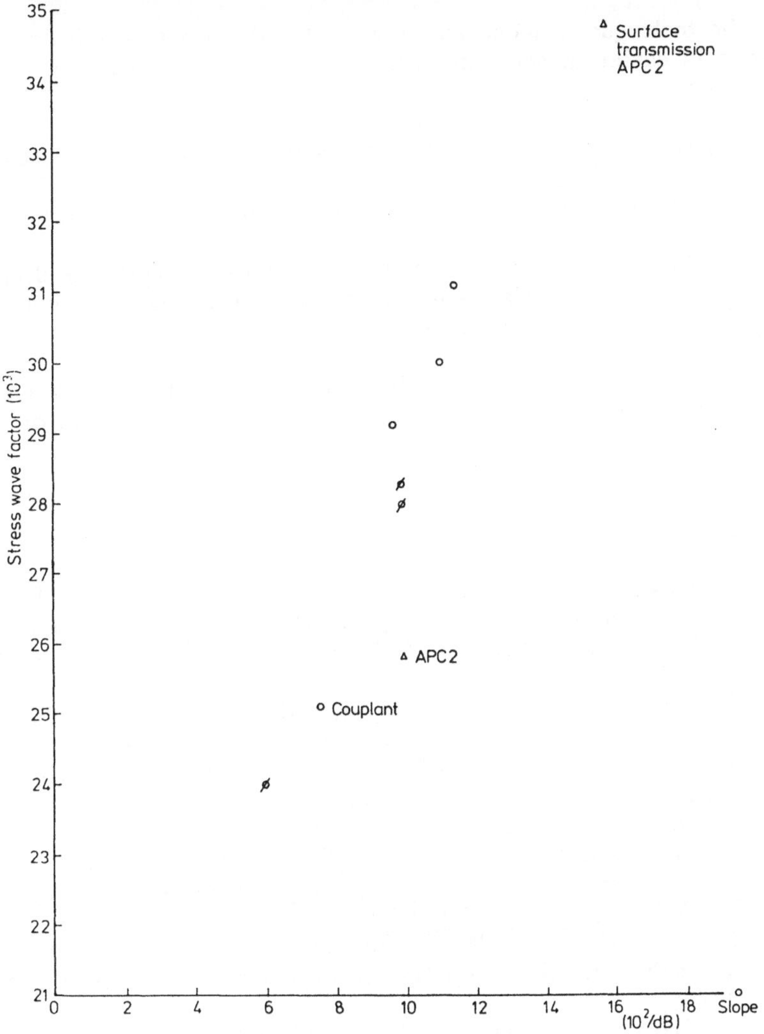

Fig. 11. Correlation between initial slope and saturation plateau within the stress wave factor-gain curve.

would then lead to problems of reproducibility, as the SWF is found to be rapidly varying with the noise background. When the same-surface readings are compared, it is found that there is a very small noise independent region, making regions I and III the dominant regions. As the stress wave factor varies rapidly with voltage within these regions, care is required to work within the region of validity. Notice also that the "region of validity" for the same surface configuration at 40 dB has a value of 1000, which is the chosen constant of reference. This implies that only the unmodified emitter signal is being picked up by the receiver in this configuration, and care is again required in interpreting results.

CONCLUSIONS

Although the stress wave factor shows potential, there are problems associated with reproducibility, and interpretation of the parameter. In particular, the technique is most useful when the through transmission method is used, as it is easier to home in on the signal that is independent of the background noise.

A region has been found where the stress wave factor may be considered to be free of extraneous noise, and reproducibility is enhanced by working at high gains. A possible source of scatter has thus been eliminated.

Sensitivity is found to be much improved, leading to detectable differences between inferior material, and that which is undamaged, provided the specimen is internally consistent in its readings.

Cross comparison between materials is difficult, giving rise to ambiguous results, suggesting that there are differing degrees of modification of the signal.

REFERENCES

1. A. K. Govada, J. C. Duke, Jr., E.G. Henneke, II, and W. W. Stinchcomb, "NASA Contractor Report 174870," NASA, Cleveland (1985).
2. J. E. Green and J. Rodgers, SAMPE National Symposium and Exhibition, San Diego (1982).
3. J. C. Duke, Jr., E. G. Henneke, II, and W. W. Stinchcomb, "NASA Contractor Report 3976," NASA, Cleveland (1986).
4. K. Ono and R. DeSpain, in: "Proceedings of the International Symposium on Acoustic Emission in Reinforced Composites," Society of the Plastics Industry, Inc., (1983).
5. K. K. Phani and N. R. Bose, J Mat Sci. 21:3633 (1986).
6. K. K. Phani and N. R. Bose, Comp Sc & Tech. (in press).
7. J. M. Rodgers, Acoustic Emission 1:322 (1982).
8. A. Sarrafzadeh-Khoee, M. T. Kiernan, J. C. Duke, Jr. and E. G. Henneke, II, "NASA Contractor Report 4002," NASA, Cleveland (1986).
9. V. K. Srivastava, J Mat Sci. 21:3638 (1986).
10. R. Talreja, A. Govada and E. G. Henneke, II, in: "Proceedings of the Review of Research in Quantitative NDE, Vol. 2" D. O. Thompson and D. Chimenti, eds., Plenum Press, New York (1982).
11. A. Vary, in: "Mechanics of Nondestructive Testing," W. W. Stinchcomb, ed., Plenum Press, New York (1980).
12. A. Vary, Matls Eval. 40:650 (1982).
13. A. Vary and K. J. Bowles, Poly Eng Sci. 19:373 (1979).
14. A. Vary and R. F. Lark, J Test & Eval. 7:185 (1979).

MEASUREMENT OF THE ENERGY CONTENT IN ACOUSTO-ULTRASONIC SIGNALS

M. J. Sundaresan and E. G. Henneke, II

Materials Response Group
Virginia Tech
Blacksburg, Virginia 24061

ABSTRACT

Important features of the techniques currently employed for the measurement of acousto-ultrasonic (AU) parameter commonly referred to as stress wave factor are briefly described. An alternate procedure for characterizing this AU parameter, in which the energy content of the received signal is used to rank the material's interaction with the propagating stress wave is proposed. This procedure employs simultaneous counting of acousto-ultrasonic signals at a number of threshold levels, suitably distributed across the amplitude range of the signals encountered in a particular test. The resulting counts at different threshold levels are given weightings according to their amplitudes and are summed up to provide a measure of energy. The accuracy of this scheme is verified by measuring the acoustical signals produced by the impact of steel spheres of different masses on a borosilicate glass block. In a related study, the same procedure was applied to measure acoustic emission signals from composite materials and was found to provide an estimate of the damage level with fair degree of success.

INTRODUCTION

The acousto-ultrasonic (AU) method developed by Vary,[1] uses the propagation character of ultrasonic waves through composite laminates to evaluate the material's quality. A schematic illustration of this technique is given in Fig. 1. A transmitting and a receiving ultrasonic longitudinal wave transducers are placed a short distance apart on the same surface of the laminate, and usually coupled to the material by a viscous fluid. The transmitting transducer sends in compressional waves into the laminate. These waves, in turn, excite lamb waves, which after propagating through the material in between the transducers are sensed by the receiving transducer. Recent studies[2] suggest that these waves are primarily transmitted in the first extensional and flexural modes. During their passage through the composite material, these ultrasonic waves undergo changes depending on such features as microstructural defects, delaminations and transverse cracks. In

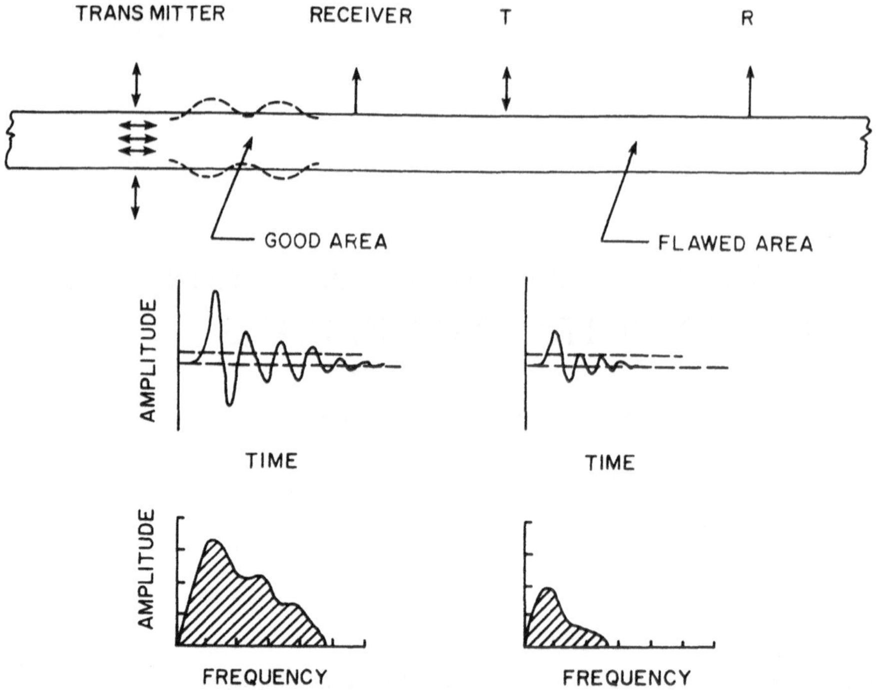

Fig. 1. Principle of Acousto-ultrasonic technique

principle, changes to each of the different characteristics of the received AU signal, namely the phase velocity, excited mode shapes, frequency spectrum etc., could be tried for possible correlation with the material's quality. But, most of the AU applications have relied on the changes produced in the amplitude or the energy of the received signal to rank the anomalies in the material. The energy contained in the AU signal could be measured by one of the several methods, all of which are directly borrowed from acoustic emission technique. It is obvious that the success of the AU technique crucially depends on the accuracy of the signal measurement technique employed. In addition, for actual industrial applications, in general large surfaces are required to be examined and this would involve several thousands or tens of thousands of stress wave factor readings for each component. Hence the measurement technique should be sufficiently fast and amenable to automation.

In this paper important features of the techniques currently employed for the measurement of the AU signal level are briefly described. An alternative scheme of measurement is proposed and its accuracy evaluated.

AU Signal Measurement

Ringdown counting. Because acousto-ultrasonic signals resemble acoustic emission (AE) signals, some of the AE measurement techniques,[4] are directly applicable for the stress wave factor measurement also. The simplest and the most commonly used method is the ringdown counting. Here the number of times the AU signal crosses a threshold level, Fig. 2, are counted. The number of threshold crossings within a single AU received signal, after normalizing with certain factors are termed the stress wave factor,[1] and can be interpreted as a measure of the material's ability to transmit stress

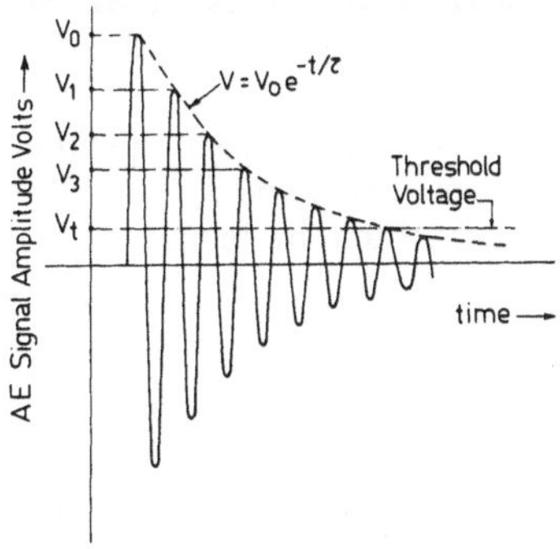

Fig. 2. Wave form of an idealized acousto-ultrasonic signal

waves. An idealized signal in the form of exponentially decaying sinusoid can be represented by

$$V = V_o\ (\exp(-t/\tau))(\sin(2\pi ft)) \qquad (1)$$

where, V_o, is the peak voltage, f, is the frequency of the oscillation of the AU signal, and τ, is the decay constant depending on various factors, like the laminate material and the type of transducer and excitation. The number of counts N registered by such an idealized waveform would be given by

$$N = (f/\tau)\log(V_o/V_t) \qquad (2)$$

Hence the SWF so obtained could be interpreted as a logarithmic measure of the peak voltage of such an idealized waveform. However, actual AU waveforms are quite complex and cannot be adequately represented by the exponentially decaying sinusoid. Further, the AU waveform changes, depending on the type and extent of defect present in the laminate. As a consequence, it is no longer possible to uniquely relate the ringdown counts measured at an arbitrary threshold level to the peak voltage or the total energy content in the AU signal. This probably is the reason why extensive calibration procedure to optimize instrument settings, namely, the gain and threshold level, are found to be required for each specimen geometry, in order to achieve a reliable predictive capability.[3]

Conventional Acoustic Emission Energy Measurement Methods. Beattie and Jaramillo[4] and Harris and Bell[5] used similar schemes for the measurement of the energy content in acoustic emission signals. This method involves rectifying the AE signal, squaring the rectified waveform and fitting an envelope by passing it through a lowpass filter. The area under this envelope is determined by feeding it to a voltage to frequency convertor. The total number of the pulses given out by the voltage to frequency convertor is assumed to represent the energy in the signal being measured. Though this method is in principle capable of measuring the energy, it has some practical limitations. How well the low pass filter would be able to track the true envelope of a rapidly fluctuating waveform is uncertain because of the frequency limitations of the filter itself.

In some of the AE applications the energy of an event is simply taken as the square of the peak voltage of the waveform. Williams and Lampert[7] used a modified stress wave factor defined as the sum of the individual oscillations contained in an acousto-ultrasonic signal.

Frequency Spectrum and RMS voltage measurement. Another AU signal processing method has been recently proposed,[8] in which the frequency spectrum of the waveform is quantized in terms of various moments defined by

$$M_r = \int_0^\infty s(f)f^r df \qquad (3)$$

where, M_r, is the rth moment of the AU signal, s(f) is the power spectral density and f is the frequency.

To evaluate these moments one may proceed by first performing a Fast Fourier Transform of the digitized AU signal and obtain these moments by integration. Of these, the zeroth moment was found to correlate well with the level of damage in the material.[8] The zeroth moment is in fact equivalent to the RMS voltage of the AU signal and measuring the analog RMS voltage directly is a faster alternative to obtaining the above zeroth moment of the AU frequency spectrum, through digitization. RMS measurement was recently used for the mapping of flaws in filament wound spherical pressure vessels.[9]

AU ENERGY MEASUREMENT BY MULTI-THRESHOLD COUNTING

In this section a new method of measurement of AU energy is proposed. This method is a modification of the conventional ringdown counting technique, and the aim of the modification is to devise a measure which is linearly related to the energy content in the AU signals.[9] Referring to Fig. 2, the peak amplitude of an AU signal is designated as V_o, and the amplitudes of individual half waves are referred to as V_1, V_2, V_3, etc. If V_1, V_2, V_3, etc., progressively reduce in amplitude, according to the

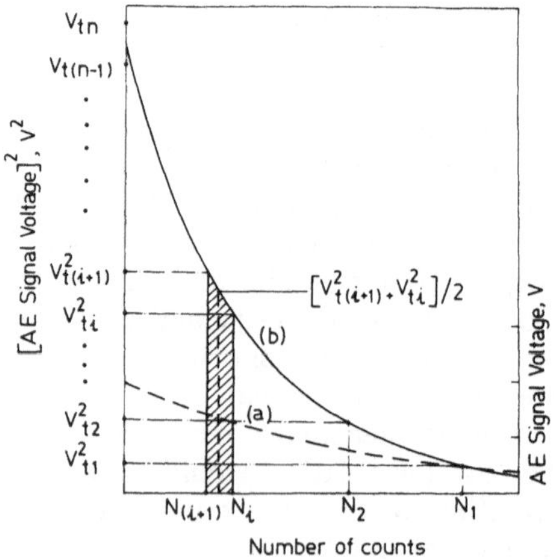

Fig. 3. Envelopes of Half-waves regrouped in the descending order of their amplitudes (a) Proportional to their voltage, (b) proportional to their voltage square

eq. (1), the squared values of the individual half waves, V_1^2, V_2^2, V_3^2 etc. would lie within the envelope given by

$$V^2 = V_o^2(\exp(-2t/\tau)) \tag{4}$$

The area under the envelope given by eq. (4) would be proportional to the energy content in AU signal. This area could be measured using multi-threshold counting, according to the following scheme.

The AU waveform is counted simultaneously by a number of counters with threshold levels set at equal intervals, ranging from V_1 to V_n, Fig. 3. The corresponding number of counts registered at those threshold levels would be N_i, 'i' ranging from 1 to n. Then the number of half waves whose amplitudes are lying between the threshold levels V_i and $V_{(i+1)}$ would be given by $(N_i - N_{(i+1)})$. The corresponding energy would be given by

$$N_{ei} = (N_i - N_{(i+1)})(V_{ti}^2 + V_{t(i+1)}^2)/2 \tag{5}$$

Summing over all the counters, we get the value of the energy counts N_e, proportional to the total energy content in the AU signal as

$$N_e = \sum_{i=1}^{n} N_{ei} \tag{6}$$

Here the maximum threshold level V_n should exceed or at least equal the maximum signal amplitude encountered in a given test. It is obvious that as the number of counters increases the accuracy of energy measurement also increases. This type of energy calculation would be equally valid for exponential decay pattern of eq. (1) as well as for non-exponentially varying and randomly fluctuating signals.

EXPERIMENTAL VERIFICATION OF THE ACCURACY

For the verification of the accuracy of the method, a stress wave source, whose energy content is measurable is necessary. In applications such as AE transducer calibration, various sources of acoustical energy have been used in the past. They included the stress waves created by glass capillary or lead break, impingement of gas molecules from helium gas jet on a plane surface and impact of spherical masses on a specimen surface. Among these sources, only the last one appears amenable to estimation of the energy input into the stress waves, in a simple manner. Hence stress waves produced by impact, is selected for the present experiments, even though the frequency content of these signals are different from those in actual AU tests.

To produce stress waves having different energy levels, the masses of spheres used in these impacts were varied. Here, it is assumed that the same fraction of the energy lost during the impact is converted into stress waves, regardless of the mass. It is hoped that this method, in spite of its limitations, could at least provide an estimate of the accuracy of the proposed scheme of AU energy measurement.

A simple instrumentation comprising a 50 kHz resonant frequency transducer, a voltage comparator LM 317 and a universal counter, was used in these

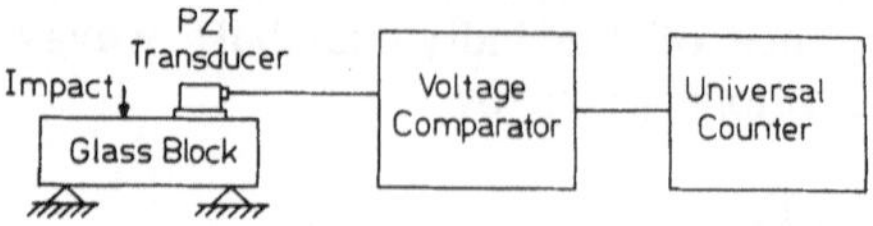

Fig. 4. Experimental setup for measuring the number of ringdown counts due to sphere impacts, at different threshold levels

experiments, Fig. 4. Any desired threshold voltage in the range of 0 to 15 V could be set in the comparator. Since the impact of steel spheres produced signals of sufficient amplitude, amplification of the signals was found unnecessary in these experiments. The masses of the steel spheres are given in Table 1. These spheres were dropped from a height of 25 cm on to a borosilicate glass block of dimensions 130 mm x 84 mm x 25 mm, so that the point of impact was at the center of the 130 mm x 84 mm face. These spheres were held back during their first rebound, so that there was only one impact each time they were released from an electromagnet. The transducer was mounted at a distance of 40 mm from the point of impact, using a constant force spring. Though the number of counts varied by a maximum of about 15% for identical impacts, the average value for any ten successive impacts was quite consistent, and was within a variation of 2%. This type of averaging was used in plotting the variation of the number of counts versus threshold shown in Fig. 5. These plots were obtained for the four different masses included in Table 1. It is clear from Table 1 that the conventional ringdown counts underestimates the higher amplitude signals.

The energy counts N_e for the impact of each of the four spheres were calculated using eq. (7). Here seven threshold levels ranging from 75 mV to 700 mV were used. Values of corresponding counts are included in Table 1.

Table 1. Comparison of Conventional and Multi-Threshold Counting Methods

Sl. No.	Threshold Voltages V_{ti}	Sphere 1 (0.130 gms)		Sphere 2 (0.253gms)		Sphere 3 (0.427gms)		Sphere 4 (1.040gms)	
		$N_i^@$	N_{ei}^*	N_i	N_{ei}	N_i	N_{ei}	N_i	N_{ei}
1.	100 mV	208	4.50	207	3.68	252	3.85	403	5.85
2.	200 mV	28	1.82	70	4.16	98	3.51	169	2.67
3.	300 mV	--	-	6	0.63	44	4.50	128	8.63
4.	400 mV	-	-	1	0.21	8	1.23	59	3.49
5.	500 mV	-	-	-	-	2	0.61	42	4.58
6.	600 mV	-	-	-	-	-	-	27	7.23
7.	700 mV	-	-	-	-	-	-	10	5.70
$N_e = \Sigma N_{ei}$			6.32		8.68		13.70		38.15

@ N_i - is counts at threshold voltage V_{ti}

* N_{ei} - is energy at V_{ti}, $N_{ei} = (N_i - N_{i+1})(V_{ti}^2 + V_{t[i+1]}^2)/2$

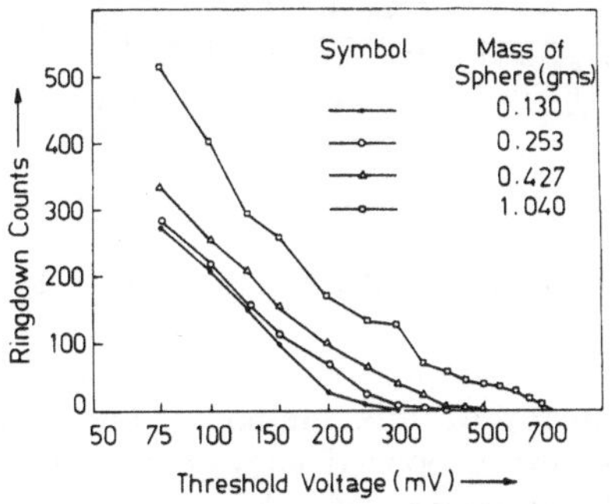

Fig. 5. Variation of ringdown counts with the threshold voltage for the four spheres

In Fig. 6, the values of the energy counts N_e are plotted against the respective masses for the four spheres. It is apparent from this figure that the proposed scheme of energy measurement is capable of giving reasonable estimations of the energy content in AU signals.

A block diagram of the instrumentation developed for the above scheme of energy measurement is shown in Fig. 7. Though this instrumentation has not been applied for the measurement of the acousto-ultrasonic signals, it has been used with fair degree of success, for the acoustic emission monitoring of unidirectional glass-epoxy composite material.[10]

In this study, the extent of damage to the composite was estimated from the rate of AE energy released as a function of strain. The damage level so evaluated was in agreement with the composite ultimate strength.

SUMMARY

Currently employed techniques for the measurement of the acousto-ultrasonic signals were described and their important features were compared. A new method of AU energy measurement was proposed. A simple experiment was

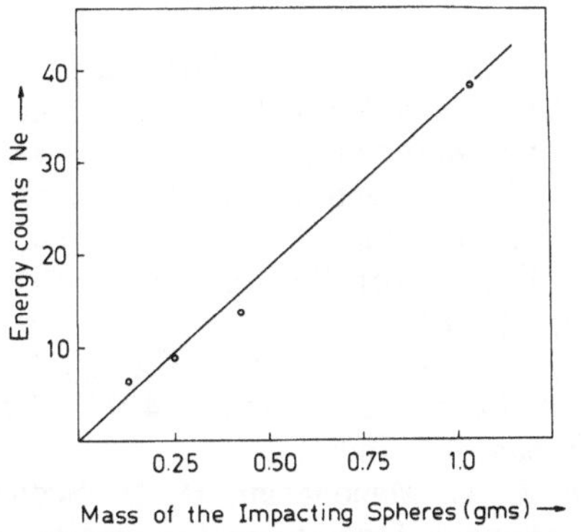

Fig. 6. Plot of the energy counts versus the mass of the spheres for the four spheres

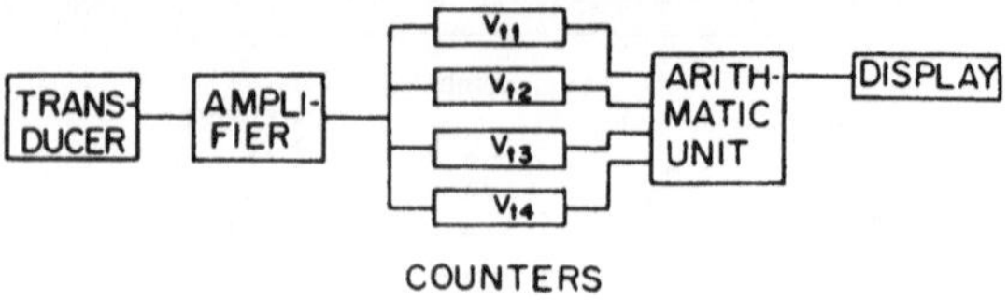

Fig. 7. Block diagram of the instrumentation for AU energy measurement

conducted to verify the accuracy of this method. The results of this experiment appear to suggest that the proposed energy measurement technique is capable of giving reasonable estimate of the actual energy contained in the acousto-ultrasonic signals.

ACKNOWLEDGEMENTS

The first author gratefully acknowledges the support and encouragement of Dr. B. R. Somashekar and Mrs. L. C. Manoharan, for the portion of the work performed at the National Aeronautical Laboratories, Bangalore, India.

REFERENCES

1. A. Vary and R. F. Clark, "Correlation of Fiber Composite Tensile Strength with the Ultrasonic Stress Wave Factor, TM 78846," NASA, Cleveland (1978).
2. J. C. Duke, E. G. Henneke, and W. W. Stinchcomb, "Ultrasonic Stress Wave Factor Technique for the Characterization of Composite Materials, CR-3876," NASA, Cleveland (1986).
3. E. G. Henneke, J. C. Duke, W. W. Stinchcomb, A. Govada and A. Lemascon, "A Study of the Stress Wave Factor Technique for the Characterization of Composite Materials, CR-3670," NASA, Cleveland (1983).
4. A. G. Beattie and R. A. Jaramillo, The Measurement of Energy in Acoustic Emission, Rev of Sc. Instr. 45:352 (1974).
5. D. O. Harris and R. L. Bell, The Measurement and Significance of Energy in Acoustic Emission Testing, Exp Mech. 17:347 (1977).
6. J. H. Williams, Jr. and N. R. Lampert, Ultrasonic Evaluation of Impact Damaged Graphite Fiber Composite, Matls Eval. 38:68 (1980).
7. A. K. Govada, J. C. Duke, E. G. Henneke, and W. W. Stinchcomb, "A Study of the Stress Wave Factor Technique for the Characterization of Composite Materials, CR-174870," NASA, Cleveland (1984).
8. M. J. Sundaresan, E. G. Henneke, K. L. Reifsnider and D. Post, "Nondestructive Evaluation of Filament Wound Pressure Vessels, CCMS-87-02," Center for the Composite Materials and Structures, Virginia Tech, (1987).
9. M. J. Sundaresan, "Acoustic Emission Energy Measurement Through an Improved Count Technique, TM-ST-407/243-81," National Aeronautical Laboratory, Bangalore, India.
10. M. J. Sundaresan, L. C. Manoharan, H. N. Sudheendra and Viveka Naik, Evaluation of Damage Growth in Composite Materials Using Acoustic Emission Energy Measurement Technique, in: "Proceedings of the 6th Intl. Symposium on Acoustic Emission," Susuno, Japan (1982).

ACOUSTO-ULTRASONICS: APPLICATIONS TO WIRE ROPE, WOOD FIBER HARDBOARD, AND ADHESION

Henrique L. M. dos Reis

Department of General Engineering
University of Illinois
Urbana, IL

ABSTRACT

Applications of the Acousto-Ultrasonic Stress Wave Factor Technique (SWF) to the nondestructive evaluation/characterization of wire rope, wood fiber hardboard, and adhesion between rubber-like materials and steel substrates are presented. It was observed that for each application the SWF technique has the potential of being used as a nondestructive testing tool. The SWF technique proved successful to monitor progressive damage in wire rope and wood fiber hardboard, and to monitor the strength of the bond between rubber layers and steel plates.

WIRE ROPE

Background

Wire rope consists of twisted elements such as filaments and strands in various constructions. Its use is common in many applications such as ship mooring, vehicular towing, hoisting and mining.[1,2] Failure of a wire rope can result in significant property damage and/or personal injury. Because wire ropes can fail with no quantifiable change in visual appearance, the development of nondestructive evaluation (NDE) techniques is very desirable to improve the safety and reliability of operations. Various NDE techniques practiced today include visual inspection, acoustic emission, radiographic examination and electromagnetic methods. Electromagnetic methods complemented with visual inspection are the most common methods of wire rope inspection. A good state-of-art critical review of wire rope inspection procedures is given by Weischedel.[3] Acoustic emission monitoring has proven very reliable in detecting wire rope breakage and the rate of breakage during fatigue tests.[4-9] However, all these procedures have drawbacks due either to inaccuracy such as in visual inspections or to economic factors such as in radiographic examinations. Furthermore, while these methods can detect broken wires and reduction of metallic cross-sectional area, they do not evaluate the residual strength or the remaining useful life of a used rope. Several articles related to the NDE of ropes have appeared in the scientific literature.[3-9]

Analytical ultrasonics implies the measurement of material microstructure and associated factors that govern mechanical properties and dynamic response. It goes beyond flaw detection, flaw imaging and defect characterization and includes assessing the inherent properties of material environments in which the flaws reside. A good review of analytical ultrasonics in materials research and testing is given in reference 10. Acousto-ultrasonics is an analytical ultrasonic NDE technique which measures the relative efficiency of energy transmission in the specimen. An ultrasonic pulse is injected with a transmitting transducer mounted on the surface of the specimen. A larger amount of damage in the specimen produces a higher signal attenuation, resulting in lower stress wave factor (SWF) readings. Traditionally, the SWF is evaluated as a number of oscillations higher than a chosen threshold in the ringdown oscillations in the output signal from the receiving transducer. In some applications the SWF has already been correlated with the mechanical strength of composite materials.[11-13] The purpose of this study is to investigate the applicability of the acousto-ultrasonic stress wave factor technique to the nondestructive evaluation of wire ropes.

Experimental Procedures

The wire rope specimens were taken from a single spool of 5/8 in. (1.59 cm) diameter, regular lay, 6 x 25F IWRC Filler rope made of improved plow steel by MacWhyte Co. in Kenosha, Wisconsin. The nominal ultimate strength of the new rope is 35,8000 lbs. (159.103N). Each of the rope specimens was 2 ft. (61 cm) long with ends splayed like a brush and fixed for a length of about 3 in. (7.6 cm) into spelter sockets with a binder. The binder used in this study was resin (Socketfast) manufactured by Philadelphia Resins Corporation.

A set of ten undamaged specimens was prepared. The specimens with controlled damage were obtained by submitting each of the ten undamaged specimens to a different number of fatigue cycles as shown in Fig. 4. Among the ten damaged specimens, the specimen with the least damage and the specimen with the most damage were obtained by submitting undamaged specimens to .125 x 10^6 and 1.250 x 10^6 fatigue cycles respectively. The specimen with the most damage showed 47 visible broken wires on the outside surface. For the undamaged specimen a new rope taken form the spool was used. Damage due to cyclic fatigue loading was applied on a MTS fatigue testing machine at a rate of 6 Hz. The applied sinusoidal force had a minimum value of 2,000 lbs. (8896 N) and a maximum of 10,000 lbs. (44,480 N).

As a measurement of damage the wire rope compliance was measured at the middle of each specimen. Strain versus load plots were obtained using an extensometer (MTS, model 632.11).[14] The rope specimens were remounted in the fatigue machine and run 200 cycles to allow the relative rotation of the mounting fixtures to stabilize. Then, each specimen was loaded with a static load of 10,000 lbs. (44,480 N). The extensometer output voltage change (i.e., strain) readings were then taken as the specimen was unloaded to 4,000 lbs. (17,790 N) in 1,000 lbs. (4,948 N) increments as shown in Fig. 2. For each specimen the strain-load relationship was quite linear over this range of applied loads (i.e., 4,000 to 10,000 lbs. (17,790 to 44,480N)). The

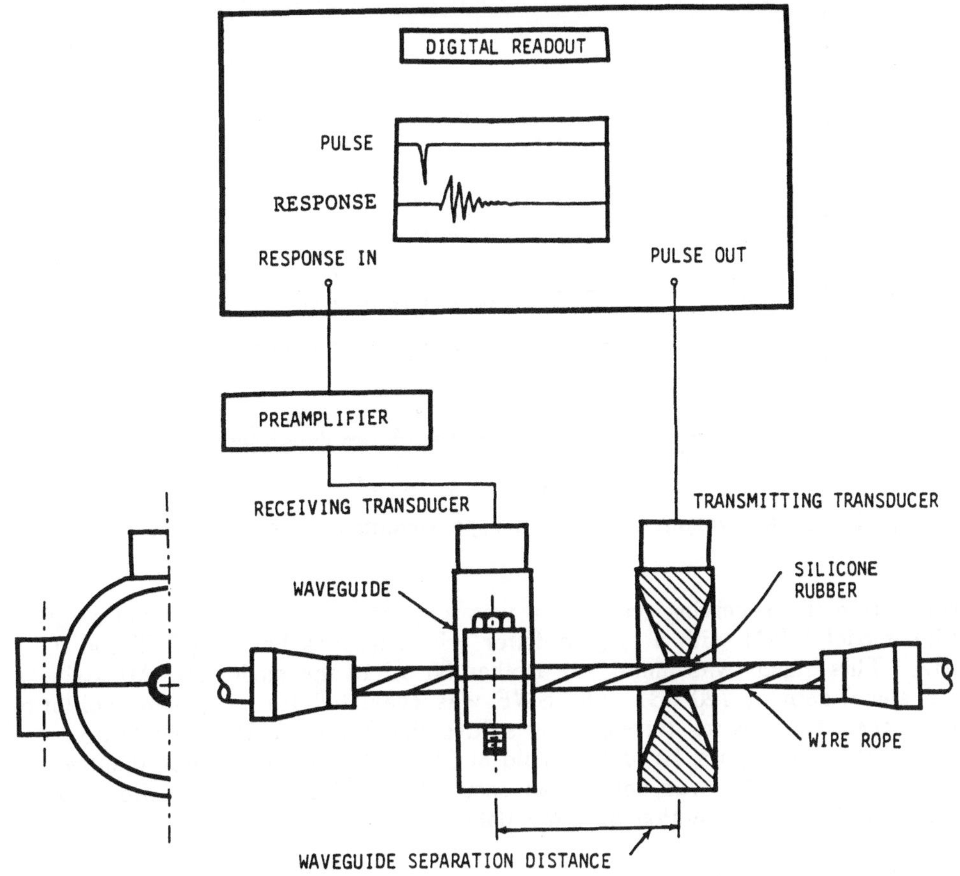

Fig. 1. Schematic diagram of the acousto-ultrasonic stress wave factor measurement system for testing wire rope.

compliance of each specimen is the slope of a straight line fitted to the strain-load data using the least squares method. Because of the strong nonlinear strain-load relationship, which exists for loads less than 4,000 lbs., it was necessary to remount the extensometer.

The stress wave factor (SWF) measurements were obtained using the portable Acoustic Emission Technology (AET) model 206 AU Instrument. A schematic diagram of the measurement system is shown in Fig. 1.[15] The broad-band transmitting transducer was the AET model FC-500 having approximately flat sensitivity of -85 dB (relative to 1V/μBar) from 0.1 to 3 MHz. The resonant receiving transducer was the AET model AC-375L having an approximate sensitivity of -65 dB (relative to 1V/μBar) at the resonant frequency of 375 kHz. Both the transmitting and receiving transducer were mounted to aluminum waveguides as shown in Fig. 1. Silicone rubber was used as couplant between the wave guides and the wire rope specimen such that the rubber would conform to the topography of the wire rope upon tightening of the waveguides. The center-to-center spacing between the waveguides was 1 in. (2.54 cm) and petroleum grease was used as couplant between the transducers and waveguides.

The transmitting transducer was excited by a -250V pulse generated by the 206 AU unit, which was set at the pulsing rate of 250 pulses/s. The output

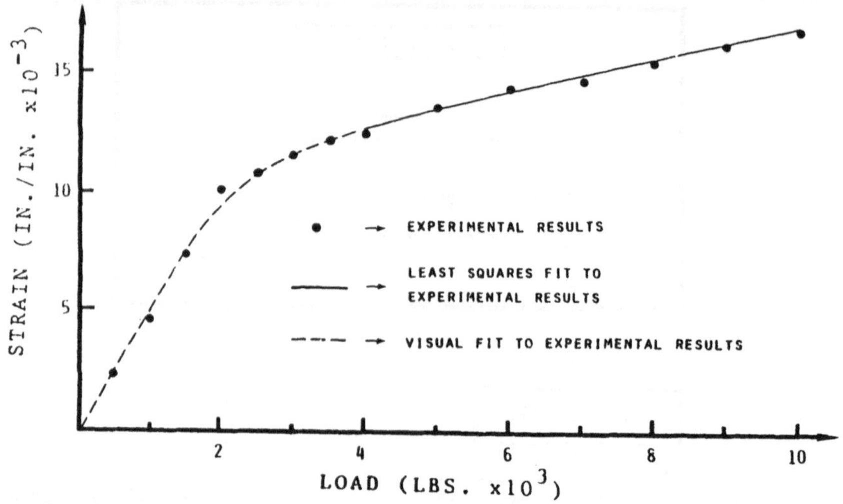

Fig. 2. Strain versus load for a rope specimen with 1,125,000 cycles.

signal from the receiving transducer was amplified by 40 dB in a preamplifier (AET model 140B) with a plug-in filter with bandpass between 0.25 and 0.5 MHz. This signal was amplified another 80 dB in the model 206 AU giving a total signal gain of 120 dB. The SWF was evaluated for ringdown oscillations that exceeded a floating threshold by setting the 206 AU unit in the "auto mode" after setting its fixed threshold at 1V. The digital counter display on the model 206 AU was set for an accumulation time of .25 s. Zero load was applied during the collection of SWF data.

Experimental Results

Figure 2 shows a typical plot of the strain versus load for a rope specimen. For the other specimens the strain-load diagrams were similar. For

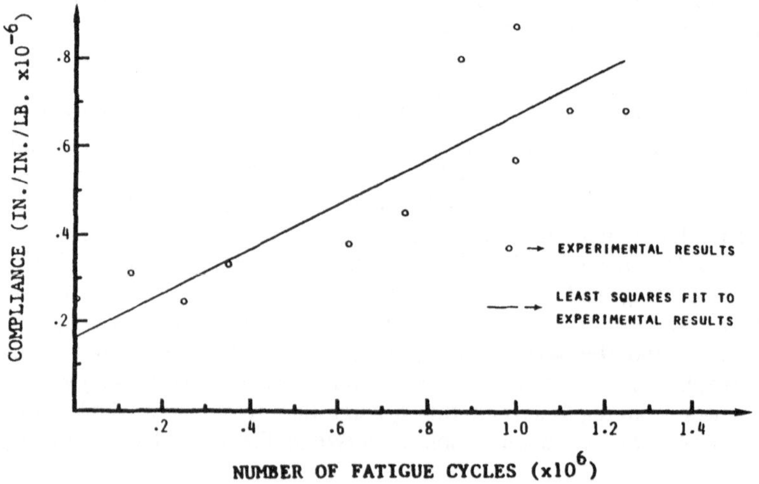

Fig. 3. Compliance versus number of fatigue cycles for the wire rope specimens.

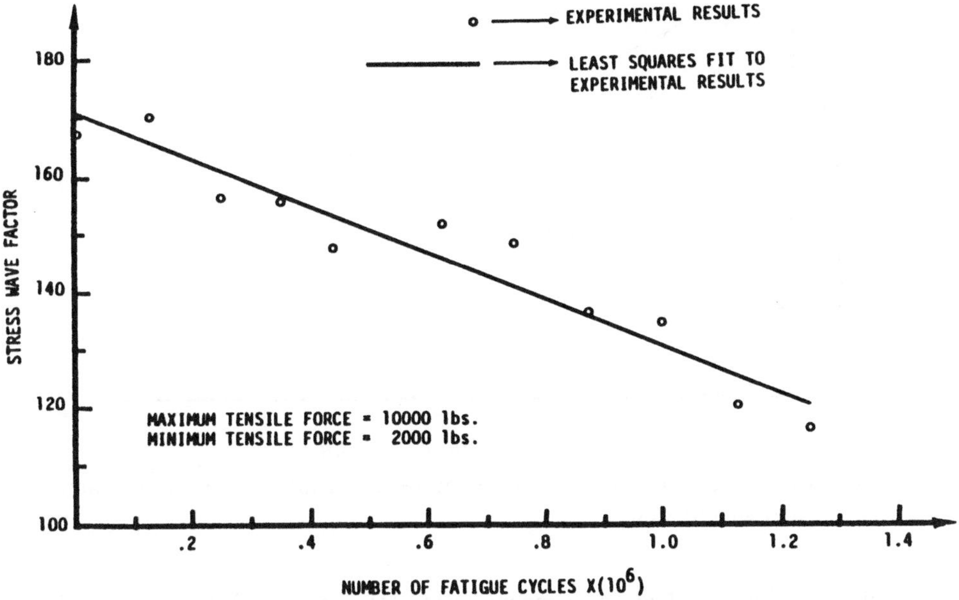

Fig. 4. Stress wave factor versus number of fatigue cycles for the wire rope specimens.

this type of rope construction the strain versus load plots appear to be quite linear for loads above 4,000 lbs. (17,790 N). For the linear regions of the strain load curve a straight line was fitted using the least squares method. The compliance of each wire rope specimen was obtained from the slope of this straight line. Figure 3 shows the compliance of the wire rope specimens versus the number of fatigue cycles. The straight line was again obtained using the least squares method. The plot in Fig. 3 shows that the compliance of wire rope has a good correlation with the number of fatigue cycles (i.e., damage). Therefore, wire rope compliance is a parameter that can be used to monitor damage.

Figure 4,[15] shows a plot of the stress wave factor for specimens that were subjected to different amounts of fatigue cycles. The straight line was also obtained using the least squares method. Figure 5 shows a plot of the compliance of the wire rope specimens versus the corresponding stress wave factor measurements. Figure 4 and Fig. 5 clearly indicate that the stress wave factor is correlated to damage in the wire rope specimens. For more effective and practical applications of the technique, a more detailed investigation should be undertaken by studying wire ropes of different diameters and different construction.

WOOD FIBER HARDBOARD

Background

Wood fiber hardboard is produced by pressing wood fibers and various additives such as petrolatum and resin into a hard high-density board. The process begins by loading wood chips into a container called "gun" and then injecting high pressure steam into the gun. This pressure, referred to as

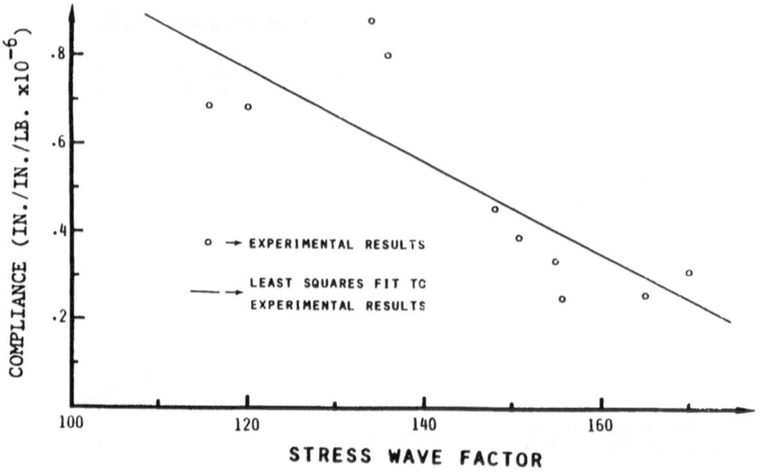

Fig. 5. Compliance of wire rope specimens versus stress wave factor measurements.

"gun preheat" is maintained for approximately 90 sec at 400 psig (2756 kPa gage) allowing the chips to become saturated with steam. The pressure is then raised to 500 psig (3445 kPa gage), a gate on the gun is open and the chips are allowed to escape. As the wood chips reach the lower pressure of the atmosphere the high pressure steam "explodes" out of the chips, thus breaking the chips into coarse fiber bundles. These fiber bundles are then sent to a disc fiber refiner where a serrated disk shreds the fiber bundles into single fibers. Using high pressures and temperatures these fibers, mixed with approximately 1% resin, 2 to 3% petrolatum and acid, are compressed to form a grainless wood fiber hardboard. The temperatures range from 375 to 400°F (191 to 204°C) while the pressure is kept at approximately

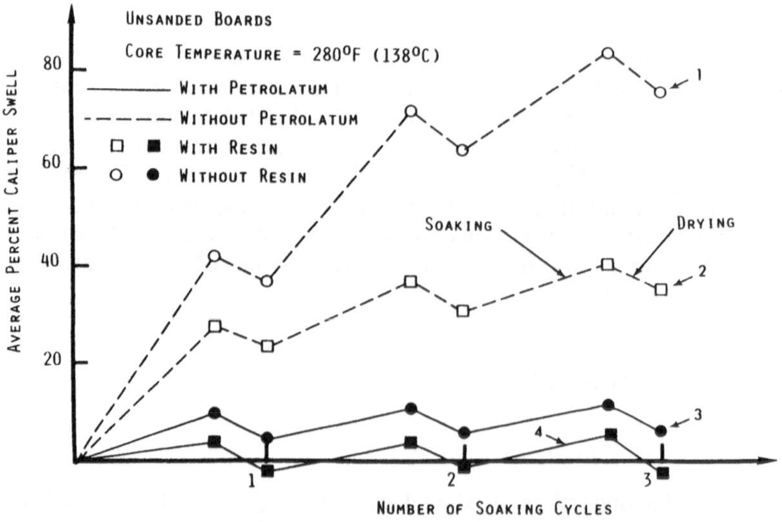

Fig. 6. Average percent caliper swell versus number of soaking cycles for unsanded boards, core temperature = 280°F (138°C).

1000 psig (6890 kPa gage) for 20 minutes. Mataki,[16] gives a good description of the internal structure of fiber hardboard, and its relation to mechanical properties.

Presently, the most commonly used tests to assess the structural functional integrity of the hardboard are the boil swell and the cyclic soak tests. In these tests, a percent caliper swell is introduced and calculated by subtracting the board's initial thickness from the board's final thickness (after the test is completed) and then dividing the result by the initial thickness. The percent caliper swell is accepted to be a good indicator of the dimensional stability of wood fiber hardboard, see ASTM Annual Book of Standards (1985). Although these tests predict how well the hardboard will perform, with the exception of visual examination, there are at present no nondestructive evaluation (NDE) techniques for the functional structural integrity of wood fiber hardboard. One of few studies in this area was performed by Beall[17] where a relationship of acoustic emission and internal bond strength of wood-based composite materials was found. The purpose of this study is to investigate the applicability of the acousto-ultrasonic stress wave factor technique to assess the structural functional integrity of the wood fiber hardboard.

Experimental Procedures

Eight 16 inch by 16 inch (406 mm by 406 mm) wood fiber hardboards with an average thickness of 0.32 inches (8.1 mm) were manufactured under controlled conditions. To obtain boards that would exhibit a different caliper swell when submitted to cyclic soak tests, each of the eight boards was manufactured with a different combination of resin, petrolatum and core temperature as shown in Table 1. All the other design variables that have influence upon the board properties, such as gun preheat, remained constant during the manufacturing process.

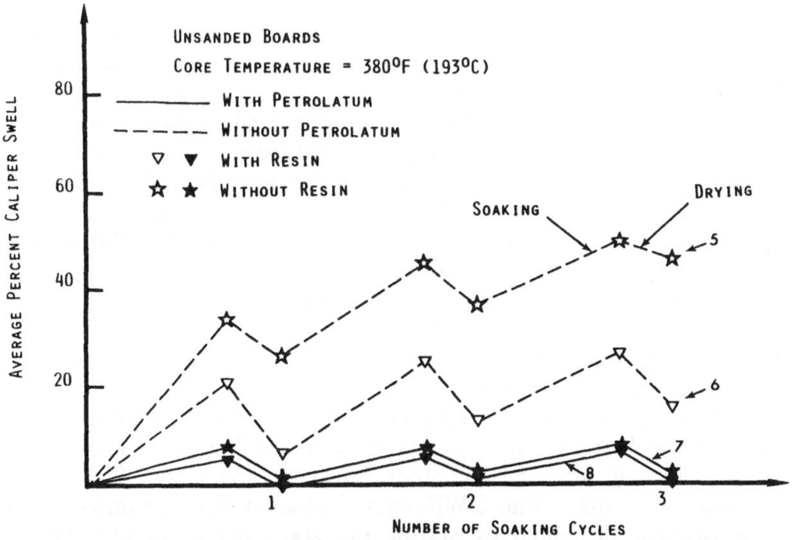

Fig. 7. Average percent caliper swell versus number of soaking cycles for unsanded boards, core temperature = 380° (193°C).

The eight boards were progressively damaged by submitting the boards to cyclic soak tests. Each cycle consisted of soaking the boards in water at room temperature for 24 hours and then drying the boards in an oven at the temperature of 220°F (104°C) for nine hours as shown in Figs. 6 and 7.

Stress wave factor measurements were obtained for the undamaged boards and then for the damaged boards at the end of each cycle (i.e., after the drying period) as shown in Figs. 8 and 9. The stress wave factor measurements were obtained using the portable Acoustic Emission Technology (AET) model 206 AU Instrument. The transmitting transducer was the AET model AC-375 L having an approximate sensitivity of -65 dB (relative to 1V/μBar) at the resonant frequency of 375 kHz. The receiving transducer was the Dunegan/Endevco model S9204 sensitive to frequencies over a range of 25 kHz to 410 kHz and having a peak sensitivity of -60 dB (relative to 1V/μBar).

TABLE 1. Controlled Content Boards

Board Number	Resin (%)	Petrolatum (%)	Manufacturing Core Temperature oF(oC)
1	0	0	280 (138)
2	2	0	280 (138)
3	0	5	280 (138)
4	2	5	280 (138)
5	0	0	380 (193)
6	2	0	380 (193)
7	0	5	380 (193)
8	2	5	380 (193)

Both the transmitting and the receiving transducers were mounted inside of aluminum load cells. A sponge foam was packed tightly into the loading cells to secure the transducers in their position and petroleum grease was used as couplant medium between the transducers and the load cells. One inch (25.4 mm) diameter silicone rubber disks with a thickness of 1/16 inch (1.6 mm) were used as a dry couplant medium between the boards and the loading cells. The rubber disks were attached to the load cells using a silicone base adhesive and the contact pressure between the silicone rubber and the boards was such that the saturation pressure was exceeded. The center-to-center spacing between the loading cells was 2 inches (50.8 mm).

The transmitting transducer was excited by a -50 V pulse generated by the 206 AU unit which was set at a pulsating rate of 1000 pulses/s. The output signal of the receiving transducer was amplified 40 dB by a preamplifier (AET model 140B) with a plug-in filter having a passband between 0.25 and 0.5 MHz. The signal was further amplified another 56 dB in the model 206 AU, giving a total signal gain of 96 dB. The SWF was evaluated for ringdown oscillations that exceed a floating threshold by setting the 206 AU unit in the "auto mode" after setting its fixed threshold at 0.6 V. The digital counter display on the model 206 AU was set for an accumulation time of 0.5 s.

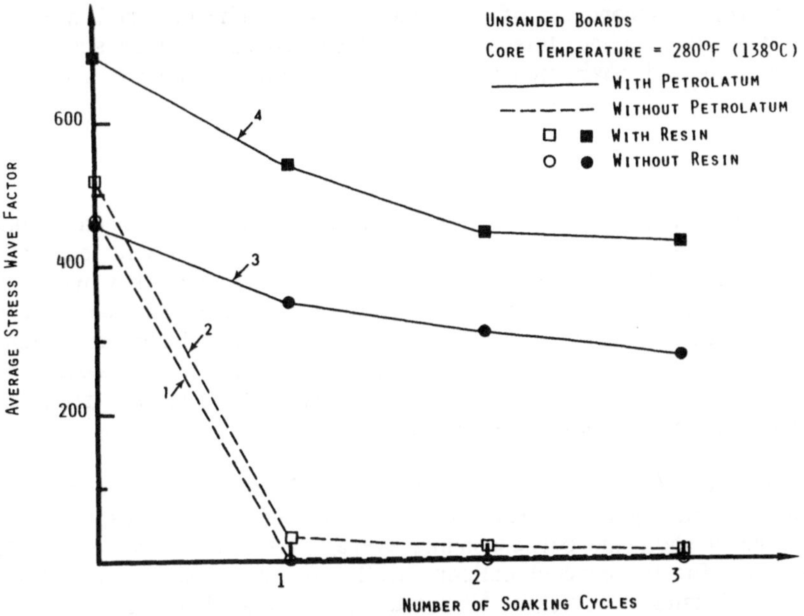

Fig. 8. Average stress wave factor versus number of soaking cycles for unsanded boards, core temperature = 280° (138°).

To investigate if the SWF measurements on the surface of the hardboard would reflect the swelling properties of the hardboard core, approximately 3/64 in. (1.2 mm) of material was removed from the surface on both sides of

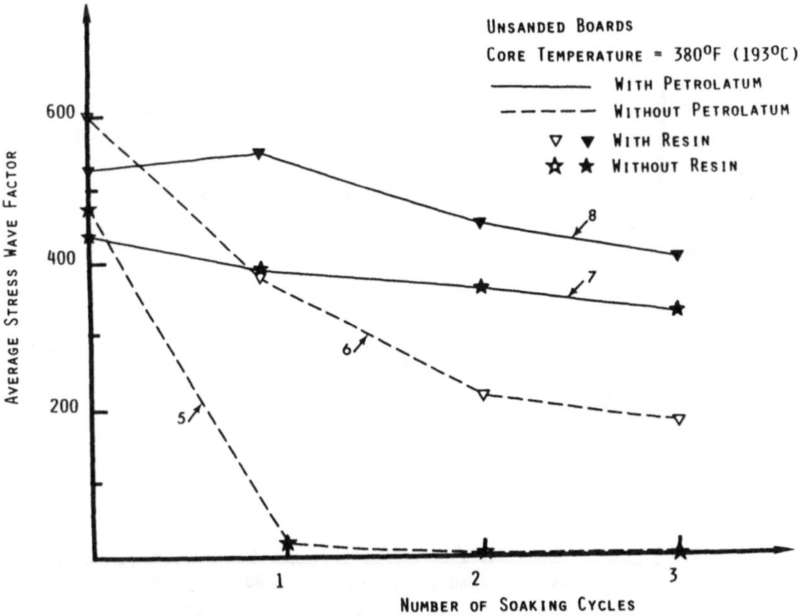

Fig. 9. Average stress wave factor versus number of soaking cycles for unsanded boards, core temperature = 380° (193°).

each board from a second set of eight hardboards. The material was removed using a belt sander. Then, this set of eight sanded boards was submitted to cyclic soak tests and SWF measurements were again obtained at the end of each cycle.[18]

Experimental Results

Acousto-ultrasonic stress wave factor measurements have been conducted on undamaged and damaged wood fiber hardboard manufactured with controlled contents. To evaluate the response of the board specimens, they were subjected to cyclic soak tests. Stress wave factor measurements were recorded at the end of each cycle (i.e., after the specimens were dried) for each board. Cyclic soak tests and SWF measurements were also conducted on boards were approximately 3/64 in. (1.2 mm) of material was removed from each surface by sanding both sides of the boards.[18]

Figures 6 and 7 show the average percent caliper swell as a function of the number of soaking cycles for the eight unsanded boards, respectively. Figures 8 and 9 show the corresponding average SWF measurements. All the averages (i.e., for the percent caliper swell and the stress wave factor) were computed using readings at 52 points equally distributed over the face of the boards. For practical applications it is recommended to use a moving average of the SWF measurements obtained using a dry-couple wheeled sensor fixture similar to the one reported by Rogers.[19] Figures 8 and 9 also show that for each board the average SWF measurement decreases as the number of soaking cycles (i.e., damage) increases. This fact may be used to evaluate boards stored in warehouses or to predict the remaining useful life of boards already in service. It was also hypothesized that the addition of petrolatum (i.e., wax) would have two effects: first, by increasing the boards' structural damping it would decrease the SWF measurements, and second, by increasing the degree of water repellency it would lower the percent caliper swell

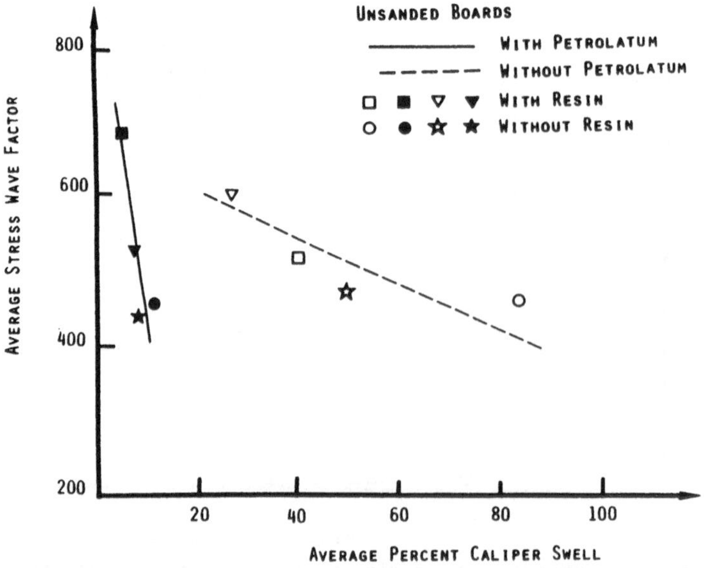

Fig. 10. Average stress wave factor of undamaged board versus average percent caliper swell at the third soak cycle.

of the boards. This hypothesis is consistent with the data reported. Therefore, SWF measurements of one board can only be compared with SWF measurements of boards within the same group, i.e., boards with or without petrolatum. Figure 10 shows the average SWF measurements for the undamaged unsanded boards as a function of the average percent caliper swell after the third soaking cycle. Similar results were obtained for the sanded boards.[18] Comparing boards within the same group, it was observed that undamaged boards with higher average SWF measurements have corresponding lower average percent caliper swell in cyclic soak tests. This observation shows the potential of using the acousto-ultrasonic stress wave factor measurement technique in quality control inspection during manufacturing of fiber hardboard.[18]

ADHESION

Background

Adhesion, or strength of the bond between rubber and other material (metal, plastic, or other rubber), is a critical requirement for the functioning of a great variety of products which are vulcanized composites of two or more materials. Usually, adhesion is measured using the destructive stripping or peeling method.[20] The need for a nondestructive testing method to evaluate the strength of the adhesive bond between rubber and other materials has long been established. The purpose of this study is to investigate the applicability of the acousto-ultrasonic stress wave factor techniques to the nondestructive evaluation of the adhesive bond strength between rubber sheets and steel plates.

Traditionally, the SWF readings depend upon several instrumentation and experimental parameters such as the threshold level. To eliminate this dependence, the signal energy has been used to define the SWF, Kautz,[21] where the signal energy is defined as the square of the amplified transducer output voltage integrated over the time of the sweep. Other alternate method for quantifying the SWF has been proposed by Govada et al.[13] This method consists in using a Fast Fourier Transform (FFT) to perform the spectral analysis of the output signal. It is based on the observation by Talreja[22] that three classes of parameters are needed to describe distribution functions (power spectrum in particular), namely location, scale, and shape parameters. Talreja[22] suggested that a convenient set of parameters to represent the frequency spectrum can be defined as follows: the location parameter can be taken to be the location of the centroid of the spectrum; the area of the spectrum forms a suitable scale parameter; and the shape parameters can be described in terms of the various moments of the power spectrum about a convenient axis.[13, 23] Following Govada et al,[13] in this study the root mean square of the power spectral density, $(M_o)^{0.5}$, is also used as an alternate method to quantify the SWF.

Experimental Procedures

Thirty-two specimens for peel strength testing were prepared.[20] Each specimen consisted of a rubber sheet with a thickness of 1/4 in. (6 mm) and dimensions of 1.0 x 5.0 in. (25 x 127 mm) which was bonded in vulcanization to one side of a steel plate with a thickness of 1/16 in. (2 mm) and dimensions of 1.0 x 2 3/8 in. (25 x 60 mm) as shown in Fig. 11. The test specimens were prepared such that the bonded area of 1.0 x 1.0 in. (25 x 25 mm)

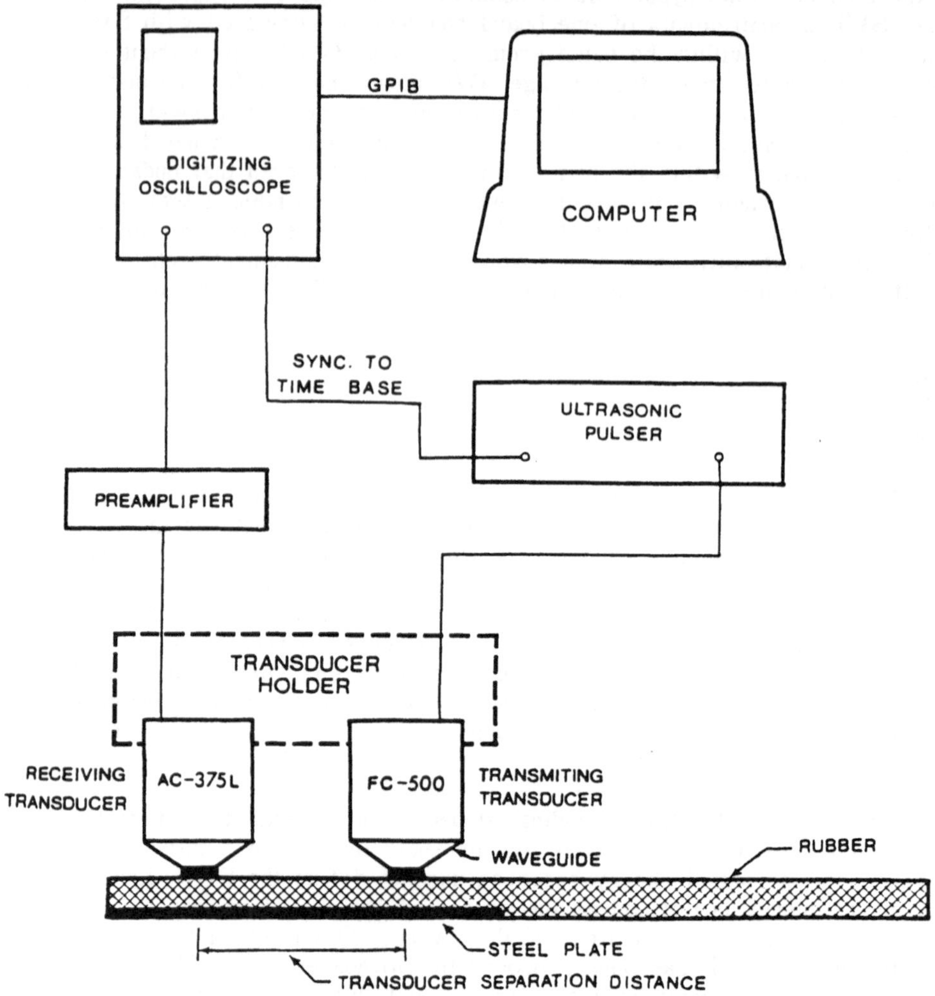

Fig. 11. Schematic diagram for the AU SWF measurement system

was located approximately in the middle of the metal member. Using a Shore A-2 durometer testing machine the rubber had an average hardness of 51. The steel plate surface was grit blasted. The thirty-two specimens were divided into eight groups of four equal specimens each. To assure specimens with different quality of bond peel strength between the rubber and the steel plate, each group of four equal specimens was prepared with a different amount of adhesive (i.e., primer cement and cover cement) and with controlled defects in the bonded surface area as described in Table 2. Three of the controlled defects were obtained by masking a portion of the bonded area prior to the application of the adhesive system. Four other defects were obtained by deleting either the primer cement or the cover cement in conjunction with contamination of the bond surface with a centrally located thumb print.

A schematic diagram of the alternate Stress Wave Factor measurement system is shown in Fig. 11. The broad-band transmitting transducer was the AET model FC-500 having approximately a flat sensitivity of -85 dB (relative to 1V/μBar) from 0.1 to 3 MHz. The resonant receiving transducer was the AET model AC-375L having an approximate sensitivity of -65 dB (relative to 1V/μBar) at the resonant frequency of 375 kHz. Both the transmitting and the receiving transducers were mounted in waveguides as shown in Fig. 11. The area of contact between each waveguide and the specimen was a circle with a diameter of 1/4 in. (6 mm). The center-to-center spacing between the waveguides was 1.5 in. (38 mm), and the contact pressure between the waveguides and the specimen was such that the saturation pressure was exceeded. The transmitting transducer was excited by the ultrasonic pulser (Parametric, model 5052 PRX) which was set at the pulsing rate of 200 pulses/s. The output signal from the receiving transducer was amplified by 20 dB in a preamplifier (Dunegan/Endevco, model 301). The acousto-ultrasonic signal waveform was then digitized in a digitizing oscilloscope (Tektronix, model 7854) using 1024 points. The calculations for the signal energy and for the power spectrum analysis were then carried out by a Tektronix computer (model 4052) equipped with a signal processing ROM. For the peel strength tests, the load was applied at an angle of 45° using a Scott tensile testing machine with a constant crosshead speed of 2.0 in/min (0.8 mm/s).

TABLE 2. Rubber-Steel Specimens with Controlled Adhesion

		Controlled Defects					
		Adhesive		Three Masked		One Masked	
Specimen Type	Number of Specimens	Primer Cement	Cover Cement	Longitudinal Stripes	Four Masked Squares	Longitudinal Stripe	Bond Contamination
1	4	Yes	Yes	---	---	---	---
2	4	Yes	Yes	Yes	---	---	---
3	4	Yes	Yes	---	Yes	---	---
4	4	Yes	Yes	---	---	Yes	---
5	4	Yes	Yes	---	---	---	Yes
6	4	Yes	No	---	---	---	Yes
7	4	Yes	No	---	---	---	---
8	4	No	Yes	---	---	---	---

Experimental Results

Figure 12 shows a full frequency range acousto-ultrasonic signal in the time domain together with the corresponding frequency spectrum for a typical specimen. The FFT algorithm,[25] used with these results produced a 0 to 1 MHz spectrum from a 500 s time record of the acousto-ultrasonic signal.

Figure 13 shows a normalized stress wave factor versus the peel strength. In Fig. 13 the stress wave factor is based upon the signal energy which is defined as the square of the amplified transducer output voltage integrated over the time of the sweep. The SWF value represents the average of twenty eight measurements for each group of four equal specimens (seven measurements per specimen) and the corresponding peel strength value represents the aver-

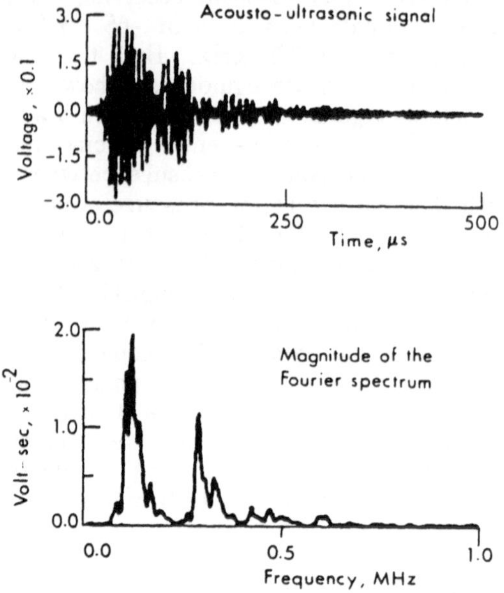

Fig. 12. AU signal and its Fourier spectrum for a typical specimen

age of four measurements for each group of four equal specimens (one measurement per specimen). The straight line was obtained using the least squares method. The corresponding correlation coefficient is equal to 0.83. Figure 14 also shows a normalized stress wave factor versus the peel strength. Following Govada et al,[13] in Fig. 14 the stress wave factor is defined as the root mean square of the power spectral density. Again, as in Fig. 13 each stress wave factor represents the average of twenty eight measurements for each group of four equal specimens and the corresponding peel strength value

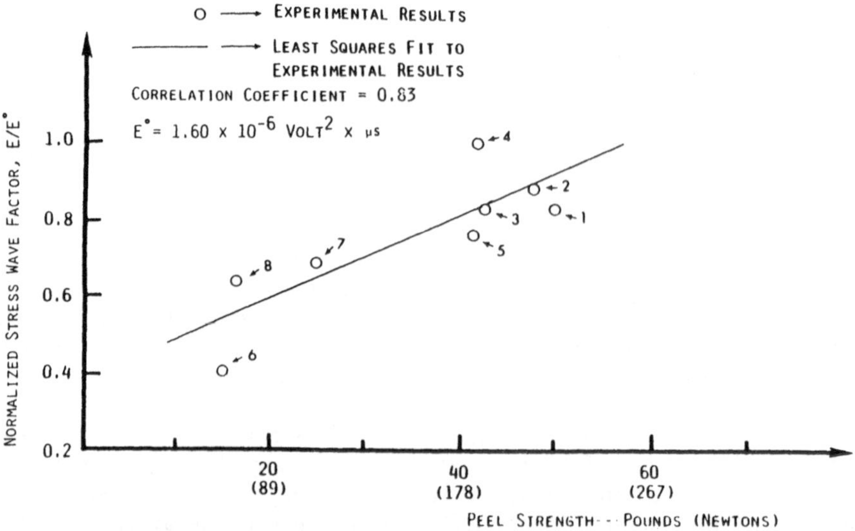

Fig. 13. Normalized stress wave factor, E/E°, versus peel strength for rubber sheet bonded to steel

represents an average of four measurements for each group of four equal specimens. The straight line was also obtained using the least squares method and the corresponding correlation coefficient is equal to 0.75.

Figures 13 and 14 clearly indicate that the stress wave factor is correlated with the peel strength of specimens made of rubber bonded to steel. Apparently, the injected signal relies on the signal path into the steel plate through the adhesive bond between rubber and steel, and re-emerges at the receiving probe through the bonded area. For weaker adhesive bonds, the transmission into the steel plate is also weaker, and the injected signal is trapped in the rubber layer and rapidly attenuates. Therefore, the feasibility of using the stress wave factor technique in non-destructive evaluation of the adhesive bond strength between rubber sheets and steel substrates has been demonstrated. Traditional SWF measurements were reported by the author.[26] To improve the correlation coefficient between the stress wave factor defined as the root mean square value of the power spectral density and the peel strength it is recommended to calculate the stress wave factor using different filter bands of the acousto-ultrasonic signal as reported by Kautz.[21] Furthermore, for practical applications where large areas are to be nondestructively evaluated, it is recommended to use a moving average of the SWF measurements obtained using a dry-coupled wheeled sensor fixture similar to the one reported by Rogers.[19]

CONCLUDING REMARKS

Acousto-ultrasonic nondestructive evaluation has been conducted on 5/8" (1.59 cm) diameter wire rope using the stress wave factor (SWF) measurement. Wire rope specimens with controlled damage were prepared by cyclic fatigue loading the specimens for a prescribed number of cycles. It was observed that the stress wave factor is correlated to the number of fatigue cycles (i.e., damage) and to the compliance of the wire rope specimens.

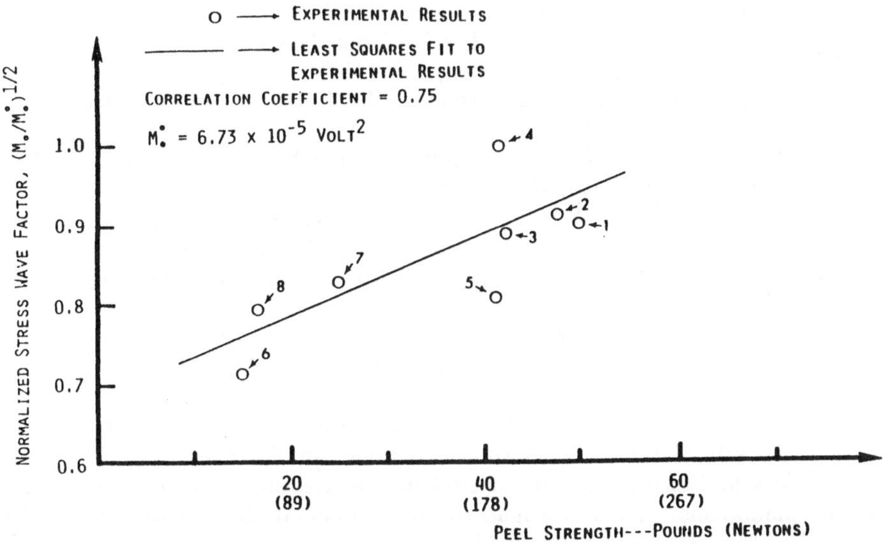

Fig. 14. Normalized stress wave factor, $(M_o/M_o{}^o)^{0.5}$, versus peel strength for rubber sheet bonded to steel

Wood fiber hardboards with controlled contents were manufactured and characterized using the acousto-ultrasonic stress wave factor (SWF) technique. The board specimens were subjected to cyclic soak tests and at the end of each cycle stress wave factor measurements were obtained. It was observed that the stress wave factor decreased with an increase in damage, i.e., higher number of soaking cycles. Furthermore, tests showed that undamaged boards with higher SWF readings would have corresponding lower percent caliper swell for the same number of soaking cycles.

Acousto-ultrasonic nondestructive evaluation has also been conducted to evaluate the adhesive bond strength between rubber and steel plates using the stress wave factor (SWF) measurement technique. Specimens with different bond strength were manufactured and tested using the SWF technique. The strength of the rubber steel adhesive joint was then evaluated using the destructive peel strength test method. It was observed that higher values of the SWF measurements correspond to higher value of the peel strength test data.

ACKNOWLEDGMENTS

The author wishes to thank Mr. Alex Vary and Mr. Harold E. Kautz, of the Structures Division at NASA Lewis Research Center, for the use of the laboratory equipment. The author is grateful to Mr. James Bauer, at the MacWhyte Company, for providing the wire rope, and to Dr. Peter Chin, at Masonite Corporation, for manufacturing the hardboards with controlled content. The author is also grateful to Professor L. A. Bergman, of the Department of Aeronautical and Astronautical Engineering at the University of Illinois, and to Mr. J. H. Bucksbee, at Lord Corporation, for manufacturing the peel strength specimens.

REFERENCES

1. H. H. Vanderveldt and R. D. Young, "A Survey of Publications on Mechanical Wire Rope and Wire Rope Systems," The Catholic University of America, Washington, DC (1970).
2. "Workshop on Marine Wire Rope," The Catholic University of America, Washington, DC (1970).
3. H. R. Weischedel, The Inspection of Wire Rope in Service: A Critical Review Matls Eval. 43:1592 (1985).
4. P. A. Laura, H. H. Vanderveldt, and P. G. Gaffney, Acoustic Detection of Structural Failure of Mechanical Cables, JASA 45:791 (1969).
5. D. H. Harris and H. L. Dunegan, Acoustic Emission Testing of Wire Rope, Matls Eval. 23:1 (1974).
6. J. W. William, Jr. and S. S. Lee, Acoustic Emission/Rupture Load Characterizations of Double-Braided Nylon Rope, Marine Tech. 19:268 (1982).
7. Y. Toda, H. Yokota, and M. Hanzawa, Detection of Wire Breakage During Tensile Fatigue Tests of Wire Rope, in: "Proceedings of the 10th World Conference on Nondestructive Testing," Moscow (1982).
8. F. Mantazano, "Axial Fatigue of Wire Rope in Sea Water, OTC 1579," ASME, Houston, Texas (1973).
9. F. R. Stonesifer, and H. L. Smith H. L., "Tensile Fatigue in Wire Rope, OTC 3419," ASME, Houston (1979).

10. "Analytical Ultrasonics in Materials Research and Testing," A. Vary, ed., NASA, Cleveland (1985).
11. A. Vary and R. F. Lark, "Correlation of Fiber Composite Tensile Strength and The Ultrasonic Stress Wave Factor, TM 78846," NASA, Cleveland (1978).
12. J. H. Williams, Jr. and N. R. Lampert, Ultrasonic Evaluation of Impact-Damage Graphite Fiber Composite, Matls Eval. 38:68 (1980).
13. A. K. Govada, J. C. Duke, Jr., E. G. Henneke, II, and W. W. Stinchcomb, "A Study of the Stress Wave Factor Technique for the Characterization of Composite Materials, CR 174870," NASA, Cleveland (1985).
14. G. A. Costello, and J. W. Phillips, "Stress Analysis of Wire Hoist Rope, UILU-ENG 83-6006," University of Illinois, Urbana (1983).
15. H. L. M. dos Reis and D. M. McFarland, On the Acousto-Ultrasonic Non-Destructive Evaluation of Wire Rope Using the Stress Wave Factor Technique, Brit J. NDT. 28:155 (1986).
16. Y. Mataki, "Internal Structure of Fiberboard and its Relation to Mechanical Properties - Theory and Design of Wood and Fiber Composite Materials, Syracuse University Press, Syracuse (1972).
17. F. C. Beall, Relationship of Acoustic Emission to Internal Bond Strength of Wood-Based Composite Panel Materials, J Acous Em. 4:19 (1985).
18. H. L. M. dos Reis and D. M. McFarland, On the Acousto-Ultrasonic Characterization of the Wood Fiber Hardboard, J Acous Em. 5:67 (1986).
19. J. M. Rodgers, "Quality Assurance and In-Service Inspection Applications of Acousto-Ultrasonics to Bonded and Composite Structures," Acoustic Emission Technology Corporation, Sacramento (1983).
20. Standard Test Methods for Rubber Property-Adhesion to Rigid Substrates, ASTM-D429-81, in: "Annual Book of ASTM Standards, 09.01," ASTM, Baltimore (1985).
21. H. E. Kautz, "Ultrasonic Evaluation of Mechanical Properties of Thick, Multilayered, Filament Wound Composites, TM 87088," NASA, Cleveland (1985).
22. R. Talreja, On Fatigue Life Under Stationary Gaussian Random Loads, Eng Frac Mech. 5:993 (1973).
23. H. L. M. dos Reis, and H. E. Kautz, Nondestructive Evaluation of Adhesive Bond Strength Using the Stress Wave Factor Technique, J Acous Em. 5:144 (1986).
24. S. O. Rice, Mathematical Analysis of Random Noise, Bell Sys Tech Jour. 23:282 (1944).
25. Tektronix 4050 Series, R08 Signal Processing ROM Pack 2 (FFT) Manual, Part No. 070-2841-00, 1980, Tektronix, Inc., Beaverton, Oregon.
26. H. L. M. dos Reis, L. A. Bergman, and J. H. Bucksbee, Adhesive Bond Strength Quality Assurance Using the Acousto-Ultrasonic Technique, Brit J. NDT. 28:357 (1986).

AN ACOUSTO-ULTRASONIC METHOD FOR EVALUATING WOOD PRODUCTS

Marcia Patton-Mallory

U.S. Department of Agriculture, Forest Service
Forest Products Laboratory
Madison, WI

Kent D. Anderson

Electrical and Computer Engineering
University of Wisconsin
Madison, WI

ABSTRACT

The application of an acousto-ultrasonics technique to detect decay in wood members is discussed. The method uses a 75 kHz pulsing transducer to introduce a stress wave into the material being tested. The stress wave is received by a 60 kHz resonant transducer. The method uses a through transmission arrangement with transducers on opposite faces of the member being tested. Measurements have been taken on members up to 9-1/2 inches measured across the grain. Measurements on the received waveform are the traditional acoustic emission waveform parameters (Fig. 1a) and additional acousto-ultrasonic parameters (Fig. 1b). Measurements in addition to the traditional acoustic emission parameters were centroid time, start time, RMS voltage, and negative and positive peak voltage times. The acousto-ultrasonic system is detailed in Fig. 2. It is best described in terms of signal flow. The pulse generator output a short pulse to the pulsing transducer, and at the same time was connected to the oscilloscope to start the oscilloscope trace. The stress wave generated by the pulsing transducer was transmitted into the wood through a couplant and was detected by a similar transducer at the other end of the wood. The receiving transducer converted the stress wave to an electrical signal which was amplified and filtered before being captured by the oscilloscope. Once captured by the oscilloscope, the signal was transmitted to a microcomputer for further processing and storage.

Two methods of signal processing were used to enhance the waveform parameter measurements. First, an ensemble averaging technique averaged multiple waveforms. This resulted in lowering the background noise and more consistent measurements of waveform parameters. Second, a moving average of the time domain waveform smoothed some of the peaks and valleys. This also

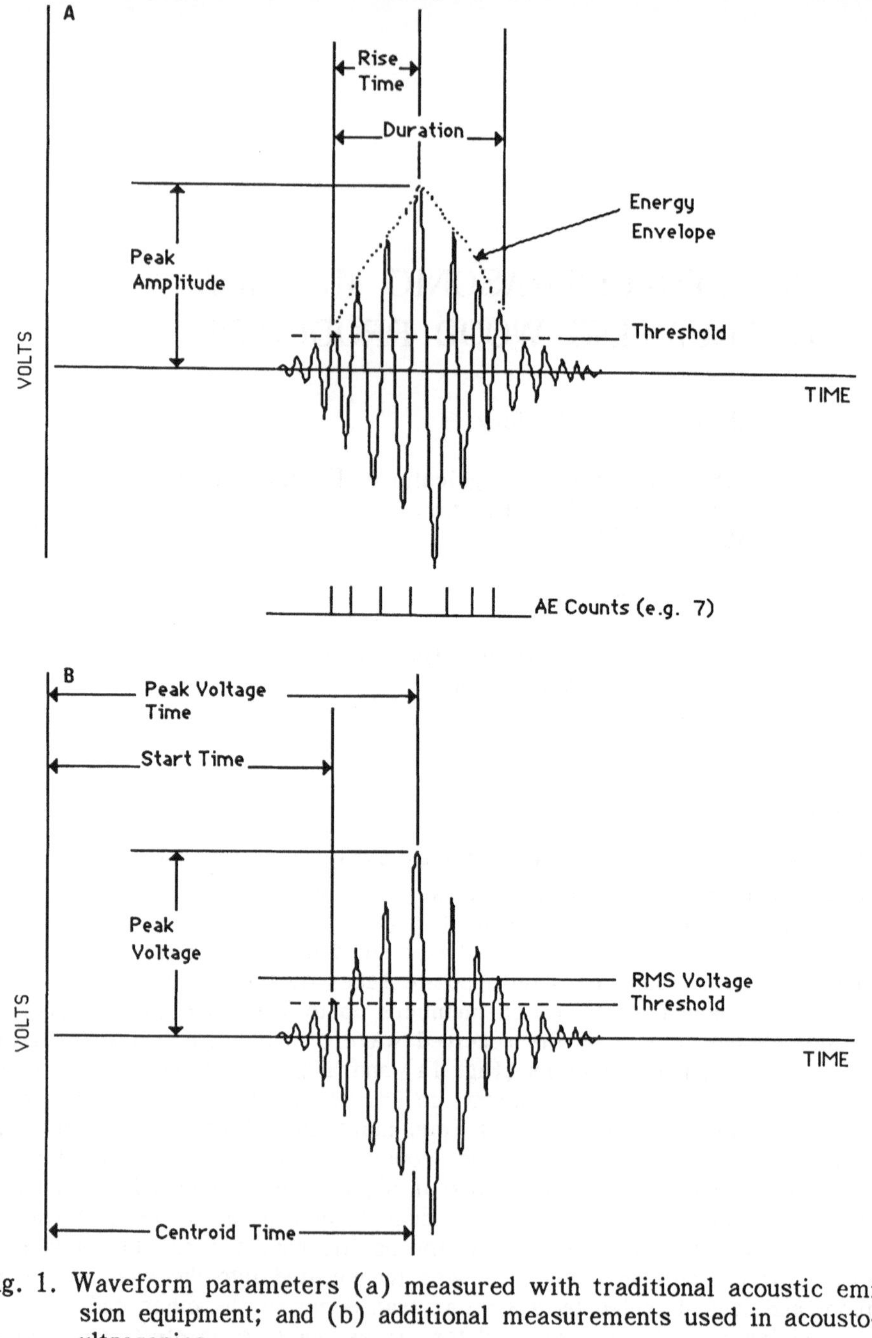

Fig. 1. Waveform parameters (a) measured with traditional acoustic emission equipment; and (b) additional measurements used in acousto-ultrasonics.

resulted in more repeatable waveform measurements. A Fourier transformation of the data gave parameters of the waveform in the frequency domain.

Combinations of the waveform parameters in the time and frequency domain will be used in future studies on partially decayed wood members. Preliminary data show centroid time and time to positive or negative peak to be sensitive to decayed wood. Higher frequencies within the bandpass (20-100 kHz) tended to attenuate more in decayed wood.

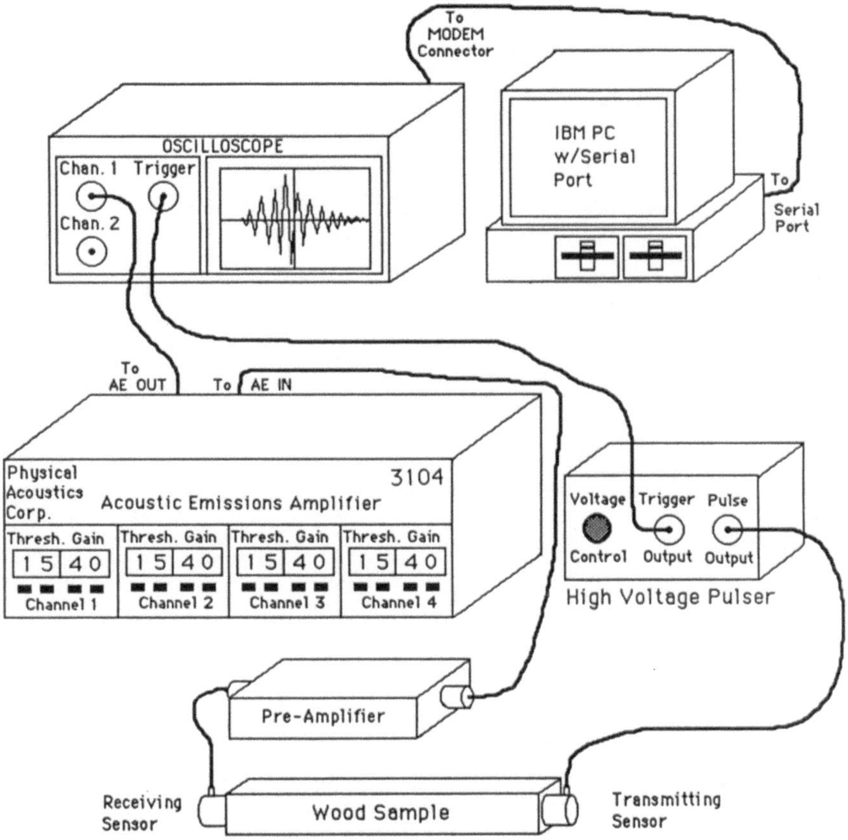

Fig. 2. Acousto-ultrasonics hardware configuration.

ACOUSTO-ULTRASONIC CHARACTERIZATION OF PHYSICAL PROPERTIES OF HUMAN BONES

Avraham Mittelman, Itzhak Roman, Arye Bivas,
Isaac Leichter, and Joseph Y. Margulies

Hebrew University
Jerusalem, Israel

Arie Weinreb

Hadassah Medical Center
Jerusalem, Israel

ABSTRACT

In this work the Acousto-Ultrasonic (AU) technique was utilized to non-destructively characterize two biological parameters of human bone specimens: the weight and mineral content densities. It has been previously shown that these two parameters have great effect on bone strength, thus by employing the non-invasive AU technique one can obtain vital information about bone strength without using any ionizing radiation. In the present study, the bones were characterized both by common techniques, e.g. γ-rays and AU, and the results of the study indicate that the AU parameter which best correlates with the weight and mineral content densities of human bones was the peak amplitude of the received AU signals. The peak amplitude of the AU signal and the weight density of the bone specimens was found to be linearly related with a coefficient of correlation $r = 0.68$. The coefficient of linear correlation between the mentioned AU parameter and the mineral content density was $r = 0.63$. This preliminary study demonstrated the potential of employing the Acousto-Ultrasonic technique as a non-invasive means to determine physical properties of human bones. The results suggest that the peak amplitude of the AU signals can be related to the bone strength.

INTRODUCTION

The availability of a sensitive though non-invasive method for the estimation of bone strength is of cardinal importance for the early diagnosis of osteopenia. In search of such a method, the commonly used techniques for the measurement of bone mineral content, BMC, and bone density, BD, were developed. The present study is an attempt to determine bone properties by AU so that the use of ionizing radiation is avoided, and AU can be employed in the clinical domain. The physical properties of bone which can be evaluated by the AU technique, differ essentially from those which are obtained by any other commonly used technique. Photon absorptiometry, computerized tomography and photon scattering, all measure the concentration of minerals or the density of the bone tissue. The parameters obtained by AU, on the other hand, are related to the structure of the bone tissue and to the manner in which the bone fibers are organized. Consequently, AU has the potential of providing insight into structure-related mechanical properties.

MATERIAL AND METHODS

Thirty bone specimens were removed from fresh femoral heads of cadavers. None of the subjects was known to have any disease which affects bone status. The specimens were taken from the trabecular region under the cortical layer (Fig. 1). Each specimen had the shape of cylinder of 1 cm diameter and 1 cm height. The symmetry axis of the specimen coincided with that of the femoral neck. In addition, thirty-two bone specimens were obtained from osteoporotic patients who had undergone hip-joint surgery. The bone specimens were air dried for 24 hours and then kept at -18°C during the period of measurements. Four tests were performed on each of the specimens:

1. The bone density was measured using the Compton scattering technique: Cs^{137} (662 keV) was used as the radiation source and the photons, which were Compton scattered by the specimen at a 90° angle, were detected by a NaI(T1) counter. A focusing collimator was used to define the scattering angle. The method has been described in details elsewhere.[1,2] The calibration procedure utilized various plastic cylinders of known densities and of the same dimensions as that of the bone specimens. The BD in grams/cm^3 was calculated using the calibration curve so obtained.

2. The BMC was determined by photon absorptiometry[3] using the Noriand-Cameron Bone Mineral Analyzer (Model 178). The specimens were scanned with their longitudinal axis parallel to the gamma-ray beam. The BMC per unit area as given by the device was divided by the sample's height in order to obtain the mineral content per unit volume (grams of mineral per cm^3).

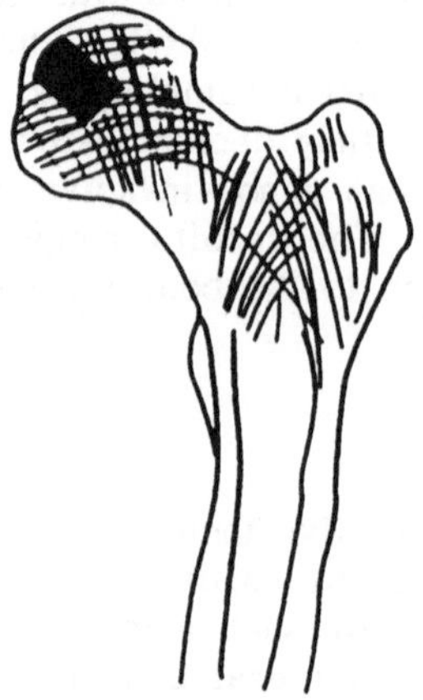

Fig. 1. The location in the femoral head from which the trabecular bone specimens were taken.

3. The average attenuation coefficient of each bone specimen was determined by computerized tomography.[4,5] The specimens were scanned by an EMI 1010A scanner for 60 seconds at 120 kV and 30 mA. The frames were transferred to a magnetic tape for further analysis on a PDP 15/76 computer. The average CT number (in Haunsfield units) was calculated for each specimen.

4. The bone specimens were placed between two ultrasonic transducers, one serving as a stimulator and the other as a detector. They were coupled to the transducers by ultrasonic coupling gel, and the coupling was maintained by application of a constant pressure. The emitting transducer was a AETC broadband FAC-500 which injected a periodically repeating series of ultrasonic pulses into the bone specimen. Each of these pulses produced simulated AU stress waves in the material. The signals arriving at the receiver resemble "burst" type acoustic emission events.[6-8] The receiving transducer was a piezoelectric acoustic emission AETC transducer with resonance at 375 kHz. After passing a 375 kHz filter the signals were analyzed by an AETC 5000A device. The analysis included an evaluation of the peak amplitude and the duration of the event. For each bone specimen 10,000 signals were analyzed and the most frequent value of each of the two parameters was chosen.

RESULTS

The bone samples examined were taken from two distinct populations. The first came from cadavers, none of which was known to have any disease, and the other group of samples were removed from patients that suffered from osteoporosis. Table 1 gives some of the physical parameters (such as BD, BMC) that were determined for all samples and which show that the two populations differ significantly. In addition, the samples were also characterized employing the AU and the results are summarized in Table 1.

Table 1. Physical parameters of bone

Parameter	Healthy	Osteoporosis	P<
Age [gr/cm^3]	63 ± 15.3	76 ± 9.2	0.0005
BD [gr/cm^3]	1.198 ± 0.20	1.143 ± 0.093	0.05
BMC [gr/cm^3]	0.332 ± 0.048	0.267 ± 0.038	0.0005
AU Peak Amplitude [dB]	28.93 ± 6.97	35.89 ± 6.54	0.003

(P is the level of significance for which the populations are different.)

The characteristics of the AU signal passing through the bone samples were evaluated, and it was observed that the level of the PA of the AU signals correlated best with the physical properties of the bone samples for the healthy population (Figs. 2 & 3). Such a correlation was not found for the osteoporotic population. The best correlation was found to exist between the level of PA and the BD of the healthy bone samples, r = 0.68, as shown in Fig. 2. Somewhat lower correlations however, still statistically signifi-

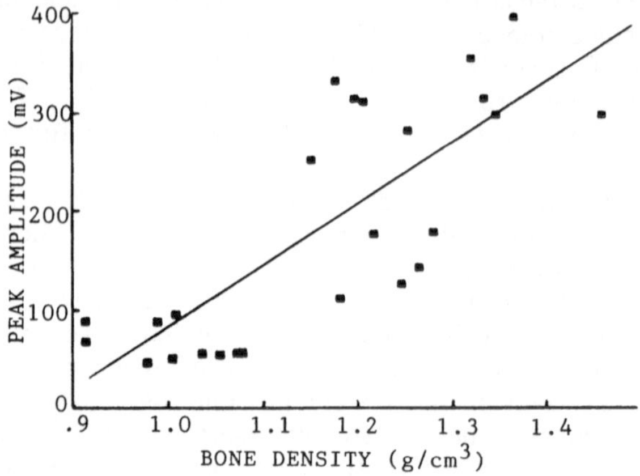

Fig. 2. The peak amplitude of the transmitted AU signals vs. BD.

cant, were noted between the level of the PA and the BMC, r = 0.63. It has to be noted that such correlation in medical results is quite significant.

DISCUSSION

This preliminary study shows that the amplitude of the transmitted AU signals, which were introduced into the bone specimens by ultrasonic pulses, are related to the physical properties of the bone. The density, mineral content, and average attenuation coefficient of the bone have already been shown to estimate its ultimate tensile strength.[9] Thus, the results of this study suggest that the AU parameters relate to the strength as well. The exact nature of this relationship can be evaluated more specifically only after destructive testing, when the bone specimens are loaded and the fracture strength is determined. Nevertheless, the correlation between the peak amplitude and the density of the specimen may indicate stronger internal reflections with decreasing bone density. The peak amplitude of the transmitted signal is related to the attenuation properties of the trabecular bone. However, the exact correlation between the attenuation mechanisms in

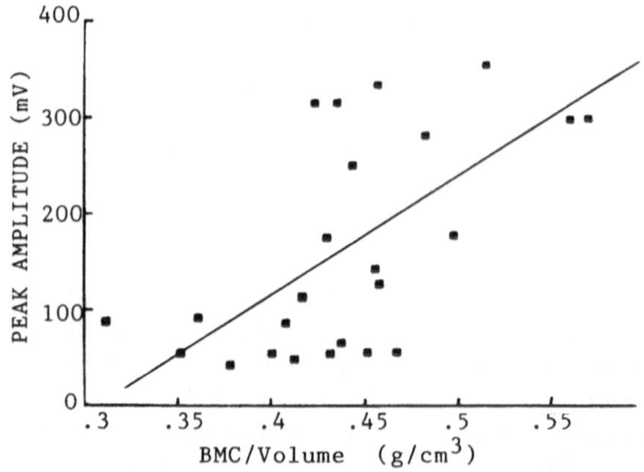

Fig. 3. The peak amplitude of the Transmitted AU signals vs. BMC.

bone and microstructural features is not well understood,[10] due to the complexity of the bone microstructure. In addition, interconnected fluid-filled pores modulate the attenuation characteristics of the bone. It is therefore not surprising that in the case of osteoporotic bones, where the volume fraction of such pores is large, no correlation between peak amplitude of AU and bone properties was obtained, as AU is mainly affected by the interconnected fluid-filled pores.

REFERENCES

1. G. Hazan, I. Leichter, E. Loewinger, A. Weinreb, and G. C. Robin, Physics in Med. Bio. 22:1073 (1977)
2. I. Leichter, G. Hazan, A. Weinreb, E. Loewinger, G. C. Robin, J. Menczel, and M. Makin, Clinic. Ortho. Rel. Research 151:232 (1981).
3. J. R. Cameron and J. A. Sorenson, Science 142:230 (1963).
4. I. Isherwood, R. A. Rutherford, B. R. Pullan, and Adams, Lancet. 2:712 (1976).
5. M. A. Weissberger, R. G. Zamenhof, S. Aronow, and R. M. Neer, J. Comput. Ass. Tomog. 2:253 (1978).
6. D. M. Egle and A. E. Brown, J. Acoust. Soc. Am. 57:591 (1975).
7. J. Rodgers in: "Acoustic Emission Trends 3," Acoustic Emission Technology Corp., Sacramento (1982).
8. A. Vary and R. F. Lark J. Testing & Eval. 7:185 (1979).
9. I. Leichter, J. Y. Margulies, A. Weinreb, J. Mizrahi, G. C. Robin, B. Conforty, M. Makin, and B. Bloch, Clinic. Ortho. Rel. Research 163:272 (1982).
10. R. Lakes et al., J. Biomed. Engr. 8:143 (1986).

MULTI-PARAMETER, MULTI-FREQUENCY ACOUSTO-ULTRASONIC FOR DETECTING IMPACT DAMAGE IN COMPOSITES

James R. Mitchell

Physical Acoustics Corporation
Lawrenceville, NJ

ABSTRACT

The inability to detect impact damage in composite material structures is a major factor blocking more extensive use of these materials. Past research utilizing a simplistic measurement of AU has indicated great potential when used in a laboratory environment. The use of AU outside of the laboratory by non-scientists has not met with the same success. This report describes the use of a more complicated approach to AU which incorporates several parameters measured simultaneously. The result is a technique, called "DCAT" (Dry Contact Acoustic Transmission; Patent Pending), which can be utilized to locate impact damage in non-laboratory environments.

INTRODUCTION

Alex Vary introduced the concepts of Acousto-Ultrasonics (AU) and Stress Wave Factor (SWF) to the AE and UT communities in the mid 70's.[1] At the time, Vary's ideas were looked upon as a gross violation of the basic definition of AE which clearly states that the source must be "generated from within the material".[2] In the 10 years that followed Vary's controversial disclosures, it has become very obvious that AU is an extremely powerful technique because the induced stress waves interact with the entire volume of material through which they travel. As a result, propagation of stress waves is related to the total damage state of the material that lies in their path[3]. A key factor in the development of AU has been the realization that advanced measurement techniques, normally associated with AE, are essential.[4]

Among the list of applications for AU are its use for measuring: fracture resistance,[1] strength properties (such as microvoids, etc.),[5] crack nucleation sites,[6] cure pressure,[7] interlaminar shear strength,[8] fracture toughness,[9] residual strength,[10] prediction of failure location,[11], loss of stiffness,[12] impact damage.[4] Along with the growing list of applications, AU has become established in the research community as an extremely promising technique for laboratory studies. For non-laboratory use the technique has met with many constraints such as: contact forces of 15 to 20 lbs. on transducers, constant couplant viscosity, constant couplant thickness, constant transducer spacing, backing plate on reverse side of sample, sophisticated calibration programs to select instrumentation settings[11] and fiber orientation[3]. Variations in the basic AU technique, which go beyond SWF, to measure other features of the received waveform, that is, RMS[3] and multiple parameters and multiple frequencies[4] hold the promise for future non-laboratory use.

IMPACT DAMAGE IN COMPOSITES

It is common knowledge that modern aircraft are subjected to a great deal of impact damage, much of which takes place while on the ground during routine maintenance and servicing. The impact damage is the result of accidentally dropped tools and inadvertent contact with a wide variety of structures such as hanger doors. The aluminum skin of conventional aircraft will show evidence of impact damage in the form of surface dents which can be seen visually and measured. Conversely, composite components of new generation aircraft may offer no visible evidence of impact damage.

When subjected to an impact force, the impacted surface of a composite material deflects elasticly and returns to its original configuration. There is no evidence of a dent even though the composite material may be severely damaged below the surface. This phenomenon is very apparent in composite materials constructed of glass/epoxy. When unpainted, glass/epoxy composites are translucent, and a bright light passed behind an impacted area will appear diffused and cloudy. Composite structures which are painted or fabricated from opaque fibers such as graphite cannot be examined in this fashion.

As illustrated in Fig. 1, damage progression through the thickness of a composite material is related to the amount of impact force. One general observation about the shape of damage progression is that it widens in the same fashion as impact fracture of glass. A common observation about impacted composite parts is that there is more damage to the back surface than to the point of impact.

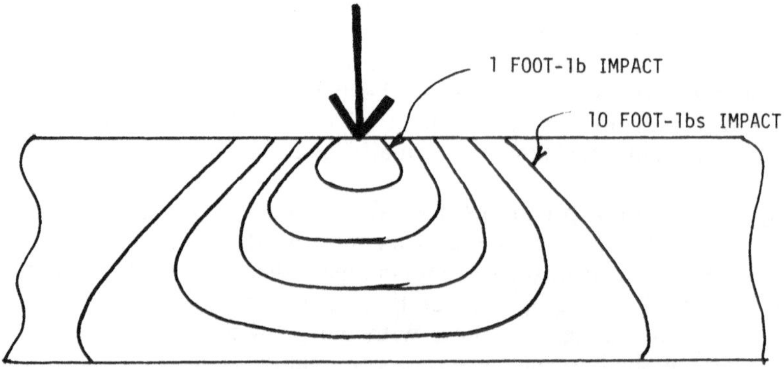

Fig. 1. Cross Section of Impact Damage.

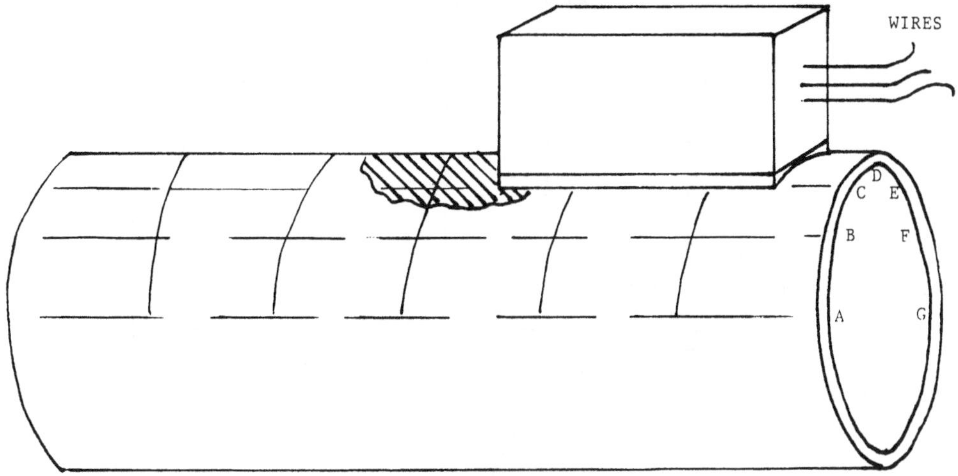

Fig. 2. Test Sample.

EXPERIMENTAL SETUP

Test samples, Fig. 2, were provided by Honeywell, Minneapolis. They were filament wound psuedo-isotropic glass/epoxy cylinders approximately 1/4 inch thick and approximately 6 inches OD. The layup was tangential and cross wound layers were 79% by weight glass content. A series of test samples were impacted with gradually increasing energy ranging from 1.5 to 57 lb-ft. Impact damage was induced by dropping a steel ball from varying heights. The samples were not painted so that crazed or cloudy areas were clearly visible.

A high performance, 4 channel acoustic emission instrument with TRA 128 transient waveform digitizer option was chosen for these experiments. This instrument is preferable to conventional ultrasonic instrumentation due to the need for simultaneous multi-parameter and multi-frequency measurements. Channel 1 is used as a trigger to time the arrival of stress waves detected on channels 2 and 3. Channels 2 and 3 used narrow band filters of 150 kHz and 400 kHz respectively. Channel 4 used a wideband filter from 20 kHz to 1 MHz and was used exclusively for digitizing the waveform. The AE instrument measured and recorded various attributes of the detected waveform for real time display and permanent storage on floppy disks. The TRA 128 option was used to digitize selected waveforms and to provide time and frequency domain plots.

The hand held probe[4] is illustrated as a cut-away at the bottom of Fig. 3. The device is a solid mass of material holding an acoustic emission sensor and pulser. The pulser, labeled R50, is used to inject a stress wave into the composite material. The sensor, labeled WD, is used to detect the signal that has traveled through the material.

Couplant is an important consideration of any acoustic emission or ultrasonic experiment, as it must efficiently transfer mechanical energy to and from transducers. Unfortunately, couplant materials are usually viscous fluids that are messy to use and often contaminate the part being tested. The data reported here were collected using a "dry coupling" technique which avoids the use of viscous fluids and eliminates contamination. As illus-

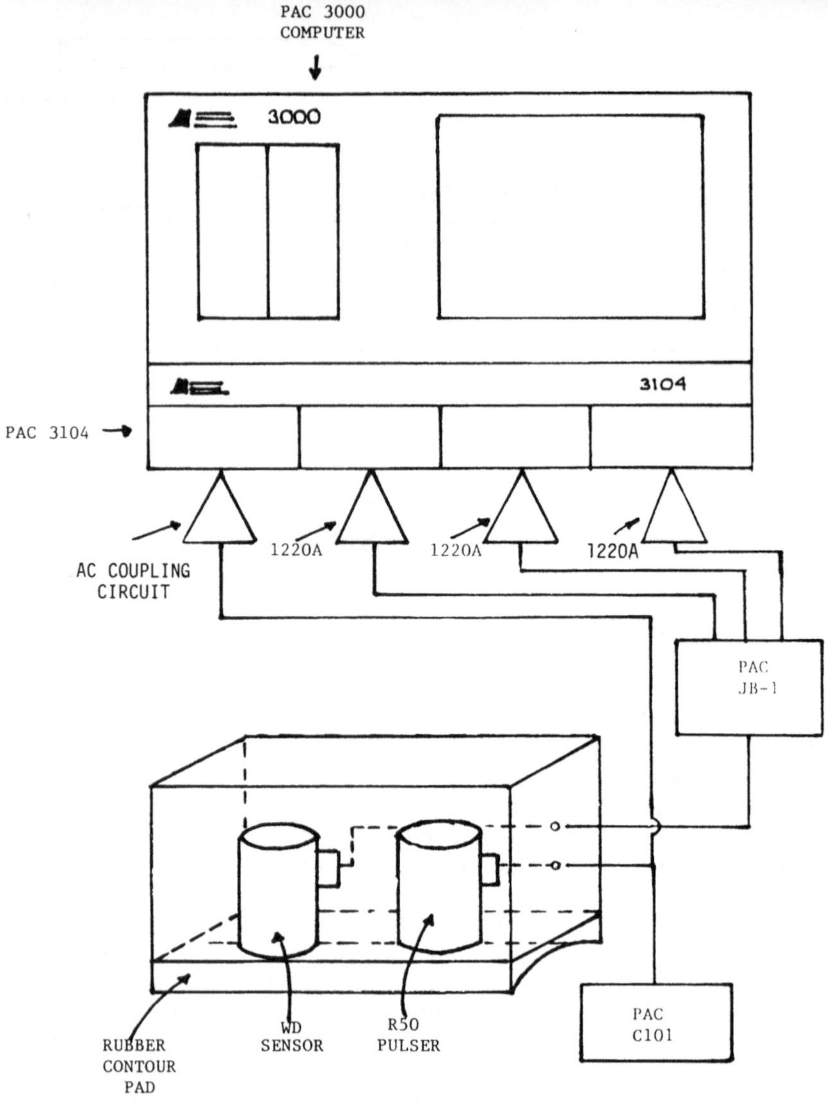

Fig. 3. Instrumentation Set-Up.

trated in Fig. 4, the interface between the sensor and surface was a flexible elastomeric material which conformed to the roughness of the surface. Though there is a practical limit to the amount of roughness that can be tolerated using this approach, the typical composite material surface is acceptable.

Dry coupling techniques are notoriously sensitive to the force that is exerted on the pulser and/or sensor. This problem and the problem of probe orientation are eliminated by normalizing attenuation measurements and comparing attenuation results with absolute time-of-flight data[4].

TEST RESULTS

Frequency plots of detected waveforms for undamaged and impacted parts are given in Fig. 5. The peak frequency detected in the undamaged part is

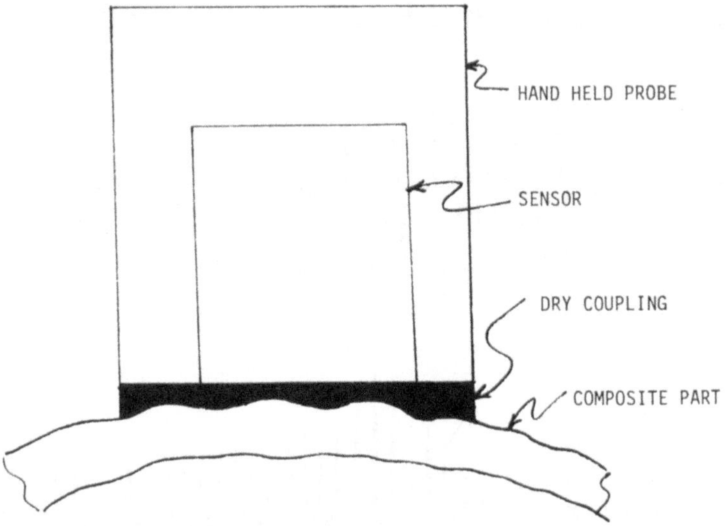

Fig. 4. End View of Probe Showing Elastomeric Dry Coupling.

390 kHz as compared to the peak frequency of 136 kHz in the impacted part. Though the peak frequency values varied from sample to sample, the trend was very repeatable. It is quite evident that the high frequency content

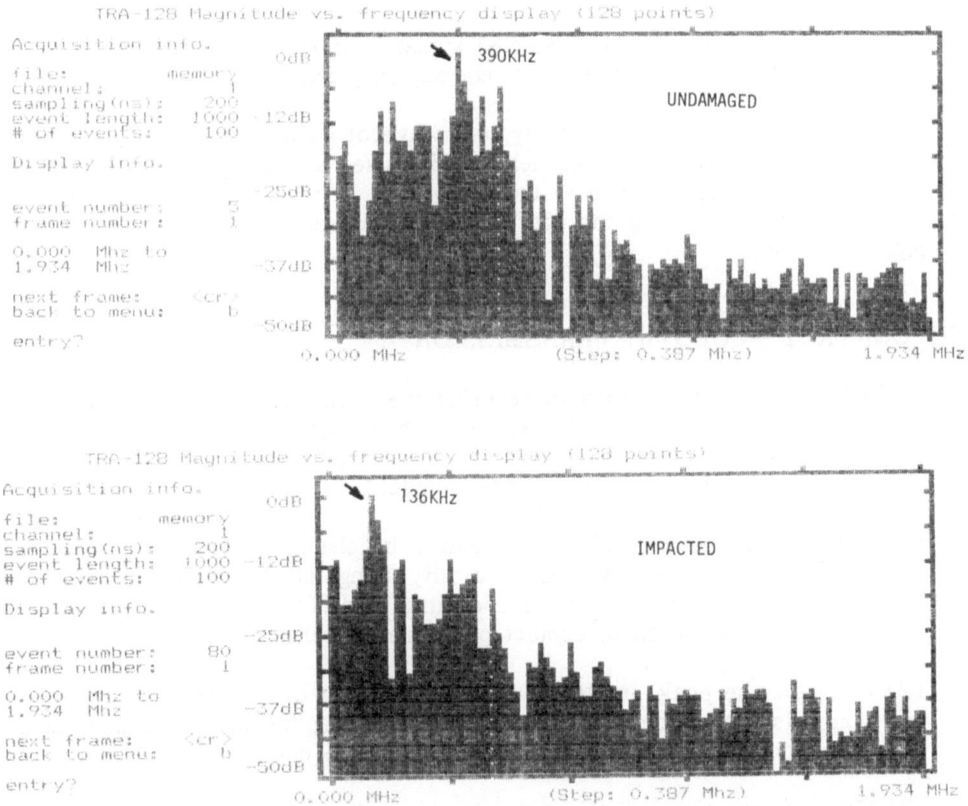

Fig. 5. Frequency Domain Plots Indicating Undamaged and Impacted Parts.

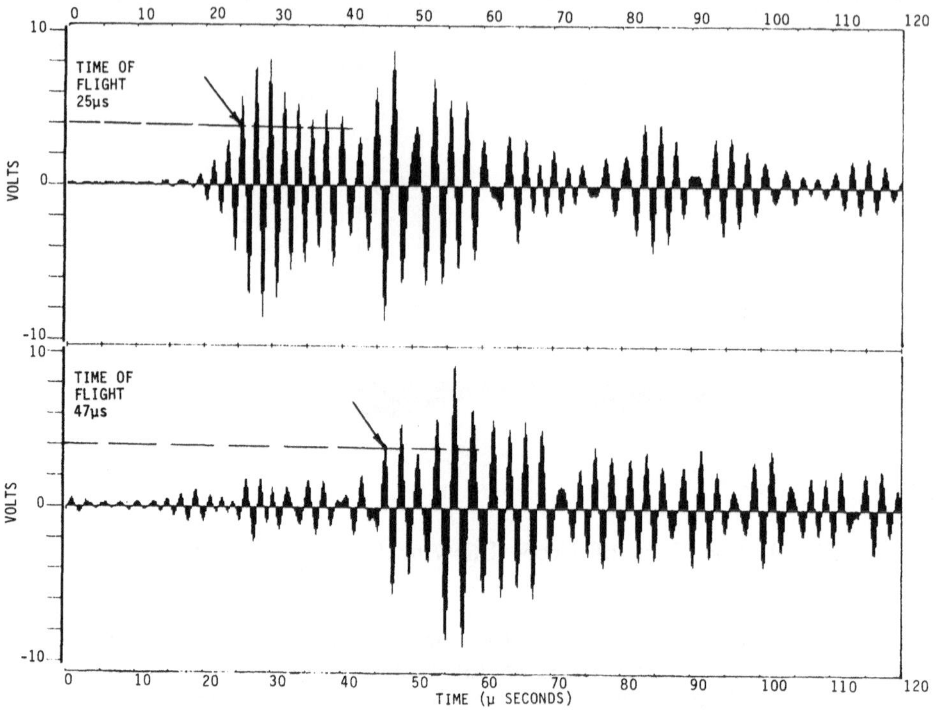

Fig. 6. Bottom Plot Used 20 dB More Gain Than Top Plot.

(approximately 400 kHz) of the detected signal is more highly attenuated by impact damage than the low frequency content (approximately 140 kHz).

Amplitude versus time plots of typical waveforms are given in Fig. 6. These plots were taken from channel 3 which uses a 400 kHz narrow band filter. Of particular significance is the difference between the time-of-flight of the stress wave traveling through an undamaged sample compared to that of an impacted sample.

DRY CONTACT ACOUSTIC TRANSMISSION

"DCAT" is the name chosen to describe the practical use of the above observations to measure impact damage and other anomalies in a composite structure. The computer is used to perform several mathematical operations on each data set that is received. These include:

(1) Subtract the peak amplitude measured by channel 3 from the peak amplitude measured by channel 2. By using the difference in peak amplitude from two channels operating at different frequencies the effects of variation in contact pressure are eliminated.

(2) Compare the peak amplitude difference to an accept/reject criteria which is determined experimentally by holding the probe on an area of the structure that is known to be free of defects.

(3) Compare the time-of-flight measurement to an accept/reject criteria which is determined as in (2) above.

(4) Compare the two accept/reject decisions from (2) and (3) above to make a comprehensive decision about the area that is currently under the probe.

CONCLUSIONS

AU is a powerful laboratory technique for identifying defects and structural anomalies in composite materials. The transition from laboratory to the field has begun with DCAT, a multi-parameter multi-frequency means of eliminating undesirable test variables. DCAT is a technique which utilizes a computer to consider several simultaneous measurements of peak amplitude and time-of-flight. The DCAT technique normalizes the affects of variation in hand/probe pressure by subtracting the peak amplitudes of the stress wave measured in two frequency ranges. A third measurement, time-of-flight, is compared to the peak amplitude difference to make a comprehensive decision about the condition of the area being evaluated. The DCAT technique has been used successfully to identify impact damage in a non-laboratory situation. Other applications will follow.

REFERENCES

1. A. Vary and R. F. Clark, "Correlation of Fiber Composite Tensile Strength with the Ultrasonic Stress Wave Factor, TM-78846," NASA, Cleveland (1978).
2. "Standard Definitions of Terms Relating to Acoustic Emission, ASTM E610," American Society for Testing and Materials, Baltimore (1986).
3. A. Govada, J. C. Duke, Jr., E. G. Henneke, II, and W. W. Stinchcomb, "A Study of the Stress Wave Factor Technique for the Characterization of Composite Materials, CR-174870," NASA, Cleveland (1985).
4. J. R. Mitchell, and R. K. Miller, Acousto-Ultrasonics Principles and Application, in: "Proceedings of the Second International Symposium on Acoustic Emission from Composites," Society of Plastics Industry, New York (1986).
5. D. E. W. Stone, and B. Clark, Ultrasonic Attenuation as a Measure of Void Content in Carbon Fiber Reinforced Plastics, NDT. 8 (1975).
6. A. Vary, Concepts and Techniques for Ultrasonic Evaluation of Material Mechanical Properties, in: "Mechanics of Nondestructive Testing," W. W. Stinchcomb, ed., Plenum Press, New York (1980).
7. A. Vary and K. J. Bowles, "Use of Ultrasonic-Acoustic Technique for Nondestructive Evaluation of Fiber Composite Strength, TM-73813," NASA, Cleveland (1978).
8. A. Vary and K. J. Bowles, "Ultrasonic Evaluation of the Strength of Unidirectional Graphite-Polyimide Composites, TM-X-73646," NASA, Cleveland (1977).
9. D. R. Hull and A. Vary, "Interrelation of Material Microstructure, Ultrasonic Factors, and Fracture Toughness of a Two Phase Titanium Alloy, TM-82810," NASA, Cleveland (1982).
10. J. H. Williams and N. R. Lampert, "Ultrasonic Evaluation of Impact Damaged Graphite Fiber Composites," Matls Eval. 38 (1980).
11. E. G. Henneke, II, J. C. Duke, Jr., W. W. Stinchcomb, A. K. Govada, and A. Lemascon, "A Study of the Stress Wave Factor Technique for the Characterization of Composite Materials, CR-3670," NASA, Cleveland (1984).

12. J. C. Duke, Jr, E. G. Henneke, II, W. W. Stinchcomb, and K. L. Reifsnider, Characterization of Composite Materials by Means of the Ultrasonic Stress Wave Factor, in: "Composite Structures 2," I. H. Marshall, ed., Applied Scientific Publications, London (1983).

TRANS-PLY CRACK DENSITY DETECTION BY ACOUSTO-ULTRASONICS

John H. Hemann and Paul Cavano

Cleveland State University
Cleveland, Ohio

Harold Kautz and Ken Bowles

NASA Lewis Research Center
Cleveland, Ohio

INTRODUCTION

The use of advanced composite materials, particularly for aerospace use, requires dimensional stability and the absence of cracks. The presence of cracks reduces mechanical characteristics and exposes the fibers and interfaces to environmental (moisture) effects. Graphite/PMR polyimide composites are candidate materials to replace metals in moderate high temperature zones (up to 316°C) in jet engines. These types of applications will introduce thermal cycling which in turn will produce trans-ply cracking unless tougher matrix materials are developed. Because these trans-ply cracks are very undesirable, NDE techniques are needed to detect the presence of trans-ply cracks. This paper will demonstrate that acousto-ultrasonics can be used to detect trans-ply cracks in graphite/PMR polyimide cross-ply and woven fabric laminates.

Adams[1], et al, have demonstrated experimentally that thermally induced trans-ply cracking occurs in graphite epoxy cross-ply laminates for various laminate configurations. Ishikawa and Chou[2] discuss the existence of trans-ply cracks in woven fabric composites. Thus it is clear that trans-ply cracks do exist both in cross-plies and in woven fabrics, due to both mechanical loads and thermal cycling.

This paper will show that the acousto-ultrasonic stress wave factor (SWF) does correlate with trans-ply crack density. The cracks were created by tensile loading of the material.

MATERIALS AND METHODS

Two different material systems were employed in this investigation, one a woven fabric reinforced system and the other a cross-ply layup made of unidirectional tape. All laminates were die molded at 3.4 MPa (500 psi) and 316°C (600°F), with a cure time of 2 hours at this temperature, none of the panels were postcured. All of the laminates had a nominal fiber fraction of 60 volume percent.

The woven fabric based laminates were prepared from prepreg obtained from the Ferro Corp., Culver City, California. The material is identified by Ferro as CPI 2237/CEL 3K, 8 HS. This designates a PMR-15 polyimide on an eight harness satin weave fabric of Celion carbon fiber. Both panels molded were six plies thick and all plies warp oriented. The panels were 15 cm wide by 30 cm long (warp direction) and .21 cm thick (6" x 12" x 0.084").

After molding, the panels were determined to be of acceptable quality by through-transmission ultrasonic (C-scan) inspection. The panels were then cut into specimens 30 cm long (12") and 1.3 cm wide (0.5") and then fitted with laminated fiberglass end tabs. These straight sided tensile specimens were then initially non-destructively inspected using the acousto-ultrasonic method.

The cross-ply laminate used a modified PMR-15 resin with a Celion C-6000 carbon fiber tow, prepared at the NASA-Lewis Research Center. The resin modification consisted of substituting pyromellitic dianhydride for the benzophenone dianhydride used in PMR-15. The construction was fourteen plies in a symmetrical $((90/0)_3\ 90)_S$ layup. The panels were 15 cm wide and 20 cm long (0 degree direction) and 0.26 cm thick (6" x 12" x 0.104"). The panels were also C-scanned and then cut into 20 cm by 1.3 cm tensile specimens. Tab bonding and initial acousto-ultrasonic readings were done as previously described for the woven fabric.

Acousto-ultrasonic measurements were made using a pair of 2.25 MHz center frequency transducers. A schematic diagram of the acousto-ultrasonic systems, transducers and specimen are shown in Fig. 1. The system is described in detail in the NASA technical memorandum by Kautz[3]. Ultrasonic coupling was achieved by using a pair of 0.05 cm x 1.27 cm (.125" x .5") RTV pads. The transducer center line separation was 3.8 cm (15") with 1.9 cm (.75") overlapping measurements done on each side of the specimens.

Crack density values reported represent an indication of cracks present, rather than an actual count of all cracks. The crack density values were assigned by counting the cracks that were microscopically apparent (200X) in the center two plies of the laminate. This was done by examining the edge of the laminate under the microscope and counting the cracks in each 1.9 cm (.75") length as established by the acousto-ultrasonic scanning procedure. The cracks that were counted were transverse to the specimen and transverse to direction of propagation of the acousto-ultrasonic wave. Figure 2 shows photomicrographs of polished sections of specimens showing these trans-ply cracks in the cross-ply laminate and in the woven fabric laminate.

The tabbed specimens were axially loaded in a tensile test machine at the loading rate of 0.002 mm/minute, held for five minutes at the preassigned load level, unloaded and microscopically examined for the presence of cracks.

For the woven fabric laminate, the loads were applied in the warp direction. One of the fabric laminate specimens was strain gaged and loaded to produce a stress-strain curve. The stress-strain curve was linear to fail-

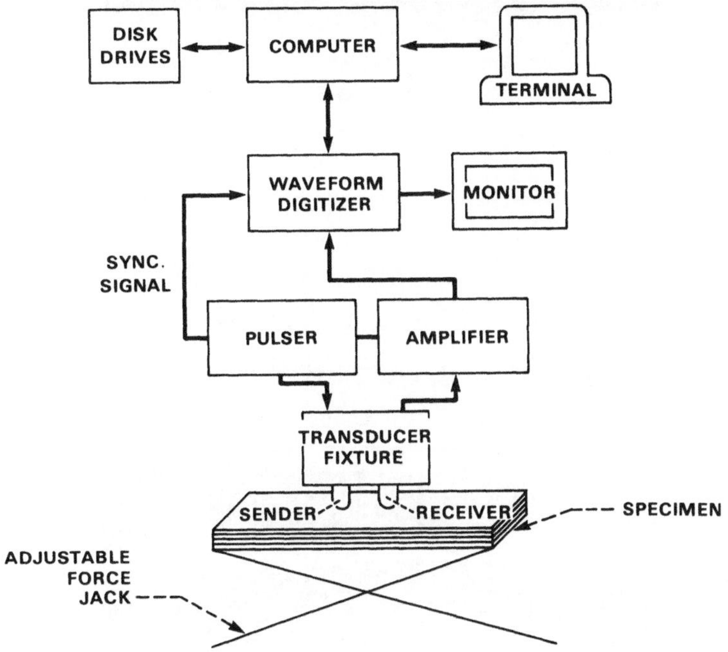

Fig. 1. Schematic diagram of the acousto-ultrasonic system

ure; the warp stiffness was 69.6 GPa (10.1 x 10^6 psi), the failure stress was 970 MPa (140 ksi), and the failure strain was 1.4%. Using the upper limit on the failure strain from the stress-strain test, a series of specimens were strained to various prescribed levels to provide a range of crack densities after which the specimens were examined microscopically and then inspected acousto-ultrasonically. The threshold strain to cause trans-ply cracking in the woven fabric laminate was measured to be 0.75%.

For the cross-ply laminate, the load was only applied in the longitudinal direction of the "0" degree plies. One of the specimens was strain gaged and used to obtain a strain-stress curve. The stress-strain curve was linear to failure; the longitudinal stiffness was 52 GPa (7.5 x 10^6 psi), the failure stress was 905 MPa (130.6 ksi), and the failure strain was 1.66%. For the cross-ply laminate, a single specimen was selected for repetitive, sequentially increased loading with microscopic crack examination and acousto-ultrasonic scanning at each load level. The threshold strain to cause trans-ply cracking in the cross-ply laminate was measured to be 0.65%.

RESULTS

Typical time and frequency domain digitized signals are shown in Fig. 3a for the cross-ply laminate. The SWF factor used in this paper is the RMS voltage over a specified time period and frequency band. At each crack density level as produced by a given load, SWF measurements were taken at 5 positions at equal intervals along the length on both sides of the specimen and averaged. This may be done over the entire time and frequency domain or the signal may be, as was done in this paper, partitioned in time and frequency. Time was partitioned in 5 μsec intervals and frequency parti-

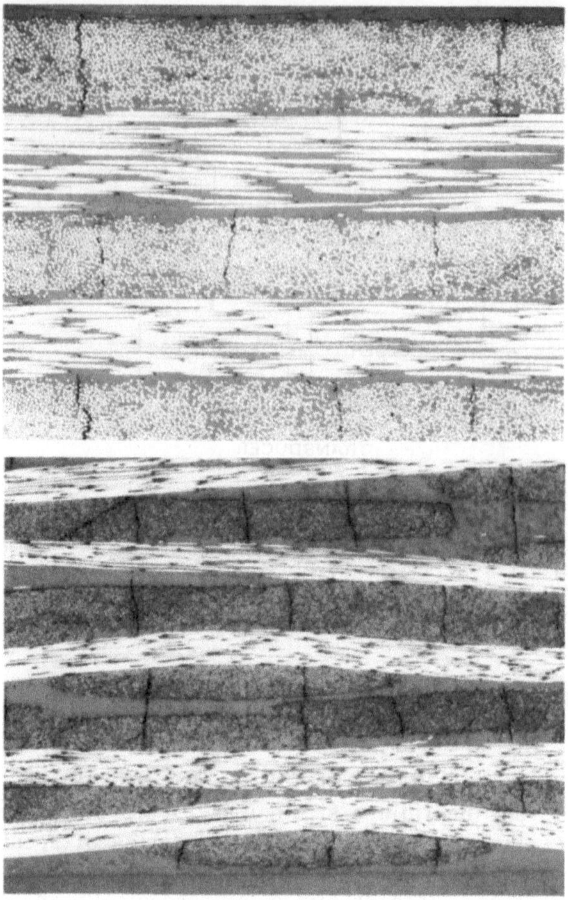

Fig. 2. Photomicrograph of polished sections of specimens showing trans-ply cracks in the cross-ply laminate (top) and in the woven fabric laminate (bottom).

tioned in 1.285 MHz bands. SWF measurements were taken at seven different crack densities.

Using a linear least squares fit to the data, correlations of SWF to crack density were determined for all time and frequencies, for each 1.285 frequency band, for each 5 μsec time interval, and for simultaneous time and frequency partitioning. Table 1 is a time and frequency partition correlation table. Figure 4a shows the SWF versus crack density data and the fit of a linear curve to the data for the 10 to 15 μsec time interval and the 0 to 1.285 MHz frequency band; these data show the best correlation, -0.907, of any partition.

The decrease of the SWF with crack density is expected because the formations of cracks are barriers to the propagation of stress and cause a decrease in amplitude of the stress wave. The good correlation occurring at the lowest frequencies and shortest times is also expected. These materials are very attenuating as is shown in Fig. 3a; thus, at low frequencies, the signal is distinct and large; at high frequencies it is small and lost in noise. At large times the signal reaching the receiving transducer is a

Table 1. Time and Frequency Partition Correlation Data - Cross-Ply

MHZ \ MSEC	0–5	5–10	10–15	15–20	20–25	25–30	30–35	35–40	40–45	45–50	50–TTL
0–1.285	.507	-.716	-.907	.156	.423	.359	.674	.485	.72	.765	.42
1.285–2.57	.583	-.626	-.824	.133	.185	-.028	-.131	-.324	-.063	0	-.65
2.57–3.855	.52	-.577	-.714	.134	.303	0	-.119	.167	-.094	.131	-.028
3.855–5.14	.084	-.624	-.85	.228	.166	-.287	-.43	-.207	-.154	-.116	-.093
5.14–TOTAL	.648	-.076	-.694	.319	.455	.274	.652	.612	.706	.785	.395

complex superposition of many Lamb modes of different speeds and types which will destroy correlation. At the shortest times only the fastest modes (or mode) arrive at the receiving transducer, and thus, constructive superpositon occurs, but decreasing in amplitude with increasing crack density.

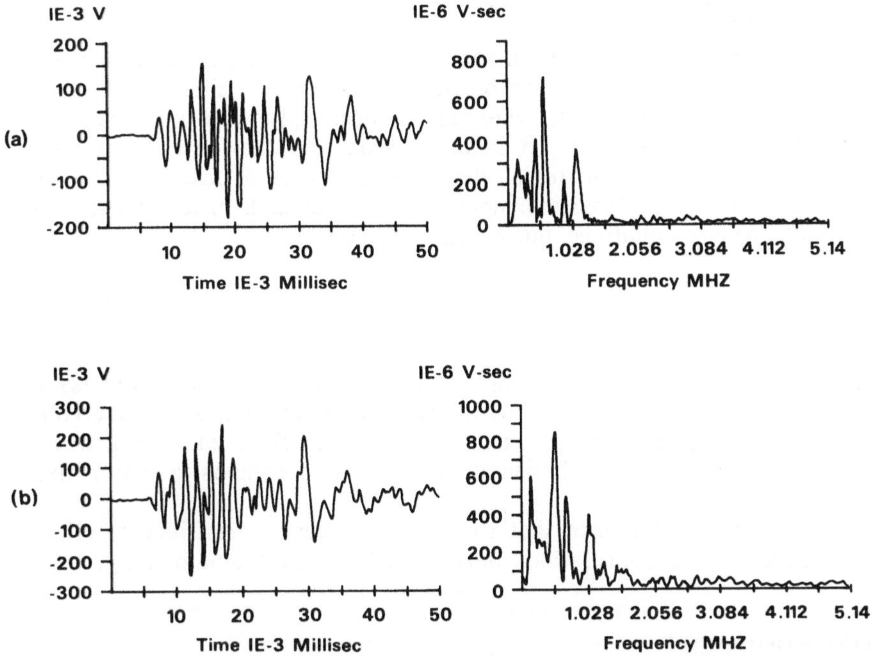

Fig. 3. Typical time and frequency domains signals a) for the cross-ply laminate, b) for the woven fabricate laminate.

Table 2. Time and Frequency Partition Correlation Data - Woven Fabric

MHZ \ MSEC	0–5	5–10	10–15	15–20	20–25	25–30	30–35	35–40	40–45	45–50	50–TTL
0 – 1.285	.374	-.924	-.863	-.899	0	-.808	-.543	.333	-.375	-.669	-.848
1.285 – 2.57	.794	-.863	-.859	-.882	-.57	-.627	-.285	-.534	-.522	-.57	-.862
2.57 – 3.855	.738	-.858	-.794	-.862	-.136	-.456	-.239	-.668	-.498	-.749	-.825
3.855 – 5.14	.069	-.868	-.695	-.826	-.415	-.533	-.522	-.734	-.578	-.838	-.767
5.14 – TOTAL	.307	-.811	-.821	-.894	.089	-.877	.104	-.76	-.539	-.673	-.852

Typical time and frequency domain signals are shown in Fig. 3b for the woven fabric laminate. Table 2 shows the time and frequency partition correlation data. Figure 4b shows the SWF versus crack density data and the fit of a linear curve to the data for the 10 to 15 μsec time interval and the 0 to 1.285 MHz frequency band. SWF measurements were taken at seven positions at equal intervals along the length on each side of the specimens and averaged. The correlation was once again better at the shorter times, but generally good at all times and frequencies.

The SWF again decreased with increasing crack density as expected indicating that cracks cause attenuation of the stress wave. The woven fabric did have cracks in all six plies, whereas the cross-ply was only trans-ply cracked in the 90 degree plies. It is clear from the data that the woven fabric is a less attenuating material than the cross-ply at low crack densities, but at higher crack densities, near failure, the high crack density in both materials nearly equalized the energy transmitted.

CONCLUSIONS

The acousto-ultrasonic method holds promise as a method to assess crack density. It is very clear from the data that the energy transmitted (SWF) decreases with increasing crack density. Thus, the SWF may be a practical measure and predictor of crack density in laminated composite materials.

There also exists evidence that time and frequency partitioning will be useful to produce good correlations between SWF and crack density.

ACKNOWLEDGEMENT

This research has been supported by grants from the NASA Lewis Research Center.

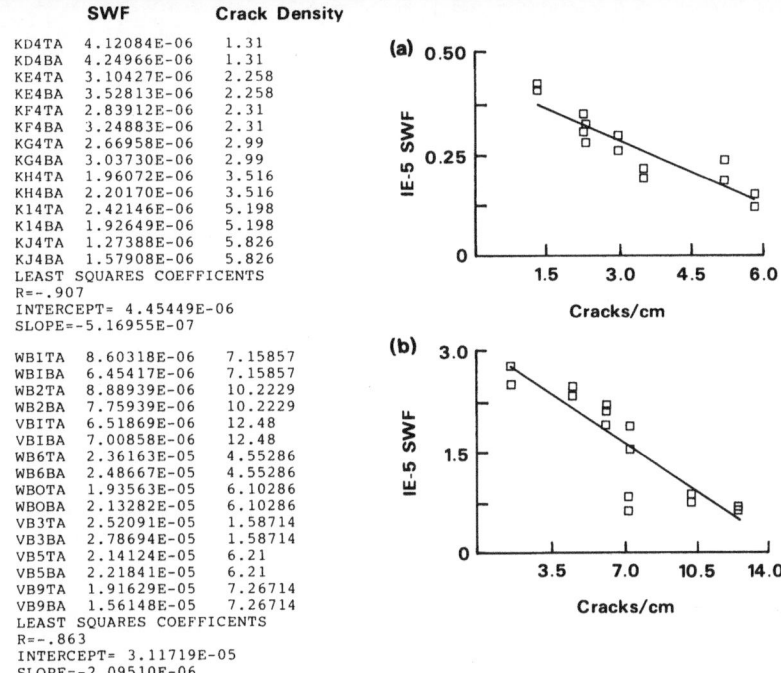

	SWF	Crack Density
KD4TA	4.12084E-06	1.31
KD4BA	4.24966E-06	1.31
KE4TA	3.10427E-06	2.258
KE4BA	3.52813E-06	2.258
KF4TA	2.83912E-06	2.31
KF4BA	3.24883E-06	2.31
KG4TA	2.66958E-06	2.99
KG4BA	3.03730E-06	2.99
KH4TA	1.96072E-06	3.516
KH4BA	2.20170E-06	3.516
K14TA	2.42146E-06	5.198
K14BA	1.92649E-06	5.198
KJ4TA	1.27388E-06	5.826
KJ4BA	1.57908E-06	5.826

LEAST SQUARES COEFFICENTS
R=-.907
INTERCEPT= 4.45449E-06
SLOPE=-5.16955E-07

WBITA	8.60318E-06	7.15857
WBIBA	6.45417E-06	7.15857
WB2TA	8.88939E-06	10.2229
WB2BA	7.75939E-06	10.2229
VBITA	6.51869E-06	12.48
VBIBA	7.00858E-06	12.48
WB6TA	2.36163E-05	4.55286
WB6BA	2.48667E-05	4.55286
WBOTA	1.93563E-05	6.10286
WBOBA	2.13282E-05	6.10286
VB3TA	2.52091E-05	1.58714
VB3BA	2.78694E-05	1.58714
VB5TA	2.14124E-05	6.21
VB5BA	2.21841E-05	6.21
VB9TA	1.91629E-05	7.26714
VB9BA	1.56148E-05	7.26714

LEAST SQUARES COEFFICENTS
R=-.863
INTERCEPT= 3.11719E-05
SLOPE=-2.09510E-06

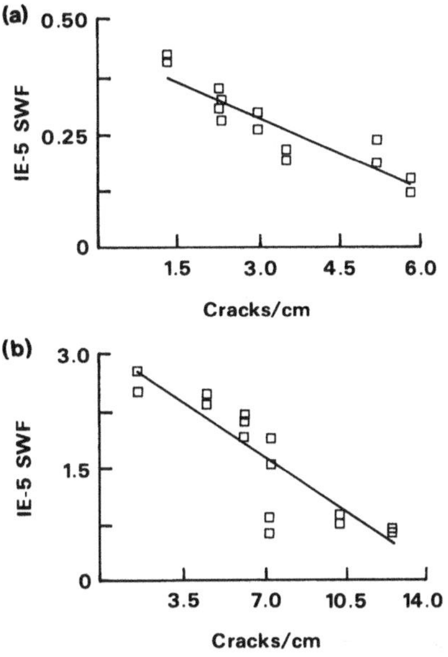

Fig. 4. SWF versus crack density data and the fit of a linear curve to the data for the 10 to 15 μsec time interval and the 0 to 1.285 MHz frequency band a) for the cross-ply laminate, and b) woven fabric laminate.

REFERENCES

1. D. S. Adams, D. E. Bowles, and C. T. Herakovich, Thermally Induced Transverse Cracking in Graphite-Epoxy Cross-ply Laminates, J of Rein Plas & Comp. 5:152 (1986).
2. T. Ishikawa and T. W. Chou, In-Plane Thermal Expansion and Thermal Bending Coefficients of Fabric Composites, J of Comp Matls. 17:92 (1983).
3. H. E. Kautz, "Acousto-Ultrasonic Verification of the Strength of Filament Wound Composite Material, TM-88827," NASA, Cleveland (1986).

APPLICATION OF ACOUSTO-ULTRASONICS FOR PREDICTING HYGROTHERMAL DEGRADATION OF UNIDIRECTIONAL GLASS-FIBER COMPOSITES

K. K. Phani and N. R. Bose

Central Glass and Ceramic Research Institute
Calcutta-700032, India

ABSTRACT

Unidirectional composites made of glass fibers of two different glass compositions, 'E' and 'N', were exposed to water at different temperatures, and the degradation process was monitored with the acousto-ultrasonic technique. The study shows that the rate of degradation is dependent on glass composition, but the degradation mechanism is independent of glass composition. The strength data of two composites exposed to boiling water for different lengths of time correlated with the normalized stress wave factor. The strength predictions made at temperatures of 60°C and 80°C, based on this correlation, show close agreement with the experimentally measured values.

INTRODUCTION

The effect of environment upon the properties of composites is of great interest in order to ensure the long-term mechanical reliability and safety of composite structures. Thermal and moisture exposure may lead to a number of undesirable effects such as polymer degradation, fiber-matrix debonding, chemical attack on the fiber surface, etc.[1-4] Reinforced plastic structures, especially those fabricated using unidirectional reinforcements, may contain defects such as voids or resin rich regions, which may cause additional defects at the microstructural level during service. Thus, assessing the accumulation of damage through non-destructive evaluation (NDE) methods and other techniques is of prime importance for assuring the safety of the structure during its service life.

Two NDE methods which have been used extensively to assess the mechanical properties of the composites in both the as-received and during service loading, are ultrasonic and acoustic techniques. Although conventional pulse-echo ultrasonics readily detects flaws, it is often difficult to correlate a detected flaw to overall performance. Additionally, evaluation of strength loss after aging may depend on sensing subtle changes that are distributed through the material rather than isolated flaws.[5] On the other hand, acoustic emission techniques require the structure to be put under stress to obtain spontaneous emissions from induced flaw growth. It is often difficult to assess the effect of this stress on the life of the structure,[6-7] especially in the case of composites. Also, the microstructure of a composite exerts a dominating influence on the observed AE signals, and no direct correlation can be made between microstructural failure events and the amplitude distribution of acoustic emission signals.[8] As a result, useful information from AE testing can be derived only if there is already a substantial background knowledge of the behavior of similar structures under known conditions.

The ultrasonic stress wave factor (SWF) technique, developed by Vary and his co-workers,[9] combines advantageous aspects of both the above techniques and at the same time overcomes the problems associated with them.[5] SWF is described[9] as a measure of stress wave energy transmission and is essentially a relative measure of the efficiency of energy dissipation in a material. If reflects the combined effects of flaws or other anomalies that exist in the volume of material being examined. Strong correlations between the stress wave factor and ultimate tensile and interlaminar shear strength in composites have already been reported in the literature.[9-12] This paper reports a study on the degradation of unidirectional glass fiber-polymer composites in water at different temperatures through this technique, and correlates the degradation process with the SWF.

EXPERIMENTAL CONSIDERATIONS

Materials

All the data reported in this paper were generated on unidirectional reinforced specimens in the form of rods having diameter 6.5 ± 0.3mm and length 120mm. The rods were prepared by drawing a bunch of rovings impregnated with resin through glass tubes. The resin used was of Isopthalic polyester type (HSR 8131 - Bakelite Hylam Ltd., India). Methyl Ethyl Ketone Peroxide and Cobalt Naphthanate were used as catalyst and accelerator, respectively. Two types of reinforcements were used --one commercially available 'E' glass rovings and the other 'N' glass rovings produced from a commercial neutral glass composition by a method described by Kumar.[13] 'N' glass fibers were treated with a silane coupling agent methacryloxy propyl trimethoxy silane (A-174, Union Carbide) compatible with polyester resin. 'N' glass fiber was chosen for its chemical resistance.[13] The composite rods were cured in laboratory atmosphere for 72 hours and then post-cured at a temperature of 100°C for 4 hours. The glass content of all the rods, as determined by burning off the resin was $73 \pm 1\%$ by weight.

Equipment

The equipment used in the experiment consisted of an Instron 1185 Universal Testing Machine (100 kN), Brooke-field constant temperature bath (Model EX 200) and an Acousto-Ultrasonic Tester (Model 206 AU of AET Corporation, USA).

The acousto-ultrasonic tester is composed of three main sections: the ulser section, the acoustic emission section and the display section. A repeated series of ultrasonic pulses from the pulser section are injected into the specimen by means of a sending transducer. Each of these pulses simulates acoustic emission events in the material.[12] A separate receiving transducer on the same side of the specimen intercepts some of the stress wave energy that radiates from the point of injection. The degree of attenuation of the ultrasonic wave as it travels through the specimen is converted to a numerical value. This numerical value, called the stress wave factor (SWF), has been defined by Vary et. al.[9] as SWF= g x r x n, where g = time before the counter is reset, r = pulse repetition rate and n = total number of counts of a single signal that exceeds a set threshold level.

This value was normalized relative to the maximum SWF value found for all specimens, i.e., $N_{SWF} = SWF/(SWF)_{max}$. SWF and N_{SWF} are both relative and depend on factors such as force applied to the transducer, signal gain, reset time, threshold voltage, repetition rate; all these factors have been kept constant in this study.

Experimental Procedure

Each specimen was subjected to measurements of the acousto-ultrasonic stress wave factor using the transducers mounted on a wheeled fixture[14]. The fixture was mounted on a spring loaded stand which applied a force of approximately 40 N to each transducer. For each specimen, two readings were taken by placing the receiving transducer at the center of the specimen and the sending transducer on either side of the center point. The mean of these two readings was assigned to that particular specimen. All acousto-ultrasonic

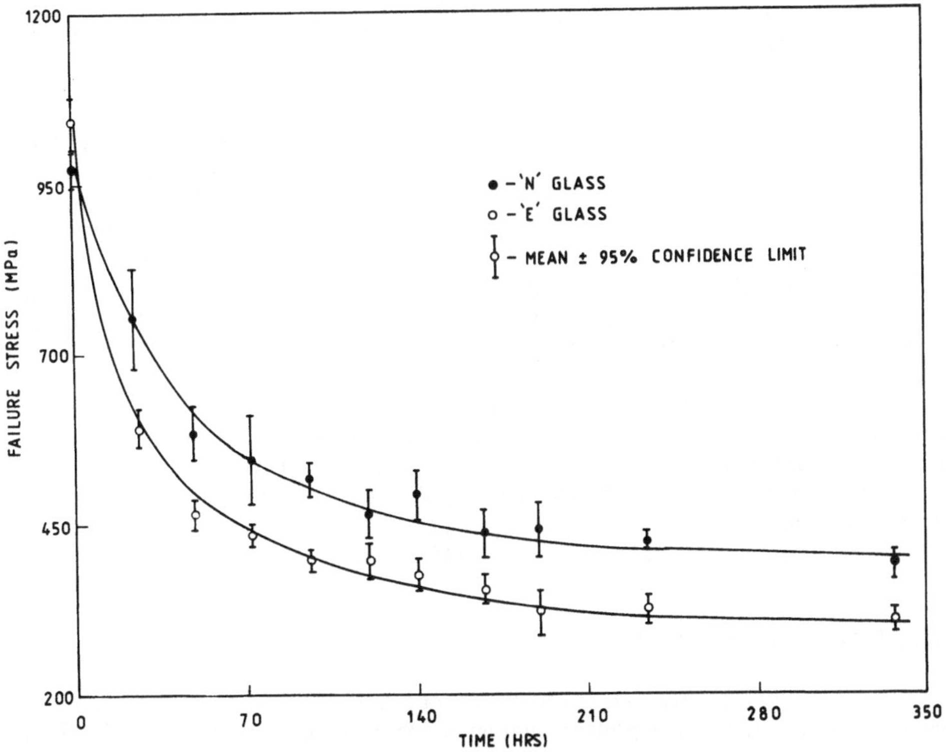

Fig. 1. Flexural Strength of 'E' and 'N' Glass Composites as a Function of Aging Time at 100°C.

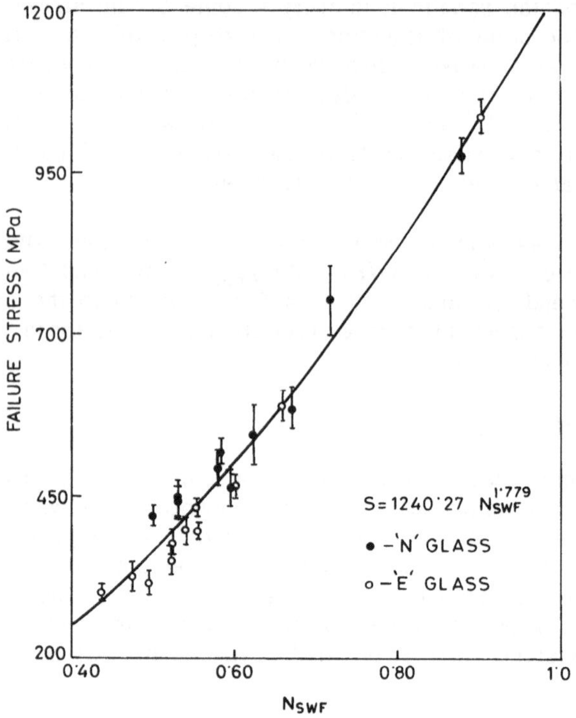

Fig. 2. Flexural Strength of 'E' and 'N' Glass Composites as a Function of Normalized Stress Wave Factor for Specimens Aged in Water at 100°C.

measurements were done by setting the instrument as follows: Threshold level: 0.5 V, Threshold mode: auto., Gate: 3 units, Trigger rate: 1 kHz, Gain: 50 dB, Sweep rate: 62.5 μsec/div., Distance between the transducers: 55 mm. Six specimens of both 'E' and 'N' glass composites were tested dry in three point flexure employing the Instron machine at a displacement rate of 2 mm/min and using a span of 100 mm. Flexural strength measurements were used in this study because they are commonly used for quality control, and 0-deg flexural strength is considered to be a filament-dominated property[15].

Sixty-six specimens were kept in boiling distilled water. Of these, six samples were drawn at random after a specified number of hours in boiling. They were again subjected to SWF measurement and flexure test after cooling to room temperature. The flexural strength was calculated by using standard beam formulae and reported as the average of six values. The same procedure was repeated for temperatures of 60°C and 80°C.

EXPERIMENTAL RESULTS

The average flexural strength values obtained after different hours of aging times at 100°C are plotted against aging time in Fig. 1 for both 'E' and 'N' glass composites. The mean dry strength of 'N' glass composites is 972 MPa which is about six percent less than that of 'E' glass composites. However, the strength of both the composites becomes equal after subjecting

TABLE 1. Comparison of strength values of composites aged at 80°C

Hours in Water at 80°C	'E' - Glass		'N' - Glass	
	Measured (MPa)	Calculated (MPa)	Measured (MPa)	Calculated (MPa)
24	545.2 ± 27.36*	603.4	798.8 ± 21.91*	790.4
75	427.2 ± 20.34	492.5	752.8 ± 44.47	716.6
141	398.4 ± 20.79	409.2	524.3 ± 35.76	550.4
189	384.4 ± 35.03	356.4	493.8 ± 38.03	493.3
213	359.8 ± 21.85	322.5	452.1 ± 42.24	418.9

* Standard Deviation

them to a boiling water enviroment for about eights hours, as shown in Fig. 1. For longer times, the strength of 'N' glass composites degrades less than that of 'E' glass. After about 280 hours, the strength of both the composites stops decreasing, at about 40% for 'N' glass and 29% for 'E' glass as shown in Fig. 1.

Fig. 2 shows a plot of the average flexural strength (S) values of both 'E' and 'N' glass composites obtained after different aging times at 100°C plotted against the average normalized stress wave factor N_{SWF}. As can be seen from the figure, the data points for both the composites fall on a smooth curve with a monotonic increase of S with N_{SWF}. A relation of the type:

$$S = K\, N^{m}_{SWF} \tag{1}$$

where K and m are the empirical constants, was fitted to the data by regression analysis. This yielded the values of K and m as 1240.27 and 1.779 respectively, having coefficient of determination of 0.95. As an additional measure of goodness of fit between the data and equation 1, a sum of squares (Q) was evaluated from the relation:

$$Q = 1 - \sum_{i=1}^{n} (S_i - \hat{S}_i)^2 / \sum_{i=1}^{n} (S_i - \bar{S})^2 \tag{2}$$

where S_i is the measured strength values, $\hat{S}_i$ is the strength value calculated from the fitted equation corresponding to measured N_{SWF} and S is the mean of the distribution. For a perfect fit the value of Q = 1. Usually a value of $Q \geq 0.95$ is considered to be a good fit. In the present case, the value of Q was calculated to be 0.972 indicating a good fit between the two parameters. Equation 1 is shown as a solid line in Fig. 2.

The average normalized stress wave factors at 80°C and different aging times were obtained experimentally, and then the flexural strength values were computed from eq. (1) obtained from data at 100°C. The calculated values as well as those measured at different aging times are given in Table 1. As can be seen from Table 1, agreement between the two strength values are quite close with a maximum deviation of about 15% to 5% from 'E' glass and 'N' glass composites respectively.

Analysis of the values of normalized stress wave factors at different aging times (t) at 60°C showed a linear relationship between $\ln(N_{SWF})$ and t with a coefficient of determination of 0.95 and 0.94 for 'E' glass and 'N' glass composites respectively. Thus a relation of the type $N_{SWF} = A \exp(-bt)$ where A and b are empirical constants, was fitted to the data following the method given by Lewis[16] yielding the relations

$$N_{SWF} = 0.768 \exp(-1.59 \times 10^{-3} t) \quad (3)$$

$$N_{SWF} = 0.850 \exp(-5.77 \times 10^{-4} t) \quad (4)$$

for 'E' glass and 'N' glass composites respectively. These equations are plotted along with the experimentally measured values in Fig. 3.

Combining eqs. 3 and 4 with eq. 1, the strength degradation with hours of aging in water at 60°C was obtained as

$$S = 775.9 \exp(-2.80 \times 10^{-3} t) \quad (5)$$

and

$$S = 928.9 \exp(-10.27 \times 10^{-4} t) \quad (6)$$

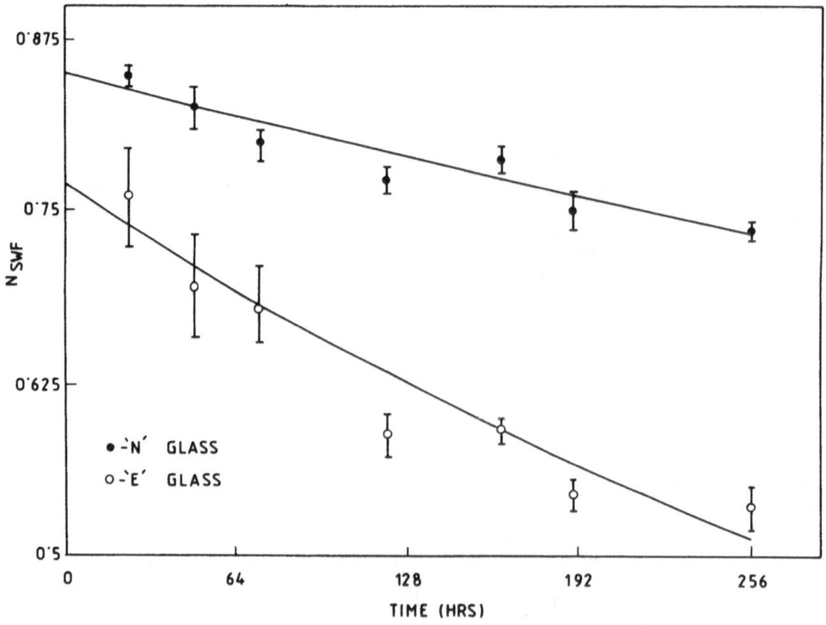

Fig. 3. Normalized Stress Wave Factor of 'E' and 'N' Glass Composites as a Function of Aging Time at 60°C.

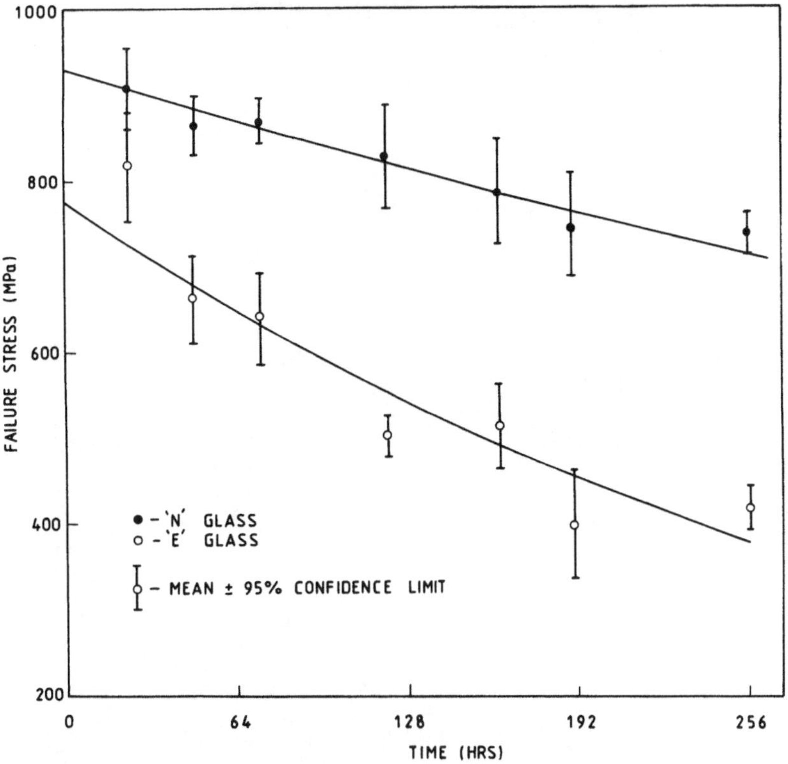

Fig. 4. Flexural Strength of 'E' and 'N' Glass Composites as a Function of Aging Time at 60°C.

for 'E' glass and 'N' glass composites respectively. These equations are plotted along with the experimental data in Fig. 4. As can be seen in the figure there is a good agreement.

DISCUSSION

Vary et. al.[9] has described the SWF as a measure of stress wave energy transmission. According to their hypothesis: decreased stress wave energy flow corresponds to decreased fracture resistance. Thus, it provides a means of rating the efficiency of dynamic strain energy transfer within a given material. If the material exhibits an efficient stress wave energy transfer, then it will have higher strength. That is, better stress wave transmission means better transmission of dynamic stress and load distribution. On the other hand, higher attenuation will give rise to a lower SWF indicating places where the dynamic strain energy is likely to concentrate and promote fracture. Fig. 2 shows that the data points for both 'E' glass and 'N' glass composites fall on the same curve indicating the dynamic strain energy concentrations which controls strength are identical in both cases. Fig. 1 also shows that initially damage occurs at a faster rate in 'E' glass composites than in 'N' glass composites. This can be explained as follows: On exposure to boiling water the fiber, the matrix, and the interfacial region of the composite are affected. The polyester component of the composite undergoes swelling and water plasticizes the resin in absence of inorganic impurities in resin.[1] In the

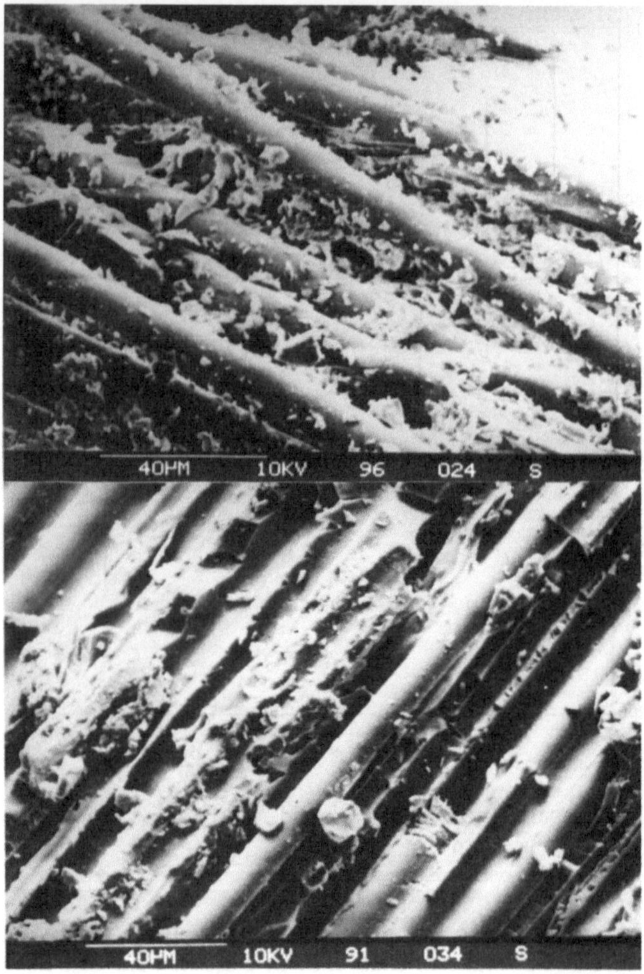

Fig. 5. Scanning Electron Micrographs of Composites Aged in Boiling Water for 25 Hours (a). 'E' Glass Composites (b). 'N' Glass Composites.

presence of traces of salts, the resin can act as a semi-permeable membrane, and the resulting osmotic pressure within the resin leads to cracking. In the case of 'E' glass composites, the resin swelling produces a radial stress at the interface which is reinforced by the osmotic pressure generated due to dissolution of leachable components of 'E' glass fiber by diffused water.[1] The osmotic pressure generated leads to fiber debonding, an increase in the shear stress transfer length, weakening the composite. In addition, 'E' glass fibers in composites are susceptible to alkali attack. Ashbee and Wyatt[1] have observed that the alkalinity of diffused water, produced by leaching of available K_2O and Na_2O from fiber, leads to the pitting of fiber surface after exposure to boiling water for a very long time. In the case of 'N' glass composites all the above phenomena will also take place but the leaching of glass constituents will be much less when compared to that of 'E' glass. Also 'N' glass has a better alkali resistance than 'E' glass.[13] Thus, debonding will be less severe in this case. This can be seen from the scanning electron micrographs of the composites aged in boiling water for 25 hours, shown in

Fig. 5; whereas, there is severe microcracking of the matrix resin in the case of 'E' glass composites when compared with the less severe damage observed in the case of 'N' glass composites. Because the matrix material is identical in both composites, the debonding at the interface will be the predominant phenomena controlling the relative strength of the composites for the temperature range and hours of immersion studied here. This explains the observed behavior shown in Figs. 1 and 2.

In the case of aging at lower temperatures, the mechanism of degradation remains the same but the process gets shifted on a time scale depending on the temperature of the water. The close agreement between the strength values predicted from SWF measurements and the experimentally measured ones only indicates that the SWF technique is a sensitive indicator of the degradation process in these composites.

CONCLUSIONS

The hygrothermal degradation of unidirectional 'E' glass and 'N' glass composites was evaluated nondestructively through an acousto-ultrasonic technique. The method involved characterization of stress wave propagation in the material. This measurement yielded the stress wave factor which was used to determine the rate of the degradation process.

Specific conclusions drawn from this study may be summarized as follows:

1. The stress wave factor is a sensitive indicator of unidirectional fiber composite strength reduction due to hygrothermal effects.

2. In unidirectional composites the hygrothermal degradation process is predominantly controlled by the interfacial phenomena, which is dependent on glass composition.

3. For a given system, stress wave factor readings can be used for the estimation of strength from the data generated under accelerated conditions.

ACKNOWLEDGEMENTS

The authors thank Dr. S. Kumar, Director of the Institute, for his kind permission to publish this paper. They also thank Mr. S. Chakraborty and Dr. S. Sen for providing the 'N' glass reinforcement and scanning electron micrographs respectively. Thanks are also due to Mr. M. Dutta for the preparation of the line diagrams.

REFERENCES

1. K. H. G. Ashbee and R. C. Wyatt, Water Damage in Glass Fiber/Resin Composites, Proc. Roy. Soc. A312:553 (1969).
2. O. Ishai, Environmental Effects on Deformation, Strength and Degradation of Unidirectional Glass-Fiber Composites, Poly. Eng. Sci. 15:486 (1975).
3. J. A. Aveston and J. M. Sillwood, in: "Advances in Composite

Materials, Vol. I," A. R. Bunsell, C. Bathias, A. Martrenchar, D. Henkes and G. Verchery, eds., Pergamon Press, Paris (1980).

4. B. J. Dewimilte, J. Thoris, R. Mailfert and A. R. Bunsell, in: "Advances in Composite Materials, Vol. I," A. R. Bunsell, C. Bathias, A. Martrenchar, D. Henkes and G. Verchery, eds., Pergamon Press, Paris (1980).
5. A. Vary, Concepts and Techniques for Ultrasonic Evaluation of Material Mechanical Properties, in: "Mechanics of Nondestructive Testing," W. W. Stinchcomb, ed., Plenum Press, New York (1980).
6. T. J. Fowler and R. S. Scarpellini, Acoustic Emission Testing of FRP Equipment-I, Chem Engrg. 10:145 (1980).
7. Idem, Acoustic Emission Testing of FRP Equipment-II, Chem Engrg. 11:293 (1980).
8. M. G. Phillips and B. Harris, in: "Advances in Composites Materials, Vol. II," A. R. Bunsell, C. Bathias, A. Martrenchar, D. Henkes and G. Verchery, eds., Pergamon Press, Paris (1980).
9. A. Vary and R. F. Clark, "Correlation of Fiber Composite Tensile Strength with the Ultrasonic Stress Wave Factor," NASA TM-78846, NASA/LeRC, Cleveland (1978). 10. A. Vary and K. J. Bowles, "Use of an Ultrasonic-Acoustic Technique for Nondestructive Evaluation of Fiber Composite Strength," NASA TM-73813, NASA/LeRC, Cleveland (1978).
11. A. Vary and K. J. Bowles, "Ultrasonic Evaluation of Strength of Unidirectional Graphite-Polyimide Composites," NASA TM-X-73646, NASA/LeRC, Cleveland (1977).
12. J. C. Duke, Jr., E. G. Henneke, II, W. W. Stinchcomb and K. L. Reifsnider, Characterization of Composite Materials by Means of the Ultrasonic Stress Wave Factor, in: "Composite Structure, 2," I. H. Marshall, ed., Applied Scientific Publishers, London (1983).
13. S. Kumar, Glass Composition for Spinning Fiber, J. Non-Crys. Solid 80:122 (1986).
14. Technical Manual, Acoustic Emission Technology Corporation, July 15, 1981.
15. J. M. Whitney and G. E. Husman, Use of the Flexure Test for Determining Environmental Behavior of Fibrous Composites, Exp. Mech. 5:185 (1978).
16. D. Lewis III, Curve Fitting Techniques and Ceramics, Am. Ceram. Soc. Bull. 57:434 (1978).

BOND QUALITY EVALUATION OF BIMETALLIC STRIPS: ACOUSTO-ULTRASONIC APPROACH

K. K. Brahma and C. R. L. Murthy

Department of Aerospace Engineering
Indian Institute of Science
Bangalore 560 012, India

ABSTRACT

Plain journal bearings used in engines such as aircraft, automobile and stationary are fabricated from bimetallic strips. In view of the performance requirements, one hundred percent nondestructive inspection of these strips is of paramount importance. But, manufacturers in general still seem to rely on destructive examination through peel-off tests for want of a suitable nondestructive evaluation (NDE) tool. To obtain a solution to this problem, in the recent past, we attempted to develop an NDE method using acoustic emission (AE) technique.[1] These attempts indicated that while the method is useful as a 'go/no go' test in the final stage of production, its applicability for one hundred percent inspection on production line does not appear to be practicable. Therefore, a program has been initiated at the Aerospace Engineering Department of the Indian Institute of Science to explore the feasibility of utilizing acousto-ultrasonic (AU) technique. To start with, two different sets of experiments were carried out. In the first set the ultrasonic and AE probes were fixed on the same side of the strip at either end. This test, though, yielded positive results; repeatability was poor. So, in the second set of tests, evaluation was carried out with the ultrasonic and AE probes mounted as in a through transmission ultrasonic test. The results of this set show that a bad bond area gives rise to higher stress wave factor (SWF) with good repeatability.

INTRODUCTION

Bimetallic strips fabricated for use as plain journal bearings are of different types (e.g. copper-lead, aluminum-tin or white metal) wherein the lining materials such as copper-lead, aluminum-tin or white metal are bonded to the base materials (mild steel) by either powder sintering or cladding or casting techniques respectively. The total thickness (base + lining) of these sheets usually varies from 0.5 to 5.0 mm. At present, to evaluate the bond quality of these strips, the technique adopted in industry consists of a manual peel-off testing and surface examination. The surface examination is carried out visually on a strip (152.4 X 152.4 mm) cut from the edges of a parent sheet approximately two meters in length and bent to expose the bond area. And, the peel test is carried out only at randomly selected small localized areas of a sheet. Thus, even after these two tests, (which are destructive in nature) it is possible that some defective strips are passed on to the final stages of production as the test is done only at selected spots of the sheet. Alternately, a good sheet may be scrapped because the random sampling indicates a bad bond. Obviously, these methods are followed even today due to the difficulties encountered in developing a reliable NDE procedure using conventional techniques. For example, it is uneconomical and unwieldy to use radiographic techniques. Ultrasonic pulse echo tests or attenuation measurements pose problems due to the small thickness involved. Recently Gilmore et al[2] have developed a novel method of evaluating the quality of diffusion bonded silicon devices by high frequency ultrasonic testing. Applicability of this method to bimetallic strips remains to be investigated. Thus, in the absence of an established and a readily available NDE procedure, in the recent past, we attempted using acoustic emission (AE) technique. The results of the acoustic emission experiments showed that specimens with good bond quality could be differentiated from specimens of bad bonds quality using cumulative counts at very small applied tensile loads. While this method would be useful as a "go/no go" test, it has two limitations viz., (i) application of a small load and (ii) inadequacy for one hundred percent inspection on production line.

So, in the current set of investigations we attempt to utilize the acousto-ultrasonic (AU) technique, as this combines the advantages of acoustic emission and ultrasonics. The technique when fully developed would be highly valuable for one hundred percent inspection on production line. In this paper, results obtained from our attempts to evaluate the bond quality by AU technique on small strips cut from a parent sheet are discussed.

SPECIMENS

The process of production of copper-lead bimetallic strips is by sintering.[2] In this process the base material in the form of long sheets is fed into a furnace wherein copper-lead powder is deposited on the sheets in the compact form continuously. Due to the low melting temperature of lead infiltration occurs. Thus, the pores in copper compact and the interstices existing due to insufficient diffusion of copper particles into the surface layer of the mild sheet strip get filled by molten lead. Figure 1 shows the infiltration of lead into copper compact (black areas identify the lead grains). Under compacting, variations in temperature, and pressure, deviations from specified atmosphere (under sintering), excessive quantity of lead and improper infiltration are some of the causes which lead to poor bonding between the base and lining materials. A typical photomicrograph of the bond zone is given in (Fig. 2). Figure 2(a) shows a neat and good bond line, and Fig. 2(b) indicates localized spots where debond exists.

The procedure in industry is to cut a two meter length piece as the bimetallic sheet continuously comes out from the rolling machine and carry out visual examination at the edges. Few such edge strips (250 mm long) of designated good and bad bond through visual examination were obtained for acousto-ultrasonic evaluation. The experimental program was carried out on 152.4 mm long, 25.4 mm wide and 2.1 mm thick copper-lead bimetallic coupons

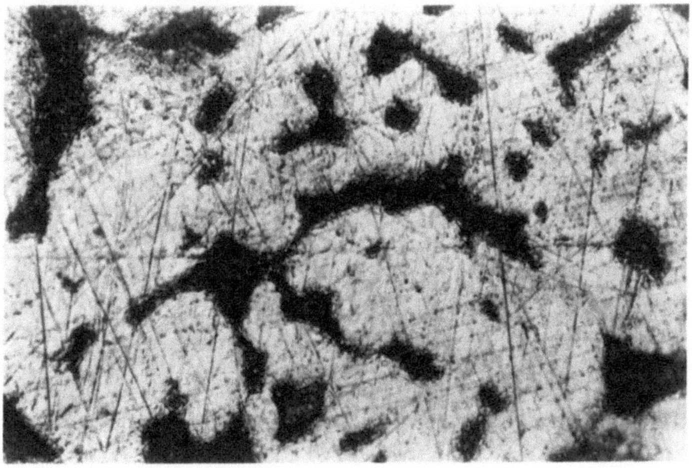

Fig. 1. Photomicrograph showing infiltration of lead into copper (magnification:500X).

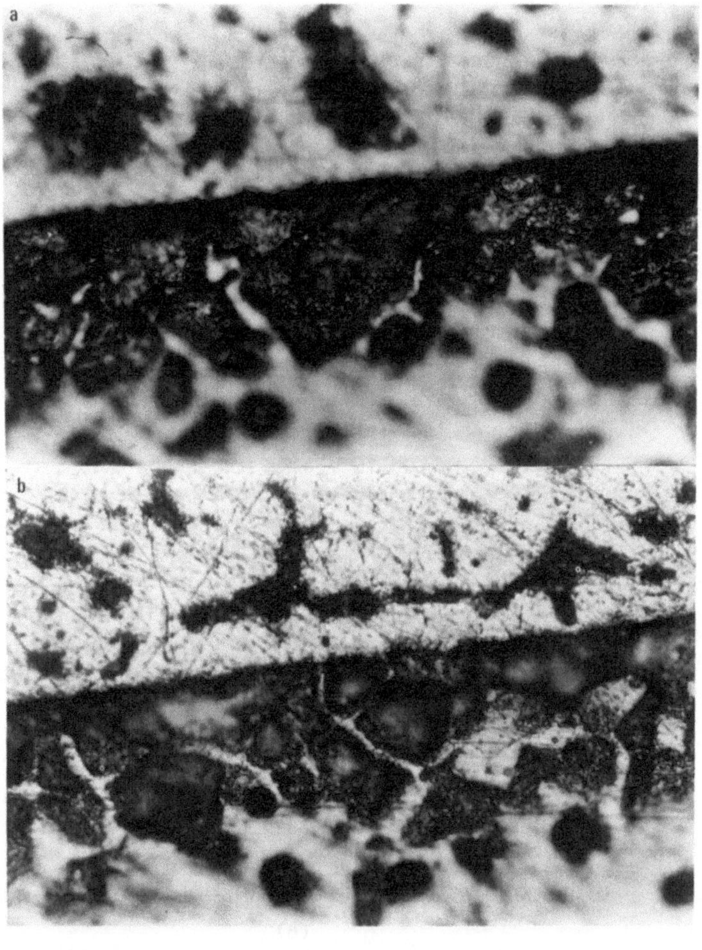

Fig. 2. Photomicrograph of a) good bond zone, b) bad bond zone, magnification:500X).

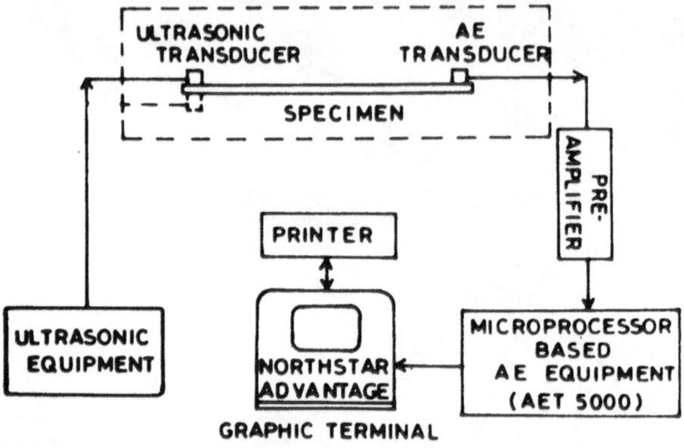

Fig. 3. Block schematic of the experimental setup.

cut from these strips. A total of twelve specimens (6 of bad bond and 6 of good bond) were utilized for the current set of investigations.

EXPERIMENTAL SETUP

The two sets of experiments planned differ in relation to the physical arrangement of the AE and UT transducers. In the first set, the UT and AE

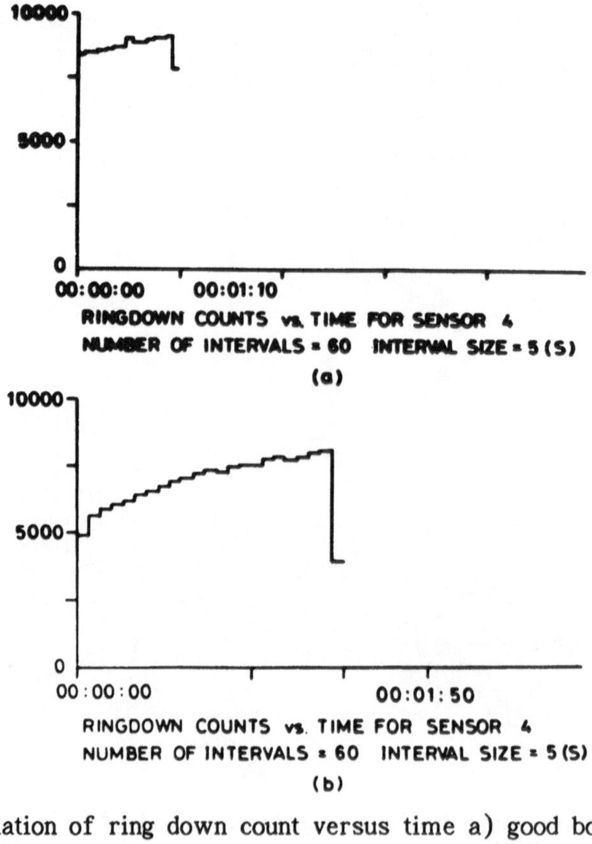

Fig. 4. Variation of ring down count versus time a) good bond, b) bad bond

probes were fixed on the same side of the strip, one at each end of the strip (Fig. 3). In the second set, evaluation was carried out with the probes mounted as in a through transmission ultrasonic test (AE transducer shown in dotted line). The AE transducer output was fed in to a preamplifier with 60 dB gain. An AET 5000 system was used for data acquisition and processing. The amplified AE signal was also displayed on a storage oscilloscope for studying the fine structure of the waveform visually.

For the experiments with the transmitter and receiver on the same side of the strip, an ultrasonic transducer of 2 MHz frequency and an AE transducer of 375 kHz resonant frequency were used. These were mounted on the same side of the bimetallic strips (on the lining material) at a fixed distance of 100 mm. The distance between the two transducers was kept constant in all the tests. In the case of through transmission mode, an ultrasonic transducer of 1 MHz frequency was coupled to one of the surfaces of the specimen. An AE transducer of 1 MHz resonant frequency was mounted on the other surface of the specimen located exactly opposite to the ultrasonic transducer. The strips were evaluated by shifting both the transducers from one area to the next or by keeping the transducers stationary and moving the strip, which would be ideal to check bimetallic sheets on production line. With this arrangement, different AE parameters such as peak amplitude, energy, and ringdown counts were examined to quantify the stress wave factor for discrimination of bond quality, because these parameters can be used to measure the energy content of the stress waveform. This arrangement also makes it possible to study the attenuation in peak amplitude of the input signal as it propagates through thickness of the specimen.

EVALUATION OF BOND QUALITY

In general, correlations with material properties are obtained by measurement of "Stress Wave Factor".[3,4] The stress wave factor may be defined as a product of G, R, and N where G is the gate width, R is the repetition rate, and N is the number of ring down counts. Stress wave factor is also defined as a measure of the efficiency of stress wave energy transmission through the material.[5] In principle and in practice, any measured AE parameter can be used for computing the SWF. In our experiments, we used ring down count as the AE parameter in accordance with the common practice. The gate width (G) was chosen as 5 seconds with a repetition rate (R) of 250 pulses per second.

Figures 4(a) and (b) show the variation of ring down count (i.e. SWF) G, and R are constant) with respect to time for strips of good and bad bond quality respectively with the transducer mounted on the same side. It can be observed from these figures that the ring down count is comparatively higher for a strip with good bond quality than that of a strip with a bad bond. The same trend repeated when the probes were shifted to the base material side (mild steel). This increase can be attributed to the higher efficiency in the transmission of the stress waves in the absence of a debond area. In other words in the case of specimens containing bad bond the stress waves are adversely effected. However, it was observed that the absolute values of the SWF varied from specimen to specimen, though the trend remained the same, which can be due to the variation in the extent of debonding in different specimens. For convenience, the ratio of the SWF for a good bond to a bond was computed, which has a value varying between 1.4

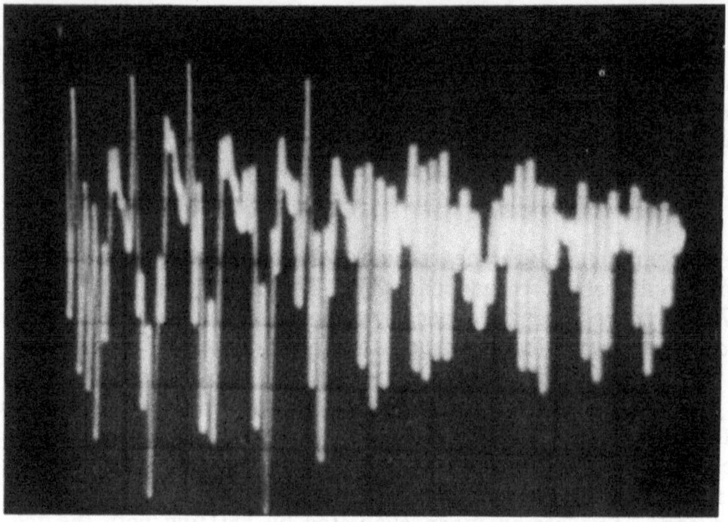

Fig. 5. Typical acousto-ultrasonic event of a good bond (x : 5 μsec/cm)

and 1.8 for different specimens. Thus, these results show that the one side arrangement of UT and AE transducers can indicate the bond quality. Further experiments were carried out which showed that this type of arrangement can also yield conclusive information regarding the quality of bond at localized areas by scanning the strip, step by step keeping the inter transducer distance constant.

Figures 5 and 6 show typical acousto-ultrasonic events for good and bad bond specimens respectively which show marginal difference in peak amplitude. A close look at these two waveforms reveals that clear differences exist in the fine structure. While the good bond specimen contains low frequency content in the first half of the event, the bad bond specimen contains higher frequency content. In the later half also the two waveforms appear to be different.

Figure 7 shows the variation of ring down count for the same good and bad bond specimens in the through transmission mode. Values of other AE parameters corresponding to a few selected tests are given in Table 1. It can be seen from Table 1 that the number of events are the same in all cases. This indicates that there is consistency in the input signal in both the cases. The events were subjected to conventional AE parametric analysis and computation of SWF. The results of the analysis showed that though all

Table 1. Comparison of several tests on "good" and "bad bonds

No. of Events	Good Bond				No. of Events	Bad Bond			
	MPA (dB)	MED (us)	MEN	RDC		MPA (dB)	MED (us)	MEN	RDC
239	92	11	90	478	226	93	29	108	3023
239	92	16	104	1195	247	93	31	108	2223
246	92	11	103	738	248	93	29	108	3070
246	90	8	98	750	245	93	25	107	2600

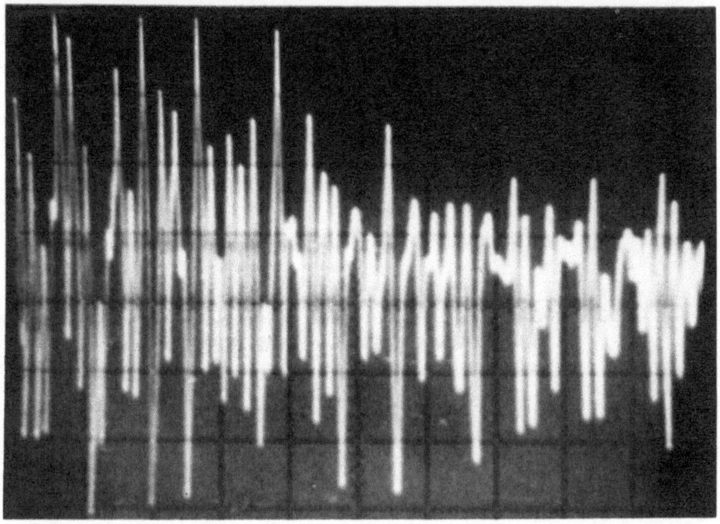

Fig. 6. Typical acousto-ultrasonic event of a bad bond (x : 5 μsec/cm)

the three parameters are derived from the same signal, while the mean peak amplitude and mean energy were almost same, the ring down count varies drastically. The ratio of the computed SWF for a bad bond to a good bond in this case is 6.3.

At this stage, we can compare the final results with the two different arrangements discussed in light of the end use viz. evaluation of bond quality of bimetallic sheets on production line. The first arrangement of placing both the transducers on the same side is applicable for strips as well as plates. However, when we consider the ratio of the Stress Wave Factors for a good bond to a bad bond (1.4 - 1.8), any local differences can further reduce this value resulting in ambiguities due to statistical scatter in the data. In contrast, the ratio obtained in the second type of arrangement viz. through transmission mode yields a value of 6.3 which is much higher and can be effectively used for evaluation of debond areas with greater confidence.

CLOSING REMARKS

Acousto-Ultrasonics appears to be a potential technique for characterization of bond quality of bimetallic strips, and Stress Wave Factor can be used to discriminate the quality of bond locally. It is found that the results obtained in through transmission mode are more consistent than those obtained by mounting the transducers on the same side. Ring down count appears to be the most appropriate AE parameter to be used for SWF when the test is carried out in through transmission mode. Further, this type of physical arrangement should be more convenient for evaluating the relative quality of bond at localized areas of a bimetallic sheet on production line. It is proposed to verify these results further by monitoring AE in the location mode during tensile tests.

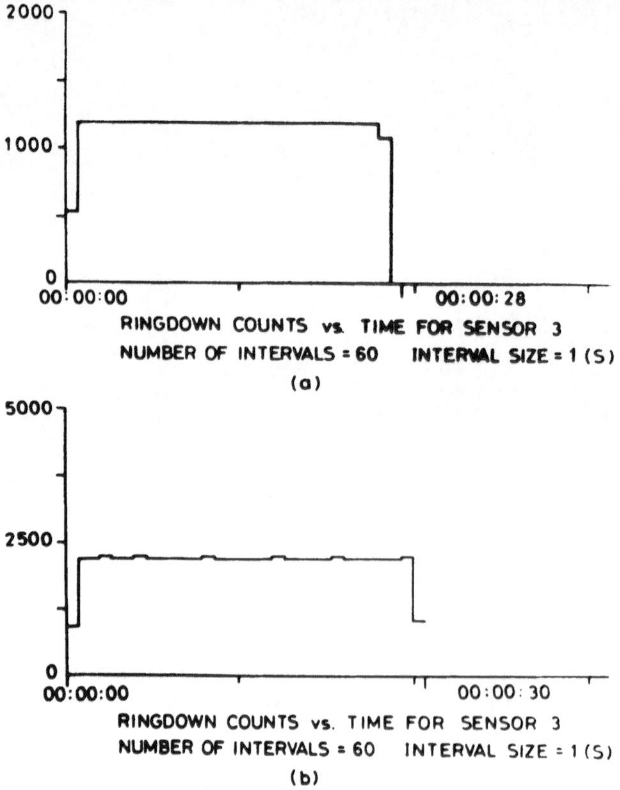

Fig. 7. Variation of ring down count versus time a) bad bond, b) good bond

ACKNOWLEDGEMENTS

M/s. Kirloskar Oil Engines Limited, Pune, have suggested the problem and provided necessary material for making specimens. The typing of the manuscript by Mr. Venkatesh is greatly appreciated.

REFERENCES

1. C. R. L. Murthy, and K. K. Brahma, Nondestructive Evaluation of Bond Quality of Bimetallic Strips Using Acoustic Emission, J. of Test & Eval. 13:371 (1985).
2. R. S. Gilmore, M. L. Torreno, G. J. Czerw, and L. B. Burnet, High Frequency Ultrasonic Testing of Bonds: Application to Silicon Power Devices, Matls Eval. 37:65 (1979).
3. A. Vary, and K. J. Bowles, An Ultrasonic Acoustic Technique for Nondestructive Evaluation of Fiber Composite Quality, Poly Eng & Sc. 19:373 (1979).
4. J. H. Williams, Jr. and N. R. Lampert, Ultrasonic Evaluation of Impact-Damaged Fiber Composite, Matls Eval. 38:68 (1980).
5. A. Vary, Acousto-Ultrasonic Characterization of Fiber Rreinforced Composites, Matls Eval. 40:650 (1982).
6. L. Alting, "Manufacturing Engineering Process", Marcell Dekker, Inc., New York (1974).

ACOUSTO-ULTRASONIC EVALUATION OF THE STRENGTH OF COMPOSITE MATERIAL ADHESIVE JOINTS

V. K. Srivastava

Mechanical Engineering Department
Institute of Technology
B.H.U., Varanasi - 221 005, India

R. Prakash

Bio-Medical Engineering
Institute of Technology
B.H.U., Varanasi - 221 005, India

ABSTRACT

The aim of the present research program is to evaluate the tensile strength of adhesively bonded lap joints of glass fibre reinforced plastic (GRP) and carbon fibre reinforced plastic (CFRP) composites by using the acousto-ultrasonic technique. The results show that depending upon the application for which the particular adhesive joints are to be used, one may choose a confidence level and then using the correlation curves, one may obtain the upper and lower bonds of strength values for a particular stress wave factor.

INTRODUCTION

The primary considerations in the design of structures using composite materials are the stiffness and strength (particularly with respect to weight) of these materials in the various forms in which they are available. However, structural components must be joined to create useful assemblies. Therefore, an equally important requirement for the complete design of practical structures is the development of attachment methods, joint designs and the problem of load introduction in composite structures. Without proper joints, it is not possible to take full advantage of the high stiffness and strength of the laminates. Adhesive joints are natural to consider for polymeric matrix composite materials because many matrix resins are also good adhesives. The primary function of a joint is to transfer load from one structural member to another. In most bonded joints, the load transfer takes place through interfacial shear. The interfacial shear gives rise to high interlaminar stresses in the adhesive layer. Berg[1] suggested that an effective way of reducing the local high stresses in the

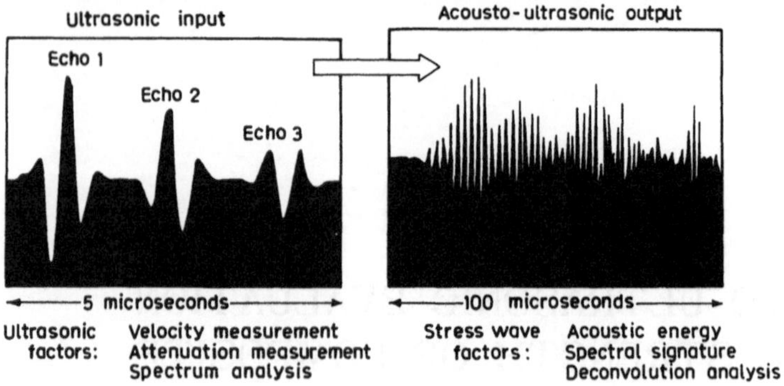

Fig. 1. Typical input and output signals in acousto-ultrasonic characterization of fiber composite laminate properties.

plies adjacent to the adhesive layer is to interleave the plies of the adherend laminates so that adhesion takes place in many layers and, consequently, stresses are distributed in many plies.

The ultimate performance of fibre reinforced plastics (FRP) has been characterized by acousto-ultrasonic technique.[2-5] It is the holistic approach, of finding out, how defects interact with the microstructural environment within the material. Acousto-ultrasonic technique is essentially an amalgamation of two separate existing techniques; acoustic emission and ultrasonics. In this acousto-ultrasonic technique, using an ultrasonic pulser, discrete ultrasonic pulses are injected into the material under test and the ultrasound is allowed to interact with the material. Due to this interaction of ultrasound with the internal features of the material (the microstructural environment within the material), the resultant waveform provides nothing but a modulated signal characterizing the material quality. An acoustic emission sensor is then used to pick-up this resultant waveform (or the modulated signal of material quality). The acoustic emission signal thus obtained is then suitably processed and digitized to provide a measure of the quality of material under test as shown in Fig. 1. The measure is usually referred to as the stress wave factor[6] (the ultrasonic pulses are produced at a repetition rate of 'g'. After amplification of the received signals are sent to a counter that counts the number of oscillations 'n' in each burst exceeding a fixed threshold voltage. The counter is reset automatically after a predetermined time interval 'r'. The product of these quantities is known as the stress wave factor. The stress wave factor is arbitrary and depends on factors such as probe pressure, coupling, signal gain, reset-time, threshold voltage, repetition rate[7-8].

The ultrasonic waves input to the material under test (FRP) is in the form of well defined discrete pulses. These ultrasonic waves are modulated/dampened differently by different materials and also by different features in the same material. A good quality material will dampen the ultrasonic pulses to a lesser extent as compared to a material having gross distributed defects. The stress wave factor measurement, which is based on the oscillation counting of the resultant waveform, shall therefore be higher for good quality materials as compared to bad specimens because for the later case the resultant waveform is a highly dampened one. Defects such as misaligned and broken fibres, micro and macro-voids, resin-cracking, cracks, etc., all reduce the ability of FRP composites to propagate ultrasonic waves and hence the resultant waveform

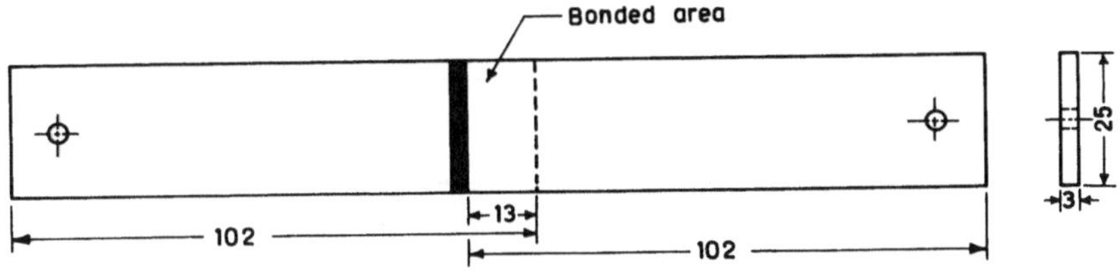

Fig. 2. Geometry of the adhesive bonded lap joints of FRP composite.

of a FRP composite containing these defects becomes a highly dampened one and the measured stress wave factor drops to a lower value for such composites.

In this paper, results are reported of recent experiments, which can be used to predict the ultimate performance of adhesive lap joints of FRP composite by the acousto-ultrasonic technique.

EXPERIMENT

Materials

The materials selected for the unidirectional FRP composites of this study were E-glass fibre, Type-I carbon fibre and epoxy resin (CY-205 Araldite and HY-951 Hardener). The composites were made by the hand lay-up technique. The glass fibre reinforced plastic (GRP) and carbon fibre reinforced plastic (CFRP) were cut into a number of pieces of standard size as shown in Figure 2.

Sample Lap Joints

The GRP and CFRP specimens were first washed with Vim Washing Powder and then immersed for 10 minutes at 70°C in a solution of Caustic Soda (2 parts by weight) and Water (8 parts by weight). All specimens were removed and washed with cold water followed by hot water before drying. Commercially available epoxy resin and hardener under the trade name Araldite was used as adhesive for simple lap adhesive bonded joints of GRP/GRP and CFRP/CFRP composites. These specimens were prepared according to ASTM D 1000-72 procedure. The glue line thickness of lap joints is approximately 0.35 mm.

Acousto-Ultrasonic Testing

The instrumentation used for evaluating the ultimate performance of adhesive bonded lap joints of GRP and CFRP composites was acousto-ultrasonic apparatus (Model AET-206 AU). It is composed of three main sections - the pulser/ultrasonic section, sensor/acoustic emission section and the display section. Using a broadband transducer (500 kHz frequency and 1 pps pulse rate) the pulser section produces stress waves that resemble acoustic emission events. The piezoelectric acoustic emission sensor probe having 375 kHz resonant frequency was used to receive the transmitted stress wave. The distance between the pulser and sensors probes is 40 mm. A broad band 40 dB preamplifier including a bandpass filter, whose frequency was matched to the AE sensor was used to obtain signal amplification before the signal is processed. The stress wave factor readings were obtained on bonded areas of

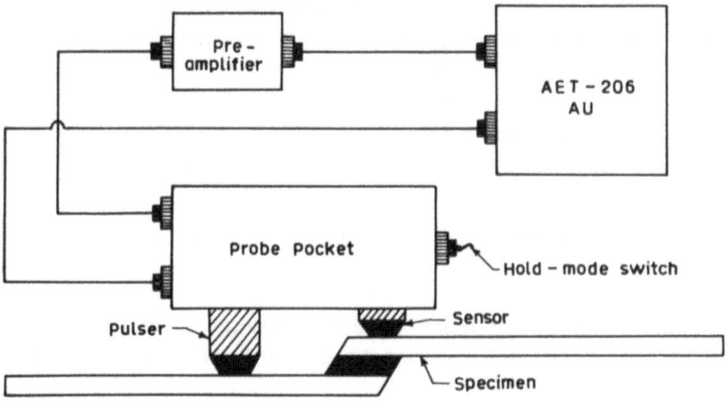

Fig. 3. Diagram of apparatus for measurement of stress wave factor.

simple lap joints of GRP/GRP and CFRP/CFRP, when the noise interference was minimum. Good acoustic contact between the probes and the surface of the specimens was obtained using high vacuum grease. Also, the pulser/sensor fixture was firmly pressed against the composite specimens to avoid changes in experimental readings with the change in applied pressure. This also eliminates friction noise caused by sliding of sensor over the composite specimens. A mass (4 kgf) was used to press the probe pocket, which was put such that the stress wave enter the adhesive layer through one adherend and wasensed through the other adherend as shown in Fig. 3. All the value of stress wave factor readings were measured at a particular setting of instrument (Gain - 62 dB, threshold voltage - 2.5 volts, Energy - 3, Rate - 1/8 sec., Trigger rate - 1 K pps, Scale - 1, Sweep - 62.5 micro sec./div., Mode pulser, counts gate, frequency, duration knobs - full clockwise). This permitted evaluation of all the layers and sublayer of the adhesive bond. Each specimen having different stress wave factors were marked by dye-pen for easy evaluation of ultimate performance of adhesive bonded lap joints of GRP and CFRP composites.

Tensile Testing

The ultimate tensile strength of adhesive bonded joints of GRP and CFRP composites were obtained by using the Hounsfield Tensometer. Each specimen having different values of stress wave factor were set in the jaws of the Hounsfield Tensometer and the load was gradually applied on the specimen. The final load at which the specimen broke was recorded and used to calculate the bond strength.

RESULTS AND DISCUSSION

The obtained values of stress wave factor for the various joints were listed with maximum and minimum value as a criterion and fed into the simple library programme of an ICL 1900 Computer. The best fit values, confidence level, and student-t values were obtained by computer results. Figure 4, shows the correlation curves indicating the relationship between bond strength of FRP composites and stress wave factor with the required degree of confidence levels. The results show that the student-t values of 9.85 of GRP and 8.75 CFRP composites, for the regression coefficient is moderately high and thereby indicating that the coefficient cannot be zero. The correlation table concludes that the greatest correlation (0.945 of GRP and 0.9325 of CFRP) exists between breaking load and the maximum value of stress wave factor. The next greatest correlation (0.752 of GRP and 0.685 of CFRP) is with the minimum value of stress wave factor. The single regression shows that satisfactory results are obtained by measuring stress wave factor. These results help in prediction of the breaking strength of adhesive bonded joints of GRP and CFRP composite with required degree of confidence level. As can be seen from the figure, the value of stress wave factor increases with increasing breaking strength of adhesive bonded joints. Because, the acousto-ultrasonic technique is sensitive enough to pick-up the influence of subtle morphological factors - microstructure, mechanical properties and flaw populations - on energy dissipation. The working hypothesis was that attenuated stress wave energy flow corresponds to decreased fracture resistance[8].

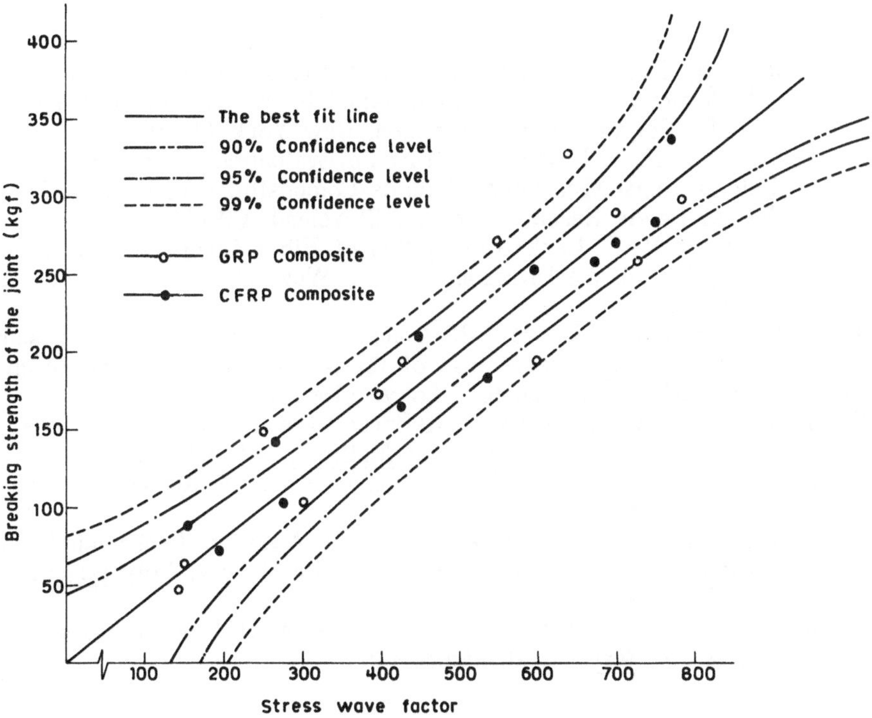

Fig. 4. Correlation curves indicating confidence levels of adhesive bonded joints of GRP and CFRP composites (at a particular instrument setting).

Apparently, the stress wave factor provides a means of rating the efficiency of dynamic strain energy transfer in a FRP composites specimens. However, the results show that a good quality adhesively bonded joints transfer more stress wave energy because minimum defects are present in the bonded area. Hence, one may predict with confidence breaking strength of newly made FRP composite material adhesive joints using acousto-ultrasonic technique.

CONCLUSIONS

The present work has provided a new and reliable technique for nondestructive evaluation of adhesive bonded joints of GRP and CFRP composites. Using the findings reported here, one may in actual practice obtain stress wave factor for adhesively bonded joints which have come for inspection and foretell the strength of the joint, since the stress wave factor has been calibrated against bond strength.

ACKNOWLEDGEMENTS

The authors wish to gratefully acknowledge the Department of Mechanical Engineering and Bio-Medical Engineering, Institute of Technology, Banaras Hindu University, Varanasi - 5, India, for providing the research facilities.

REFERENCES

1. K. R. Berg, Problems in the Design of Joints and Attachments, in: "Mechanics of Composite Materials," F. W. Wendt, H. Liebowitz and N. Perrone, eds., Pergamon, New York (1980).
2. A. Vary and K. J. Bowles, "Ultrasonic Evaluation of the Strength of Unidirectional Graphite/Polyimide Composite, NASA TX-X-73646" NASA, Lewis Research Center, Cleveland, (1977).
3. A. Vary and R. F. Lark, Correlation of Fiber Composite Tensile Strength with the Ultrasonic Stress Wave Factor, J. of Test & Eval. 7:185 (1979).
4. V. K. Srivastava, NDE of FRP Composites Using Acousto-Ultrasonic Technique, in: "Proceedings of the International Conference on Computational Mechanics," Tokyo (1986).
5. V. K. Srivastava, Correlation of Burst Pressure of GRP Cylinders with the Stress Wave Factor, J. of Matls Sc. 21:3628 (1986).
6. R. H. Wehrenberg, New NDE Technique Finds Subtle Defects, Matls Engrng. 12:59 (1980).
7. A. Vary and K. J. Bowles, An Ultrasonic-Acoustic Technique for Nondestructive Evaluation of Fiber Composite Quality, Poly Engrng and Sc. 19:373 (1979).
8. A. Vary, Acousto-Ultrasonic Characterization of Fiber Reinforced Composite, Matls Eval. 40:650 (1982).

INDEX